Semi-Insulating III-V Materials

Nottingham 1980

Semi-Insulating III-V Materials

Nottingham 1980

Edited by G J Rees

Plessey Research (Caswell) Ltd

Shiva Publishing Limited

SHIVA PUBLISHING LIMITED
9 Clareville Road, Orpington, Kent BR5 1RU, UK

IMPRINT EDITIONS INC.
Suite 1104, 415 South Howes, Fort Collins, CO 80521, USA

British Library Cataloguing in Publication Data

Semi-Insulating III-V Materials *(Conference)*,
   Nottingham, 1980
   Semi-Insulating III-V Materials.
   1. Gallium arsenide — Congresses
   2. Semiconductors — Congresses
   3. Chromium — Congresses
   I. Rees, G J   II. Institute of Physics.
   Solid-State Physics Sub-Committee
   III. Institute of Physics. *Electronics Group*
   537.6′22   QC611.8.G3
   ISBN 978-1-4684-9195-1   ISBN 978-1-4684-9193-7 (eBook)
   DOI 10.1007/978-1-4684-9193-7

Typeset and printed by Devon Print Group, Exeter, Devon

CONTENTS

ACKNOWLEDGEMENTS

The Organizing Committee is glad to acknowledge generous financial support from the Department of the US Navy Office of Naval Research; European Research Office of the United States Army (USARSG) and the European Office of Aerospace Research and Development of the United States Air Force (EOARD); GEC Ltd; Hirst Research Centre; Philips Research Laboratories; Plessey Research (Caswell) Ltd; and Standard Telecommunications Laboratories Ltd.

This work relates to Departments of the Navy Grant N00014-80-G-0043 issued by the Office of Naval Research. The United States Government has a royalty-free license throughout the world in all copyrightable material contained herein.

The views, opinions and/or findings contained in this report are those of the author(s) and should not be construed as an official Department of the US Air Force position, policy, or decision, unless so designated by other documentation.

Delegates to the Semi-Insulating III—V Materials Conference, Nottingham, 14—16 April 1980.

PREFACE

The study of deep levels in semiconductors has seen considerable growth in recent years. Many new techniques have become available for investigating both the electronic properties of deep levels and the chemical nature of the defects from which they arise. This increasing interest has been stimulated by the importance of the subject to device technology, in particular those microwave and opto-electronic devices made from GaAs, InP and their alloys.

While previous conferences have covered specialist areas of deep level technology, the meeting described here was arranged to draw together workers from these separate fields of study. The following papers reflect the breadth of interests represented at the conference.

For the sake of uniformity we have chosen the English alternative where an American expression has been used. We have also sought to improve grammar, sometimes without the approval of the author in the interests of rapid publication.

The Editor wishes to thank the referees for their ready advice at all stages, Paul Jay who helped with many of the editorial duties and Muriel Howes and Lorraine Jones for rapid and accurate typing.

G.J. REES
Plessey Research (Caswell) Ltd

July 1980

Section 1:
Invited papers

WHITHER CHROMIUM IN GALLIUM ARSENIDE?

A.M. WHITE
Royal Signals and Radar Establishment, Great Malvern,
Worcs. WR14 3PS, UK

Abstract

The principal empirical properties of the GaAs:Cr system were established nearly 20 years ago. Despite its great technological importance and much investigative work, progress in understanding has been, until comparatively recently, painfully slow. The reasons for this are examined, setting the scene for consideration of the more exciting developments of the last three or four years.

Introduction

The history of chromium-doped gallium arsenide is short. It is littered with abandoned models and, one could say, still bound by models soon to be abandoned. Because of its technological importance, there has been a huge investment of effort to reach an understanding of its behaviour beyond the empirical facts. However, the most disturbing fact of all is that until three or four years ago no major progress in that understanding had really been gained since the very first measurements of thermal activation energy. Of course, our appreciation of the complexities of the GaAs:Cr system has grown enormously, accompanied by really creditable advances in material quality. But the lack of basic understanding has remained, indeed a nightmare to the quality control engineer. Theoretical studies of deep levels and lattice interactions are now subjects of major activity and, coupled with advances in EPR and luminescence not to mention the quality of the material to which these techniques are applied, allow us to review our positions and prejudices with a little more confidence than was formerly possible.

The problem of GaAs:Cr

The problem has always been to grow reproducibly semi-insulating GaAs of good crystal quality, and then to be able to characterize the material with confidence. The requirements here are circular, since knowledge of the second matter is to some extent required for progress in the first. The intrinsic carrier concentration at room temperature is about 3×10^7 cm^{-3} so we can safely assume that production of intrinsic material at an early date is unlikely. Further, the *control* of defect and impurity concentrations at levels near 10^{15} cm^{-3} is similarly unlikely to be such as to achieve so perfect a compensation to place the Fermi energy at the intrinsic level. One needs an excess of deep donor levels at mid-gap *above* the amount necessary to deal with uncompensated acceptors (hence oxygen doping) or, similarly, an excess of deep acceptors (electron traps) over uncompensated donors (hence chromium doping). Bridgman grown material doped with chromium is sometimes semi-insulating, sometimes p-type, often with different regions spatially segregated. Liquid encapsulated Czochralski (pulled) material is much more reproducibly semi-insulating. The chemical circumstances of the techniques are different, with more oxygen but less silicon present in the latter. Type conversion on heat treatment (such as when used as substrates for epitaxial growth) is surely related to these different conditions.

The particular problems that have dogged the characterization of chromium-doped gallium arsenide over the years are now easily recognized. The first difficulty is a conceptual one, associated with the view that Cr on a Ga site is said to be a double acceptor. The relevant electron shells for a Ga atom are $3d^{10}4s^24p^1$. A simple acceptor such as Zn has $3d^{10}4s^2$, loosely binding a hole. Chromium has $3d^54s^1$. However, rather than leading to two hydrogenic holes in the shallow acceptor picture, one can simply imagine electrons from the d core making up the deficit so that one has $3d^34s^24p^1$, i.e. the holes are bound in the d shell rather than bound hydrogenically. Capture of electrons leads successively to core states $3d^4$ (singly negatively charged with respect to the lattice, referred to in the EPR convention as Cr^{2+}) and $3d^5$ (doubly negatively charged, Cr^{1+}). Chromium which is neutral (the neutral 'double acceptor') is then Cr^{3+}. It would be better to regard Cr^{3+} as an isoelectronic trap.

An electron may be captured from the valence band to convert Cr^{3+} to Cr^{2+} (Figure 1). The energy absorbed in this process can be represented by a level at the appropriate position above the valence band (despite the one-electron connotation of the band diagram, for this multi d-electron situation). Naturally, the same level represents the reverse transition, where the electron is returned to the valence band or removed to the conduction band. We have the result that a single level is used to represent both Cr^{3+} and Cr^{2+}, depending on which of electron capture or release is under consideration [1]. Similarly, a second 'level' is used to describe the energetics of conversion between Cr^{2+} and Cr^{1+} leading to the further result that now two different levels relate to Cr^{2+}! Of course, the whole picture would be more consistently described in terms of the energy levels of the first electron to bind, and then the second electron, but this is infrequently done. These levels are not eigenvalues of some Hamiltonian, but represent transition energies — for example, the difference between the energies of a core $3d^3$ with a conduction band electron, and a $3d^4$ core. This practice has led to unforgivable confusion in the literature — the first hole ionization energy level is often referred to as Cr^{3+} in the context of p-type or semi-insulating matters (tacitly assuming the non-existence of Cr^{4+}) but frequently as Cr^{2+} in circumstances involving n-type material. The higher level (closer to the conduction band, corresponding to the conversion between Cr^{2+} and Cr^{1+}) is generally called Cr^{1+}, leading to discussions on the point of 'the existence of the Cr^{1+} level within the gap', or the possibility that 'the Cr^{1+} level may be resonant in the conduction band'. These questions are more accurately stated in terms of 'can Cr^{2+} bind an electron?'. A logical and less confusing scheme [4] would be to represent the first hole ionization energy of a deep centre by a level 1, the second ionization by a level 2 and so on (see Figure 1). In circumstances with negligible lattice relaxation following ionizations, the levels would then recede in numerical order away from the appropriate band edge. Within this picture, the level found by Haisty and Cronin [5] is hole level 1 (its position, 0.79 eV, nevertheless referred to the conduction band edge, a natural procedure if one regards Cr as an electron trap rather than as an acceptor). In the event that Cr^{3+} can *capture* a hole (i.e. electron ionize) to create Cr^{4+}, etc. [2] one obtains a similar sequence, with one new level for each ionization [4].

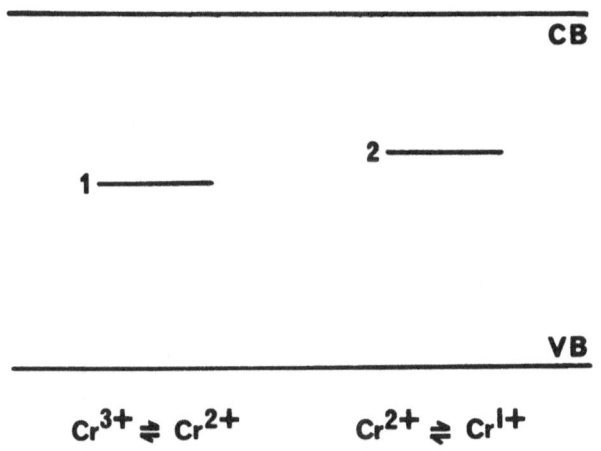

Figure 1   A representation of the hole ionization levels discussed in the text. Level 1 is approximately 0.78 eV above the valence band (VB) with scarcely any lattice relaxation. Level 2 has a zero phonon energy of approximately 1.1 eV but a Stokes shift of approximately 0.2 eV (1). It seems that the level corresponding to the electron ionization $Cr^{3+} \rightarrow Cr^{4+}$ (2) lies below hole level 1 at about 0.45 eV above the valence band [2, 3] — about the same distance as level 2 below the conduction band!

The difficulty with thermal equilibrium activation experiments is principally associated with parametrization of the model. The slope of the Arrhenius plot necessarily gives an energy value extrapolated to absolute zero. One can improve on this by attempting to fit data over a wide range to a model which includes the presence of donors and possibly acceptors as well. The simplest approach is to treat Cr as a simple acceptor, i.e. to assume that at all temperatures covered in the experiment only two charge states of chromium can exist. In semi-insulating material these have been taken to be Cr^{3+} and Cr^{2+}, and in favourable circumstances this can be a useful approximation.

In reality, the situation is far more complicated. All charge states of the chromium must be allowed, and each state has multiplicity with ground state splittings. The well-known Fermi function does not represent the occupancy of charge states of multiply ionizable impurities [4, 6]. The band gap changes with temperature, and its variation may (or may not) contribute to the observed activation depending on how the chromium ionizations vary with temperature. The result is that, depending on the circumstances of the doping and on the characteristics of the chosen model, the delivered parameter 'energy depth' can vary within a range of 0.2 eV (Figure 2). It is, of course, impossible to include the effects of unknown numbers of donors, acceptors, traps and other multiply ionizable centres of unknown identity. For complicated centres like Cr, one concludes that there is no way of determining ionization energies with any accuracy using such equilibrium type activation experiments [7]. At best, parameters of a model applied to different samples may be compared.

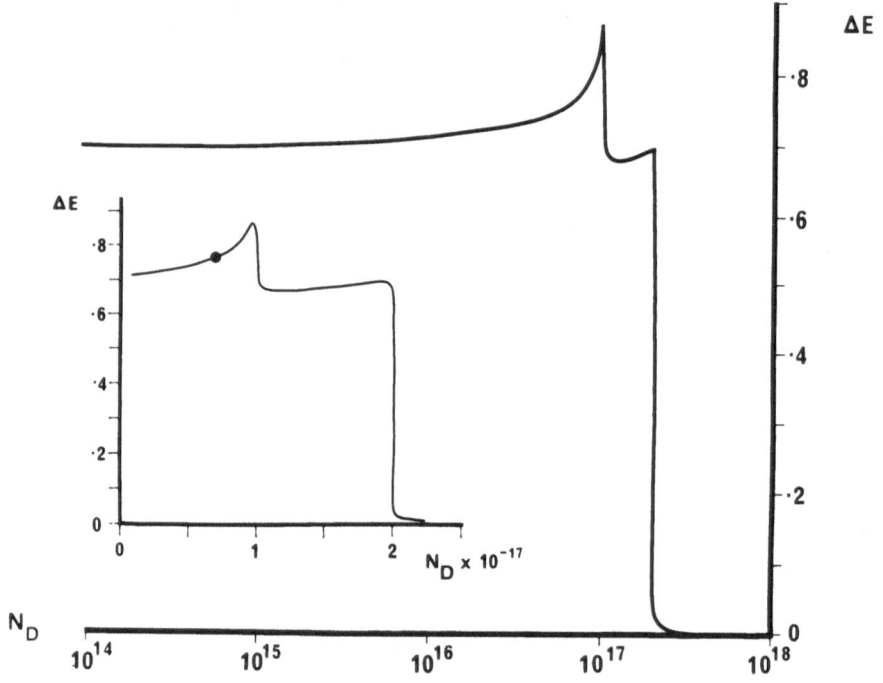

Figure 2   Computed apparent conduction activation energy (eV) as a function of shallow donor concentration (cm^{-3}), incorporating three charge states Cr^{3+}, Cr^{2+} and Cr^{1+}, excited state splittings and temperature effects on ionization energy and band gap. (From Clark [7].)

The position of transient capacitance type experiments is rather different, since the thermally induced release time constant of carriers on traps is a very sensitive function of trap depth. One still has the temperature extrapolation problem, and the temperature dependence of ionization energies and, for accuracy, the effects of temperature variable capture cross-sections must be compensated for at all points. Measurements of capture cross-section are not conducted under the same electric field conditions as the thermal emission experiments. Finally, the trap emission energy may include barriers due to lattice relaxation which are not detected by thermal equilibrium techniques, or in optical measurements.

Optical excitation experiments have been just as confusing. Such experiments rely on the photon energy of exciting radiation to exceed some threshold, generally of some ionization but possibly also of some resonant transition. We can cite photocapacitance, photoconductivity, photo-luminescence excitation and optical absorption. The chromium-gallium arsenide system presents us with what is almost a worst-possible-case situation. Consider the case of gallium arsenide containing only chromium (which is therefore neutral to the lattice, i.e. Cr^{3+}). It appears that the ionization energy of the first hole is such that level 1 is close to mid-gap (recent measurements [1, 8] place it ~ 35 meV above mid-gap). Thus, except in the more refined experiments, light of energy below $E_g/2$ has no effect, whereas light of energy above $E_g/2$ not only ionizes Cr^{3+} to create Cr^{2+}, but also photoneutralizes a proportion of the Cr^{2+} thus created, and is already well above the threshold for the further ionization $Cr^{2+} \rightarrow Cr^{1+}$ (level 2, 0.5 - 0.6 eV below the conduction band). This state of affairs exists in material where there is a sufficient number of shallow acceptors (i.e. shallower than Cr, or level 1) to compensate residual donors. In p-type material, light of energy below $E_g/2$ would serve to populate Cr^{3+} at the expense of Cr^{4+} (see comment below on the significance of [3]). Even where there is partial compensation of the chromium, the onset of these effects will initiate the observation of all manner of phenomena involving trapped carriers, with the added complication of difficulty of deconvoluting the cross-sections of all different absorptions involving the two band edges. In the matter of optical cross-sections one should note that one expects to find *dominant* optical transitions to the higher density of states regions of the bands although, of course, the thresholds for the transitions are to the Γ band edges.

A celebrated optical feature of chromium in gallium arsenide (with a corresponding phenomenon in gallium phosphide) has been a peak which occurs at about 0.9 eV [9, 10]. This has been observed in photoconductivity, absorption and photocapacitance. It could be that a persistent luminescence peak near 0.8 eV [11] is the corresponding Stokes-shifted transition (Figure 3). The peaked nature of the cross-section implies some kind of localized transition. This transition is generally regarded to be basically the ^5E (upper)—$^5T_2$ (lower) levels of the Cr^{2+} core ($3d^4$), which are split by the tetrahedral crystal field (and further split by other effects, which introduce enormous difficulties, as we shall see). Its threshold energy ($E_g/2$) compared with level 1 ($Cr^{3+} \rightarrow Cr^{2+}$) seems ominous but could be another sad coincidence. Persisting with the Cr^{2+} picture as the simplest possible explanation, one reads of the possibility that 'the excited state of Cr^{2+}(i.e. the

^{5}E state) may be a resonant state in the conduction band'. Translated, this means that the added electron may autoionize (i.e. the excited Cr^{2+} may neutralize by electron release to the bottom of the conduction band). Depending on the degree of lattice coupling, this picture may be only half true since the peak energy of the transition in absorption differs from its value in emission [12].

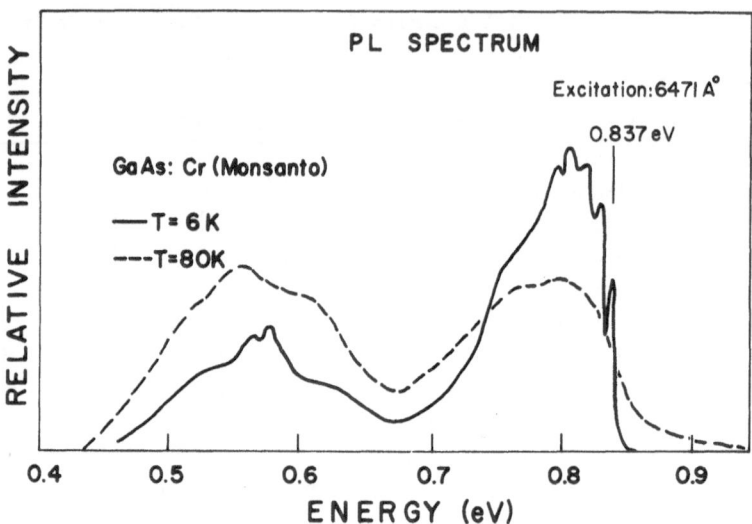

Figure 3   Photoluminescence spectra of GaAs:Cr with laser excitation, showing at the same time the principal emission bands and indications of the no-phonon lines associated with both. (From Koschel, Bishop and McCombe [11].)

The nature of the second emission band at 0.56 eV [11] is very obscure. The amount of experimental work on this band is relatively meagre so that it is difficult to choose between assignments which identify it variously as band to chromium, ^5E—$^5T_2$ transitions of Cr^{2+} or even transitions between the Jahn-Teller split components of the $^5T_2$ ground state of Cr^{2+}, or indeed anything else.

The observation of the sharp no-phonon line at 0.839 eV in both luminescence and absorption in recent years has turned attention away from the broader band, although possible correlation with it has been hinted at. There appears to be a correlation in different materials between the energy of this line and the photoconductivity threshold [13]. However, the assignment of this sharp line has caused trouble. Following initial assignments to transitions between Cr and the bands (level 1), the most popular view now is that this is the no-phonon transition of the ^5E—$^5T_2$ splitting. Further high-resolution studies [14] have shown considerable substructure within this line so that, for the first time, one is in a position to test this assignment in detail. One can make a comparison of the level splittings deduced from the luminescence with those found directly by EPR for the different charge states of Cr [15, 16]. The conclusion is most disappointing, for there is no correspondence to be found. The EPR data for Cr^{2+} show that the centre is axial, with all the characteristics of a <100> Jahn-Teller distortion corresponding to a relaxation of several hundred meV, considerably larger than might be expected from the Stokes shift of the phonon-coupled luminescence associated with the 0.839 eV no-phonon line. As a result of these observations it is not at all clear how the ^5E_T–$^5T_2$ assignment can be correct, and an alternative model has been proposed (see below).

In view of the specialist paper by Blakemore in this volume, it would be inappropriate to take space here to discuss the theoretical difficulties raised by GaAs:Cr. They are not in principle unique to chromium, although in some ways GaAs:Cr does seem to be particularly perverse. For example, the d–d transitions of Fe in InP and both Fe and Cr in the II—IV compounds seem to be reasonably well understood, at least in terms of lattice coupling and the consequential nature of the Jahn-Teller phenomena as manifested in a considerable degree of experimental detail.

The progress of GaAs:Cr

Having reviewed the problems of the GaAs-Cr system and defined our terms, the broad history of the subject can be summarized quite simply. Unlike much spectroscopic work on more tractable systems there has, until recently, been no sharply structured feature to investigate, leading to the proposal and acceptance of simple models which can satisfactorily explain the bland optical data. For many years the most interesting feature, namely the peak near 0.9 eV, has been the subject of discussion, leading to a number of proposals as to its nature, the most important being the ^5E—$^5T_2$ transition or some type of complex involving Cr. Complexes involving combinations of shallow impurities and vacancies were proposed many years ago to account for a series of bell-shaped luminescence peaks in GaAs, and the transition metal impurities have not been exempt from this idea [17]. The more recent high-resolution work [14] has led to a proposal [18] that an *excitonic* transition is involved in the transition at 0.839 eV, involving a complex of Cr and a donor (Figure 4). It is interesting that the corresponding line in GaP was classified as clearly excitonic, at a time when its relationship to Cr was unsuspected [19]. The shape of the spectrum is closely similar to other known excitonic features in GaAs and GaP. Important Zeeman experiments, both in the near and far infra-red, present us with excellent opportunities for fine scrutiny of all such models.

Due to the position of the 'levels' it has taken capacitance measurements a number of years to achieve what may be classed as reliable data, and then only with difficulty. The same comment applies to photoconductivity and photo-Hall measurements. At no time has there been much difficulty in finding consistency with the latest thermal activation measurements! Evidence that Cr^{2+} can bind an electron (level 2 within the gap) is quite meagre. This is because in many cases it is possible to fit data either way — assuming it cannot, and again assuming it can. Probably the best recent evidence for level 2 being within the gap can be found in photocapacitance measurements [1], which show unambiguously the level positions and concentrations in each state and comparisons of carrier concentrations with accurate atom counts of chromium and donors [20]. Measurements of transient EPR data [3], supposedly on Cr^{1+}, may have been confused by the presence of Cr^{4+}[2], in which case the first electron ionization level of neutral Cr would be 1.05 eV below the conduction band. In principle, the solution to this problem is also contained in the data of capacitance impurity profile experiments on n-type material, since the Schottky barrier height to electrons is greater than the depth of level 1 below the conduction band. Therefore, there should be distinct regions where the Cr is present successively as Cr^{3+}, Cr^{2+} and Cr^{1+} as can be found in the above-mentioned atom-count experiments involving critical balances of donors and chromium atoms which finely control the Fermi energy. Again, one must admit that it would not be too difficult to fit almost any model to the data!

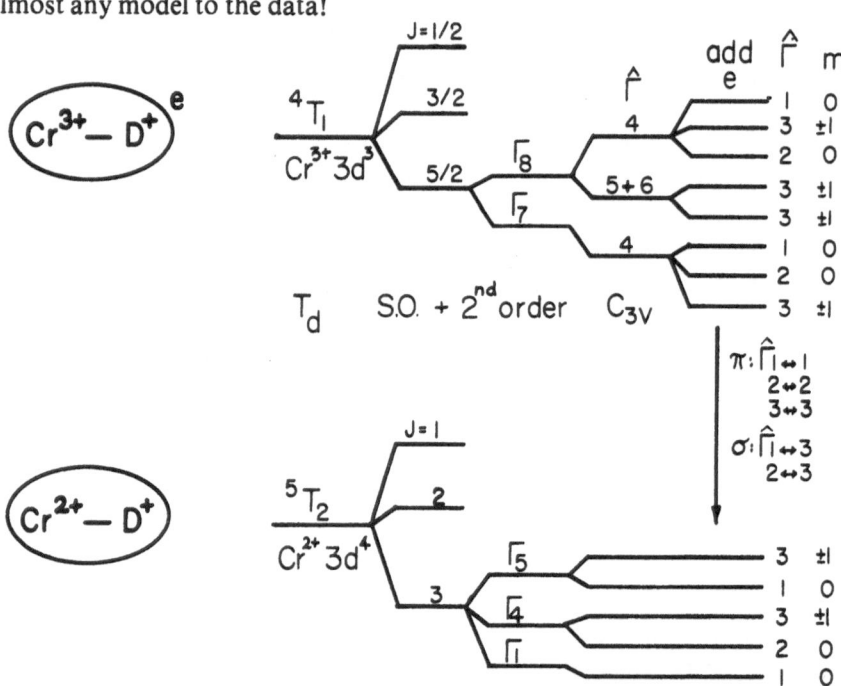

Figure 4   A suggested model of excitonic recombination at a complex of Cr and a nearby donor, in this case in the <111> direction, which matches the 0.839 eV high-resolution data of Lightowlers and Henry [14]. There is a similar high degree of splitting for the <100> direction, with more stringent selection rules. Similar considerations apply to alternative schemes which involve relatively large crystal fields but strongly quenched spin-orbit effects. The recent proposal [2] of a trapped hole state (Cr^{4+}, $3d^2$) raises the additional possibility of an inverse centre comprising an associated acceptor, with even greater opportunities for high multiplicity. The electrostatic field of an associated ionized donor centre reduces the ionization energy of holes on the Cr by up to~0.5 eV. In terms of Figure 1, this produces new ionization levels correspondingly closer to the valence band [3]. (From White [18].)

Thermal equilibrium activation measurements have progressed little in effect. Progressively more complicated features have been recognized and incorporated into models. The present state is that there are now so many disposable parameters that the model is unwieldy, suitable only for numerical processing and useful only for comparing the parameters of one sample against those of another. There is no great accuracy in either atomic concentrations or in energy levels [7]. Schmidlin and Roberts [21] showed that for the case of an arbitrary collection of any number of deep centres of a singly ionizable nature, over any restricted temperature range, two at most of the levels dominate the fixing of the Fermi energy. This principle can be extended to multi-ionizable centres, but there is greatly increased difficulty, particularly with the multiple level structures in each charge state, of determining which levels are important. Even so, with the restricted model due to Schmidlin and Roberts, it is a most interesting result that the 0.69 eV activation energy, a commonly obtained result, was shown *not* to be the energy of a trapping level as is frequently assumed [22]. In this particular case the Fermi level was phenomenologically pinned between two controlling trap levels, 0.40 and 0.98 eV, below the conduction band. It is even more interesting [17, 22, 23] that most of the Cr present was *not* involved in these centres, which were tentatively identified, for a number of reasons based on the chemistry of growth, as Cr-O complexes.

Transient capacitance experiments [24] have advanced steadily with the growing recognition of all corrections that have to be made to convert the observed activation energy to a real trap depth. This aspect, a fairly tortuous path between experiment and result, is the principal drawback. Recent data [8] on level 1 only show, in agreement with the more recent photocapacitance measurements [1], that it is about 35 meV above mid-gap.

On the growth aspect, one must mention the production of LEC material from pyrolytic boron nitride-lined equipment [25, 26]. This material is semi-insulating at intrinsic carrier concentrations *without* having any added chromium. The concentrations of all impurities are very much reduced, and it is not yet clear to what extent the semi-insulating nature is controlled by a residual excess of deep levels, or by automatic compensation by mass-action inclusion of impurity or defect centres during growth.

In contrast, with the possible exception of the type conversion phenomenon, the most unpopular discovery in the last few years has been that of the fast diffusion of chromium [27, 28] (many microns in distance at typical temperatures and times associated with epitaxial growth). We hope to see in the discussions here to what extent these two phenomena are inter-related, for the subject is a matter of great activity and urgency [29]. For many years now it has been the custom to characterize epitaxial layers for devices by recording Hall data from layers grown simultaneously on chromium-doped substrates. However, the diffusion of chromium into the control layer must destroy one's confidence in the result. The same applies to InP layers grown on InP:Fe-doped substrates. To what extent are design parameters for devices computed *ab initio*, or by extrapolation from measurements of this kind?

Conclusion

It remains to be seen whether the control of chromium doping by provision of buffer layers, or growth at low temperatures from the vapour phase using special organic compounds, will be sufficient to allow the use of GaAs:Cr to continue in the more demanding devices. In my view it will not be, and the improvements in undoped yet compensated materials will be continued. Research will then be redirected to the reasons for such accurate compensation, and the nature of the levels responsible. Back to square one, in fact!

References

1. Szawelska, H.R. and Allen, J.W. (1979). *J. Phys. C.*, **12**, 3359
2. Kaufmann, U. and Schneider, J. (1980). *Appl. Phys. Lett.*, in press
3. White, A.M., Krebs, J.J. and Stauss, G.H. (1980). *J. Appl. Phys.*, in press
4. Shockley, W. and Last, J.T. (1957). *Phys. Rev.*, **107**, 392
5. Haisty, R.W. and Cronin, G.R. (1964). *Proc. 7th Int. Conf. on Physics of Semiconductors, Paris*, p. 1161. Paris; Dunod
6. Landsberg, P.T. (1956). *Proc. Phys. Soc. (London)*, **B69**, 1056
7. Clark, M.G. (1980). *J. Phys. C*, in press
8. Martin, G.M., Mitonneau, A., Pons, D., Mircea, A. and Woodard, D.W. (1980). *J. Phys. C*, in press
9. Allen, G.A. (1968). *J. Phys. D*, **1**, 593
10. Bois, D. and Pinard, P. (1974). *Phys. Rev. B*, **9**, 4171
11. Koschel, W.H., Bishop, S.G. and McCombe, B.D. (1976). *Solid St. Commun.*, **19**, 521
12. Chantre, A. (1979). Thesis, Université Scientifique et Médicale de Grenoble, ch. 4
13. Eaves, L., Englert, T., Instone, T., Williams, P.J. and Wright, H.C. (1980). This volume
14. Lightowlers, E.C. and Henry, M.O. (1978). *Inst. Phys. Conf. Ser.*, **43**, 307
15. Krebs, J.J. and Stauss, G.H. (1977). *Phys. Rev. B*, **16**, 971
16. Wagner, R.J. and White, A.M. (1979). *Solid St. Commun.*, **32**, 399
17. Demberel, L.A., Popov, A.S., Kushev, D.B. and Zheleva, N.N. (1979). *Phys. Stat. Sol. (a)*, **52**, 341
18. White, A.M. (1979). *Solid St. Commun.*, **32**, 205
19. Dean, P.J. (1973). *J. Lumin.*, **7**, 51
20. Brozel, M.R., Butler, J., Newman, R.C., Ritson, A., Stirland, D.J. and Whitehead, C. (1978). *J. Phys. C*, **11**, 1857
21. Schmidlin, F.W. and Roberts, G.G. (1974). *Phys. Rev. B*, **9**, 1579
22. Mullin, J.B., Ashen, D.J., Roberts, G.G. and Ashby, A. (1977). *Inst. Phys. Conf. Ser.*, **33a**, 91
23. Adlerstein, M.G. (1976). *Elec. Lett.*, **12**, 297
24. Lang, D.V. and Logan, R.A. (1975). *J. Elec. Mat.*, **4**, 1053
25. Swiggard, E.M., Lee, S.H. and Von Batchelder, F.W. (1977). *Inst. Phys. Conf. Ser.*, **33b**, 23
26. Swiggard, E.M., Lee, S.H. and Von Batchelder, F.W. (1979). *Inst. Phys. Conf. Ser.*, **45**, 125
27. White, A.M., Porteous, P. and Dean, P.J. (1976). *J. Elec. Mat.*, **5**, 91
28. Tuck, B., Adegboyega, G.A., Jay, P.R. and Cardwell, M.J. (1979). *Inst. Phys. Conf. Ser.*, **45**, 114
29. Debney, B.T. and Jay, P.R. (1980). This volume

KEY ELECTRICAL PARAMETERS IN SEMI-INSULATING MATERIALS; THE METHODS TO DETERMINE THEM IN GaAs

G.M. MARTIN
Laboratoires d'Electronique et de Physique Appliquée,
3, avenue Descartes, 94450 Limeil Brevannes, France

Abstract

Assessment of insulating bulk GaAs is proposed following three directions that all throw some light on the electrical properties of the material. First, attention is focussed on deep levels, needed to obtain highly resistive materials. Conditions of detection, methods of measurement of their concentrations in insulating materials, experimental determination of their acceptor or donor nature and, especially, of their Fermi functions, are discussed. The study of compensation mechanisms in GaAs doped with Cr, with O or undoped has been based on the knowledge of the electrical properties of these different levels. In the case of insulating Cr-doped crystals, for instance, both the Cr concentration and N_D-N_A, the residual donor concentration to be compensated by Cr, can be determined. These two parameters have an important influence on the electrical properties of layers obtained by direct implantation in these substrates and subsequent capped annealing. This last treatment reveals basic properties of the starting substrates and completes their characterization.

Introduction

The aim of electrical characterization of insulating materials is to determine the concentration of shallow donors or acceptors to be compensated by deep levels. Thus, methods to obtain the spectroscopy of deep levels and their concentrations will be discussed first. From this study, it is easier to analyse Hall data and to extract the value of residual shallow levels. A correlation has been established between the results of this analysis and those obtained after direct ion implantation in the same substrates. This will be shown in a later section.

Spectroscopy of deep levels

SPECTROSCOPIC METHODS

Both thermally stimulated current (TSC) [1] and optical transient spectroscopies (OTCS) [2—4] detect deep levels in insulating materials; Figure 1 shows the corresponding spectra on the same material. In TSC, d.c.

current, which corresponds to the thermal release of carriers from traps, is recorded as a function of temperature after only one optical excitation at low temperature. The presence of dark current at high temperature prevents the detection of very deep levels such as the Cr level HL1 [5]. The OTCS method consists in studying the transient current after repetitive light excitation as a function of temperature. It offers two advantages: deeper traps can be observed, as evidenced by the comparison of the two spectra of Figure 1; and it allows determination of the emission rate of levels, given by:

$$e_n = 1/t_1 \tag{1}$$

if $t_2 = 10t_1$, t_2 and t_1 being the two times at which the transient current is measured. The labels of traps detected in bulk GaAs, insulating or not, are given in Table 1, together with their parameters E_{Ta} and σ_{na} [5—8]. The characterization of levels from TSC is more difficult and thus we prefer to compute the peak shape and its peak temperature T_M for each known level, the thermally stimulated current being proportional to [1]:

$$I(TSC) \propto e_n \exp\left(-\int_0^t e_n dt\right)$$

where $T = \beta t$, and β is the heating rate of the sample.

The theoretical and experimental TSC peak temperatures are given in Table 1, corresponding to the experimental heating rate which is also reported.

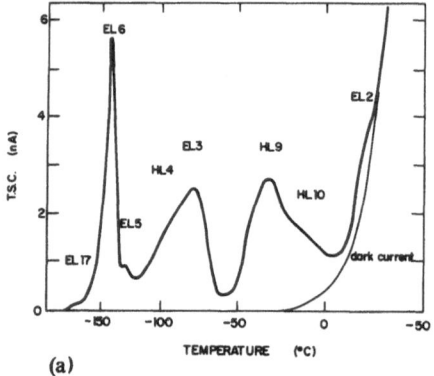 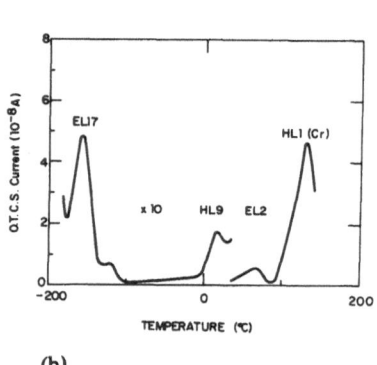

(a)   (b)

Figure 1   Spectra recorded on the same, 350 μm thick, insulating material, doped with Cr, prepared with an ohmic contact and a semi-transparent Cr evaporation. Same applied bias: 25 V. (a) TSC after 3 min white light illumination at 80 K. Heating rate given in Table 1 as a function of temperature. (b) OTCS using 30 ms, Nd Yag repetitive pulses. Sampling times $t_2 = 10t_1 = 340$ ms.

CONCENTRATION OF LEVELS

Thermal spectroscopy
The concentration of deep levels can be deduced from the amplitude of TSC or OTCS peaks if, besides the initial and final states of filling of traps, we

Table 1 List of the different levels observed in bulk, Bridgman or Czochralski GaAs, giving their electrical characteristic parameters, the temperature at which they are observed in TSC, and their initial state of filling with electrons after a 30 ms long pulse of illumination.

Position in band gap	Activation energy of the Hall constant (eV)	Label of levels	Origin	Range of concentration (cm^{-3})	Emission parameters from DLTS E_{Ta} (eV)	σ_a (cm^2)	Peak temperature in TSC Theoretical (K)	Experimental (K)	β (K s^{-1})	Sensitivity to light excitation. Percentage of the level filled with electrons $h\nu > E_G$ 10 mW (%)	$h\nu = 1.17$ eV 100 mW (%)
	$E_C-0.18$	Intermed. donor									
Upper half		EL11			$E_C-0.17$	3×10^{-16}	89				
		EL17			$E_C-0.22$	1.9×10^{-14}	96	93–95	0.68		
		EL14			$E_C-0.215$	5.2×10^{-16}	106				
	$E_C-0.35$	EL6	Complex defect	$10^{14}-10^{16}$	$E_C-0.35$	1.5×10^{-13}	137	132–136	0.60	$\simeq100$	100
	$E_C-0.42$	EL5		$10^{14}-10^{16}$	$E_C-0.42$	10^{-13}	157	144–148	0.57		
	$E_C-0.60$	EL3	Ni (?)	$10^{13}-10^{15}$	$E_C-0.575$	1.2×10^{-13}	212	197–206	0.48	$\simeq0$	$\simeq10$
Middle	$E_C-0.75$	HL1	Cr	$0-3\times10^{17}$	$E_V+0.886$	10^{-14}	321	280–289	0.33	$\simeq75$	85
		EL2	Complex defect	$5\times10^{15}-3\times10^{16}$	$E_C-0.825$	1.2×10^{-13}	288		0.36		25
Lower half		HL10			$E_V+0.83$	1.7×10^{-13}	277	262–267	0.40		
		HL9			$E_V+0.69$	1.1×10^{-13}	235	228–235	0.45		
		HL3	Fe	$\leqslant10^{15}$	$E_V+0.59$	3×10^{-15}	228				
		HL4	Cu or Ni		$E_V+0.42$	3×10^{-15}	169	172	0.53		100
		HL12	Zn associated		$E_V+0.27$	1.3×10^{-14}	109				$\simeq10$
	$E_V+0.145$ [41]		Cu		$E_V+0.14$						

Part of the data comes from [3, 5, 6−8, 26, 42] . β is the heating rate of sample used in TSC experiments.

know the active thickness W in which the released charges are swept by the electric field. The initial filling state of levels is given in Table 1 for two typical optical excitations (above band gap or below band gap). In insulating materials after illumination the space charge is no longer negligible, leading to a non-uniform electric field in the sample. Thus, W may not correspond to the sample thickness. This can be checked by varying the sample thickness (it has been observed that samples a few microns thick give a signal as large as those 100 times thicker) or the applied bias. This last measurement on very thin samples can be a way of determining the concentration of traps, as shown in Figure 2. OTCS spectra, recorded on a 5 µm thick sample for increasing values of applied bias, show that the amplitude of peaks also increase up to a plateau region. It seems likely that the plateau is reached when W becomes equal to the thickness of the sample. The concentration of traps can thus be computed from the plateau value. It has been found that [EL2]$\geqslant$2.4×10^{16} and [EL6]$\geqslant$7.5×10^{15} cm^{-3} in this case. But, the measurement of concentrations remains difficult in insulating materials, and the range of deep level concentrations in bulk GaAs, reported in Table 1, has been mainly deduced from capacitance measurements in low conductivity crystals.

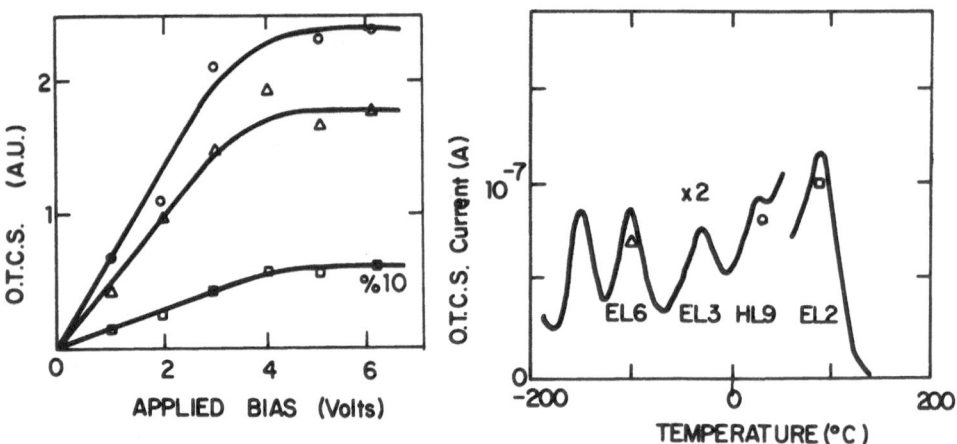

Figure 2 OTCS spectrum on a 5 µm thick insulating sample and variation of the amplitude of some peaks as a function of applied bias. Sample prepared with two Cr semi-transparent contacts.

Main levels

Besides the Cr level HL1, two electron traps are predominant, i.e. EL6 and more especially EL2 which was shown recently [9] not to be related to oxygen, contrary to what has often been suggested in the literature. It now appears that the role of oxygen is to reduce Si contamination during growth [10]. The presence of EL6 and EL2 in the highest concentration in undoped bulk GaAs is confirmed by Hall measurements (see Table 1) in high resistivity undoped materials. Undoped materials sometimes show relatively high doping levels ($n \leqslant 5 \times 10^{15}$ cm^{-3}) at room temperature corresponding to the thermal exhaustion of one or more 'intermediate donor' levels [11, 12] lying at 0.14—0.18 eV below the conduction band. A similar level has also been observed, by time of flight measurements, to control the drift mobility in certain materials [13].

Two hole traps, HL10 and more especially HL9, have often been detected by TSC and OTCS. Since we do not yet know their sensitivity to optical refilling, their concentration cannot be determined precisely. They have been observed in undoped material and are not related to Cr, nor to vanadium [14], but their origin remains unknown.

Optical spectroscopy

Optical absorption measurement appears the most reliable technique for obtaining the concentration N_T of deep levels in insulating bulk crystals. The absorption coefficient α is given by the expression:

$$\alpha(h\nu) = \gamma N_T (\sigma_n^0 f + \sigma_p^0 (1-f)) \tag{3}$$

where $\sigma_n^0 (\sigma_p^0)$ is the photo-ionization cross-section for electrons (holes) and f, the Fermi function, gives the occupancy of the level. If there exists an energy $h\nu_0$ for which $\sigma_n^0 = \sigma_p^0$, then $\alpha(h\nu_0)$ is directly proportional to N_T, whatever the occupancy of the level.

Figure 3 shows the shape of σ_n^0 and σ_p^0 measured recently for the deep donor EL2 [15, 16] and for the deep Cr acceptor HL1 [17] by a new experimental technique. The relative amplitude of σ_n^0 and σ_p^0 has been deduced from optical DLTS measurements using 1.06 μm excitation. At this energy, σ_n^0/σ_p^0 was determined to be 3.3 for EL2 [18] and 0.185 for Cr [5]. It is clear that between 0.75 and 1.4 eV, all the transitions between each of the two levels and the two bands are possible, which renders the photo-Hall and photoconductivity measurements extremely difficult to analyse in terms of the relative presence of each level. As a matter of fact, the experimental determination of optical capture cross-sections and their theoretical interpretation is still to be done for most of the deep levels in GaAs and this should help to clarify many data, such as photoluminescence results [19, 20] or photo-induced EPR signals [21].

In Cr-doped crystals, it has been shown [22] that the optical absorption between 0.8 and 1.1 eV corresponds to transitions from the Cr deep level, and the absorption coefficient at 0.92 eV has been calibrated as a function of the chemical concentration of Cr. According to Figure 3, $\sigma_p^0 \simeq 2 \times \sigma_n^0$ at this energy, which means that such a calibration is rather more sensitive to Cr^{3+} than to Cr^{2+}, the two charge states corresponding to the deep Cr acceptor level [5].

In undoped crystals, the residual absorption seems to be mainly due to the presence of EL2, as is seen in Figure 4. This figure shows good agreement between σ_n^0(EL2) and the shape of $\alpha(h\nu)$ measured in many undoped crystals. When calibrated, α should give a reliable determination of [EL2] in these crystals.

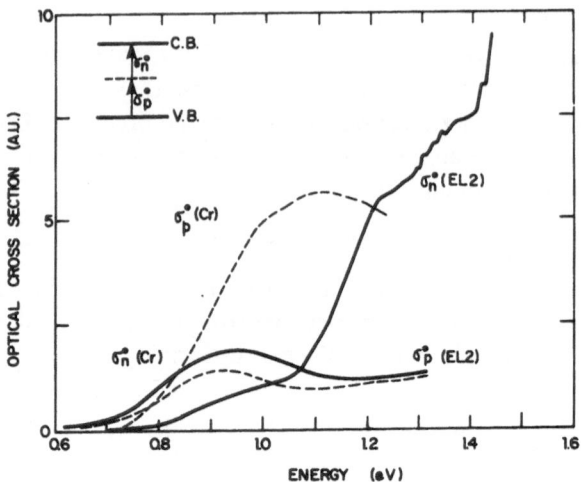

Figure 3 Variation of photo-ionization cross-sections as a function of photon energy for the Cr deep acceptor and the deep donor EL2 (after [15—17]). The relative intensity σ_n^0/σ_p^0, measured at 1.06 μm, has been determined from optical DLTS using Nd Yag laser pulses. The relative intensity of Cr and EL2 optical cross-sections should not be very different, but are not known precisely. Except for σ_n^0 (EL2) which was measured at 85 K, all other curves correspond to T=300 K.

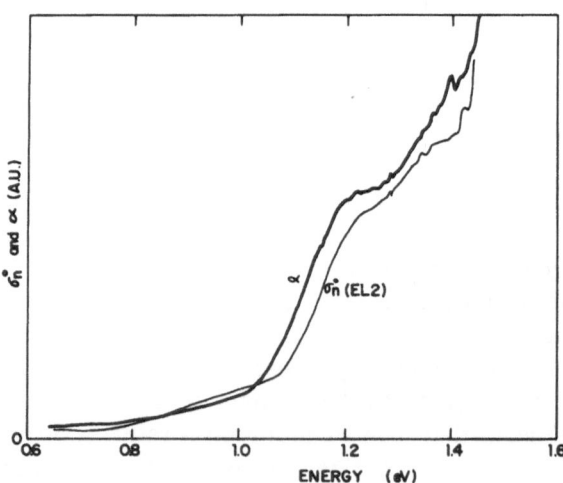

Figure 4 Comparison of the shape of the absorption coefficient recorded on bulk high resistivity undoped GaAs (5 mm thick sample) and the electron photo-ionization cross-section σ_n^0 of EL2 (T=85 K).

Fermi function of the main levels

The knowledge of this function is essential for analysing Hall data. It can be determined experimentally from the separate measurement of the emission rate, e, and of the capture cross-section σ of the level, since f can be expressed by:

$$f = \left(1 + \frac{e}{N_c \langle v \rangle \sigma} \exp\left(+\frac{E_c - E_F}{kT}\right)\right)^{-1} \tag{4}$$

This expression is easily deduced from the detailed balance equation. Thus, the measurements of e(T) and σ (T) lead to the determination of the total free energy of ionization:

$$E_T' = E_T(T) + kT \ln\frac{g_0}{g_1} = \Delta H(T) - T\left(\Delta S + k \ln\frac{g_0}{g_1}\right) \tag{5}$$

but do not permit separation of the contribution of lattice and electronic entropy ΔS and $k \ln(g_0/g_1)$, respectively.

EL2 is only detected as an electron trap ($e_n \gg e_p$) and it has been shown [23, 24] that

$$E_c - E_T'(EL2) = 0.759 - 2.37 \times 10^{-4} \times T \quad eV. \tag{6}$$

Furthermore, it has been definitively proved, from exo-diffusion studies of the level, that EL2 is a donor [25].

The behaviour of the Cr deep acceptor level is more complex, since it can be observed as a hole trap as well as an electron trap, according to initial refilling [5, 26]. Besides σ_n and σ_p, e_n and e_p can be determined in this case, which means that the free energy of ionization is known relative to the two bands. This is the only case of this type known in GaAs. The energy relative to the valence band has been obtained with better accuracy and is equal to [5]:

$$E_T' - E_v = 0.81 - \frac{3 \times 10^{-4} \times T^2}{T + 204} + 0.07 \quad kT \tag{7}$$

Furthermore, the same Cr level has been observed in GaAs grown by different methods and its concentration is very close to the chemical concentration of chromium [5]. This confirms the idea that this level corresponds to the most simple configuration, i.e. a Cr atom on a Ga site.

As far as the locations of the Cr level and the deep donor are concerned, different models were proposed [27—29]. The experimental data given by equations (6) and (7) show that the Cr level is located just above EL2 at high temperatures (T>150 K).

The existence of two deep levels, associated with three charge states of Cr, is controversial. It has been postulated to reconcile Hall and chemical analysis data [30] and is a challenge for those involved in deep level spectroscopy. Our measurements have been shown to be consistent with the presence of only one Cr deep level, in all the bulk or epitaxial Cr-doped GaAs studied, but do not exclude the existence of a second one close to the conduction band. Such a location of this second level would still be consistent with the data of Brozel *et al.* on n$^+$ material, but unfortunately, not with their results on the higher resistivity ingot. Could the results in this latter case be explained by the presence of parasitic excess shallow or deep (may be HL4 or HL9) acceptors?

Global assessment of semi-insulating GaAs

The purpose of such an assessment is to determine the residual concentration of shallow levels $N_D - N_A$ and the concentration of compensating deep levels. It will be shown that this aim can be achieved using both Hall and optical absorption measurements.

In insulating materials, the position of the Fermi level E_F depends only on the *relative* concentration of the different acceptor and donor levels. Thus, when E_F is known, the absolute concentration of the different levels can be obtained if one can determine the concentration of only one of them or, more precisely, of *the deep level* which fixes the Fermi level (all the shallower traps are fully ionized and cannot be distinguished in the measurement).

E_F can be obtained by solving the equation of charge neutrality:

$$n + N_A^i = p + N_D^i \tag{8}$$

which may be rewritten in the following form, when both the deep donor EL2 and the deep acceptor Cr are taken into account, and n and p neglected:

$$N_A + [Cr]\ f(Cr) = N_D + [EL2]\ (1 - f(EL2)) \tag{9}$$

where N_A and N_D are the concentrations of shallower levels (compared to EL2 and Cr) fully ionized at equilibrium. E_F is the solution of the equation and can be easily computed for each set of values of [EL2], [Cr] and $(N_D - N_A)$, since the Fermi functions for the two deep levels are known (equations (6) and (7)).

But the Fermi level position cannot be determined from Hall data in a straightforward way in the case of high resistivity GaAs in which mixed conductivity occurs due to the large difference between electron and hole mobilities and effective masses. Thus, the apparent Hall mobility μ_H and Hall constant R_H are given by:

$$qR_H = r_H \frac{p - nb^2}{(p + nb)^2} \tag{10}$$

$$\mu_H = r_H \mu_n \frac{n - p/b^2}{n + p/b} \tag{11}$$

where $b = \mu_N/\mu_P$ has been studied [31] as a function of temperature and ionized impurity concentration N^i. It has been shown that at 400 K, b=14 for $N^i = 3.10^{16}$ cm^{-3}, and 16 for $N^i < 10^{15}$ cm^{-3}. In the following, we will take b=15. The variation of $1/qR_H$, μ_H and the resistivity ρ calculated using the above mixed conduction model and a true electron mobility of 4000 cm^2 V^{-1} s^{-1}, is given in Figure 5 at 400 K. This temperature has been chosen for two reasons: it is high enough to minimize errors of measurements related to surface effects [31] and low enough to allow neglect of the concentrations of free electrons and holes relative to ionized deep levels.

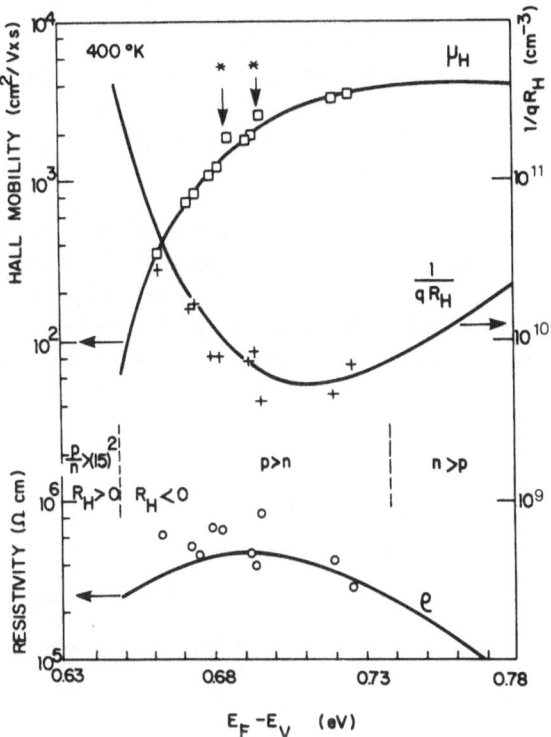

Figure 5   Theoretical curves (mixed conductivity model) giving the variation of resistivity and Hall parameters (apparent mobility μ_H and Hall constant R_H) as a function of the position of the Fermi level. Experimental points (from [32]), correspond to Cr-doped ingots and to ingots not doped with Cr (indicated by *).

Figure 5 demonstrates that of μ_H, R_H and ρ, μ_H is the parameter most sensitive to the Fermi level position, when E_F ranges between 0.65 and 0.72 eV above the valence band. Experimental data reported previously [32], have also been plotted on this figure showing that this is exactly the usual range of variation in most of the cases of Cr-doped material. Some results obtained on Bridgman materials not doped with Cr are also plotted, proving that these materials can be as resistive as those doped with Cr. It can also be noticed that the experimental data reported in this figure correspond to crystals in which p>n even if all appear n type ($R_H<0$). The value of μ_H has been drawn using a true electron concentration μ_{HO} equal to 4000 cm² V⁻¹s⁻¹. At 400 K, μ_{HO} is mainly phonon controlled and it has been measured [32] to vary only slightly as a function of ionized impurity concentration N^i.

Typically, μ_{HO} varies from 4800 when $N^i = 10^{16}$ cm^{-3} to 3400 cm^2 V^{-1}s^{-1} when $N^i = 2 \times 10^{17}$ cm^{-3}. Thus, μ_H (400 K) can be used practically to determine E_F (400 K). Conversely, knowing E_F and thus the true electron mobility, one can calculate μ_H using equation (11). More generally, starting from a given set of values of [Cr], [EL2] and ($N_D - N_A$), one can compute E_F using equation (9), and then μ_H via equation (11). Since the Fermi functions f are known, the only assumption, justified above, is that b=15.

The calculated apparent Hall mobility μ_H has been plotted [32] in Figure 6 as a function of the concentration of Cr, for various concentrations of ($N_D - N_A$) and a constant concentration [EL2]= 10^{16} cm^{-3}, which is the mean value generally observed in bulk GaAs:

1. If $N_D - N_A$ is kept constant, μ_H decreases when [Cr] increases, which corresponds to a shift of the Fermi level towards the valence band.
2. If [Cr] remains constant, μ_H decreases when $N_D - N_A$ also decreases, corresponding to a similar displacement of E_F towards the valence band. This means that the purest materials offer the lowest mobility at a given concentration of Cr!

Experimental data have been plotted in Figure 6.

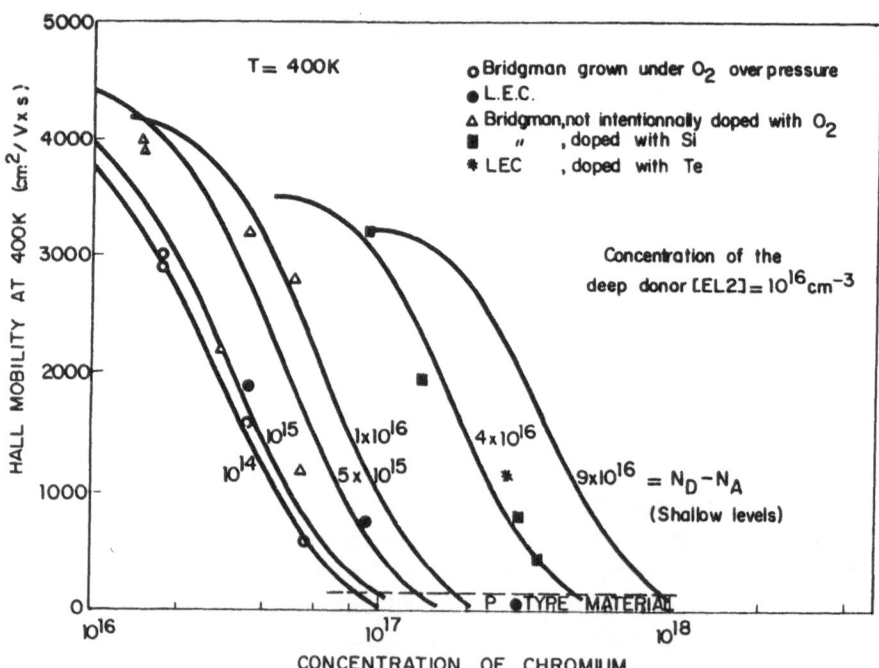

Figure 6   Theoretical (solid line) and experimental variation of the apparent Hall mobility μ_H at 400 K as a function of Cr concentration and $N_D - N_A$ (shallow levels).

The corresponding crystals have been analysed by different techniques in order to check the values deduced from reference curves. In the case of high values of N_D—N_A, good agreement has been observed with results of chemical analysis or intentional co-doping [32], as well as with EPR measurements [33]. In fact, only the EPR signals attributed to Cr^{2+} and Cr^{3+} have been observed in the corresponding crystals, while the signal assigned to Cr^{1+} was never observed. This seems to confirm our analysis of these materials based on only one deep Cr level, the second one being unoccupied. On the other hand, low values of N_D—N_A on other crystals are confirmed by assessment of materials grown under the same conditions, but not doped with Cr. In such conditions, N_D—N_A has been observed to lie below 3×10^{15} cm^{-3}.

In these crystals, EL2 remains the main donor to be compensated, with a concentration around 10^{16} cm^{-3} in good agreement with the above spectroscopic measurements. The complete analysis of data seems to show that N_D—N_A varies mainly as N_D and that N_A, the total concentration of acceptors shallower than Cr, is lower than about 5×10^{15} cm^{-3} [32].

Figure 7 corresponds to highly resistive GaAs not doped with Cr, which shows a large activation energy of the Hall constant, near 0.75 eV. It gives the variation of μ_H at 400 K as a function of [EL2] for various concentrations of N_A—N_D, calculated in the same way as for Figure 6, using the mixed conductivity model. It is well established that N_A is larger than N_D and is compensated by the deep donor EL2 in these materials. The determination of EL2 from optical absorption should allow deduction of N_A—N_D using these reference curves.

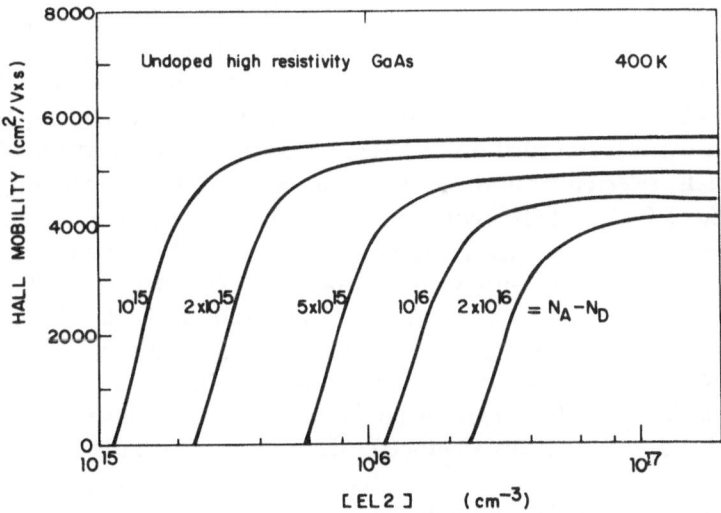

Figure 7 Theoretical variation of the apparent Hall mobility at 400 K as a function of the concentration of the deep donor EL2, valid for material which shows an activation energy of R_H close to 0.75 eV.

Figure 6 is based on the presence of both the deep donor EL2 and the Cr deep acceptor as the main deep traps, which corresponds to most practical cases. Obviously, it is not impossible that another deep trap (for instance, the level HL10)) near the middle of the gap is also predominant in some materials, or that the deep donor EL2 is negligible in certain crystals, which would mean that Figure 6 or Figure 7 is not valid to analyse these particular cases. This still emphasizes that the origin of EL2 remains to be determined in order to be able to master its concentration in insulating GaAs, and more especially in the materials not doped with Cr.

Quality requirements for semi-insulating substrates

As far as the technology of GaAs high frequency devices is concerned, semi-insulating material is used either as a substrate for the growth of the epitaxial active layer or as a material in which the doped layer is made by ion implantation.

According to the literature [34, 35], shallow acceptors (very probably Mn) [36, 37] accumulate near the surface of the substrate when heat-treated under conditions similar to those at the start of VPE growth. This is responsible for parasitic p-type thermal conversion in many cases. It arises when the concentration of these shallow acceptors N_A becomes larger than the total concentration of all shallow and deep donors (including EL2). This means that this conversion is made more difficult in materials with large concentrations of EL2 or N_D, such as those reported in Figure 6 and doped with Si or Te.

On the other hand, the presence of shallow donors and of the deep Cr acceptor in too large concentrations has been shown to affect the electrical properties of active layers prepared by direct Se implantation in insulating substrates [38]. The main data are summarized in Figure 8, giving the profile of free carriers recorded on different Cr-doped or undoped semi-insulating materials, selected from those shown in Figures 5 and 6, after implantation with Se (280 kV, 3.5×10^{12} cm^{-2}) and annealing (870°C, 15 min under a silicon nitride cap). Figure 8a corresponds to materials doped with low Cr concentrations, and with various values of $N_D - N_A$. A similar gaussian profile shape is observed in Cr-doped materials with $N_D - N_A$ lower than 10^{16} cm^{-3} as well as in undoped materials. A tail manifests itself as soon as $N_D - N_A$ becomes larger than 10^{16} cm^{-3} which seems to correspond to excess electrons compared to the results obtained on the previous materials. On the other hand, using heavily Cr-doped substrates, in which the concentration of Cr lies around 2.5×10^{17} cm^{-3}, equal to the value of the peak free carrier concentration, quite different results can be obtained, as seen in Figure 8b. If $N_D - N_A$ is large, excess free carriers are observed over the whole investigated thickness, with respect to the reference profile corresponding to undoped material. In contrast, if $N_D - N_A$ is low, lower than 3×10^{15} cm^{-3} in the case of

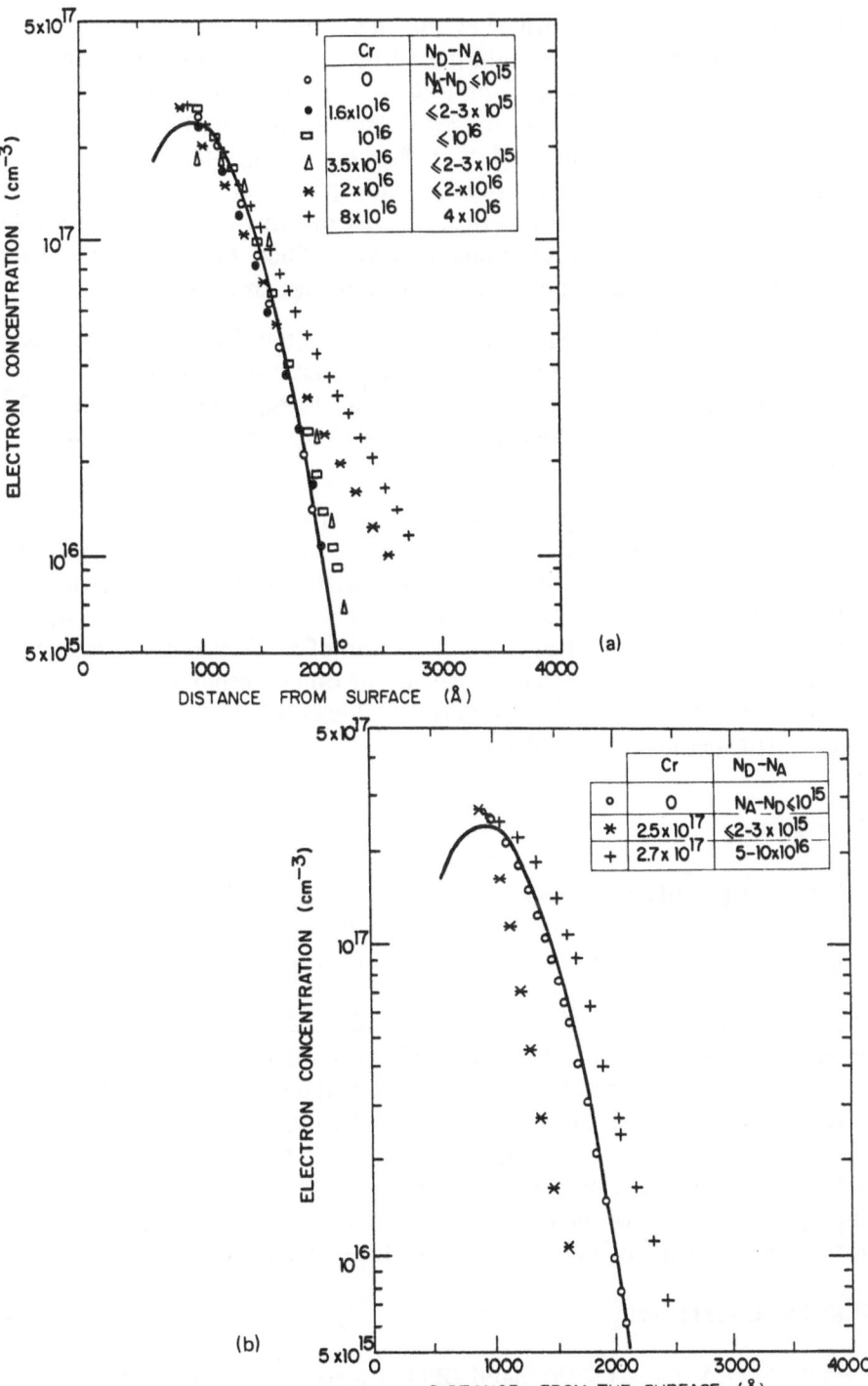

Figure 8 Profiles of the free carrier concentration obtained after 3.5×10^{12} cm^{-2}, 280 kV Se implantation in different insulating substrates and annealing at 870°C under a silicon nitride cap. The solid line corresponds to the LSS profile for a dose of 2.5×10^{12} cm^{-2}.

Figure 8b, the doping profile lies distinctly below the reference. These different observations can be interpreted by a fast, but not complete, exo-diffusion of Cr under capped annealing. This exo-diffusion, which has already been shown to occur by SIMS measurements [39], would leave uncompensated the residual level of shallow donors N_D-N_A, assumed not to diffuse at the same speed. This process can explain many results reported in the literature [40, 41] concerning the influence of insulating substrates on the electrical properties of GaAs implanted layers. Thus, the values of at least two parameters are important to control in starting substrates:

1. The concentration of N_D (shallow donors) minus N_A (all the acceptors except Cr), which is in practical cases very similar to the value of N_D-N_A (including the relatively low concentration of deep donors), given in the insert of Figure 8 and deduced from global characterization of Figure 6.
2. The concentration of chromium.

These two parameters induce opposite effects since they are respectively responsible for an increase and a decrease of free carrier profile. If the importance of each effect could be evaluated, for instance by knowing the concentration of Cr, low dose Se implantation annealed under encapsulation would appear as a method of characterization of semi-insulating substrates. But more work is still needed to clarify all the other mechanisms, such as in- or out-diffusion of impurities and defects, which probably also influence the electrical properties of implanted layers.

Conclusion

A survey of the different spectroscopic methods used to study deep levels in insulating GaAs has been given. Some of them are particularly fitted to the characterization of electrical properties of traps, others to the measurement of their concentrations. Many levels have been detected in GaAs and, although the deep Cr acceptor and the deep donor EL2 have been observed to be the predominant ones and thus extensively studied, a lot of work remains to be done in order to determine their origin and even the acceptor or donor nature for other levels. From the analysis of compensation mechanisms in different crystals and of ion implantation results on these same materials, some correlations have been established between the electrical properties of implanted layers and of their substrates; this helps to clarify the problems of quality requirements for substrates needed for high frequency devices.

ACKNOWLEDGEMENTS

I would like to thank G. Poiblaud (RTC Caen) for supplying most of the studied crystals, and also R. Ware (Metals Research), E.M. Swiggard (NRL), and W. Ford (Hewlett Packard) for the gift of some samples. This work has been pursued in close collaboration with many of my colleagues, A. Mircea,

A. Mitonneau, J.P. Hallais, G. Jacob, J.P. Farges, A. Mircea-Roussel, S. Makram-Ebeid, M. Berth and C. Venger, to whom I am particularly indebted. It is also a pleasure to thank R. Duchesne for his skilful technical assistance.

References

1. Buehler, M.G. (1972). *Solid St. Electron.*, **15**, 69
2. Mitonneau, A., Martin, G.M. and Mircea, A. (1977). *Inst. Phys. Conf. Ser.*, **33a**, 73
3. Martin, G.M. and Bois D. (1978). *J. Electrochem. Soc.*, **78**, 32
4. Hurtes, C., Boulou, M., Mitonneau, A. and Bois D. (1978). *Appl. Phys. Lett.*, **32**, 821
5. Martin, G.M., Mitonneau, A., Pons, D., Mircea, A. and Woodard, D.W. (1980). To be published in *J. Phys. C: Solid St. Phys.*
6. Martin, G.M., Mitonneau, A. and Mircea, A. (1977). *Electron. Lett.*, **13**, 191
7. Mitonneau, A., Martin, G.M. and Mircea, A. (1977). *Electron. Lett.*, **13**, 666
8. Mircea, A. and Bois, D. (1978). *Inst. Phys. Conf: Ser.*, **43**, 210
9. Huber, A.M., Linh, N.T., Debrun, J.C., Valladon, M., Martin, G.M., Mitonneau, A. and Mircea, A (1979). *J. Appl. Phys.*, **50**, 4022
10. Woods, J.F. and Ainslie, N.G. (1963). *J. Appl. Phys.*, **34**, 1469
11. Basinski, J. (1966). *Can J. Phys.*, **44**, 941
12. Blood, P. and Farges, J.P. Private communication
13. Robert, J.L., Pistoulet, B., Raymond, A., Dusseau, J.M. and Martin, G.M. (1979). *J. Appl. Phys.*, **50**, 349
14. Mircea-Roussel, A., Martin, G.M. and Lowther, J.E. Submitted to *Solid St. Commun.*
15. Chantre, A. (1979). Thesis, Grenoble
16. Bois, D. and Chantre, A. To be published in *Rev. Phys. Appl.*
17. Nouailhat, A. Private communication
18. Mitonneau, A. Private communication
19. Koschel, W.H., Bishop, S.G. and McCombe, B.D. (1977). *Solid St. Commun.*, **19**, 521
20. Lightowlers, E.C. and Penchina, C.M. (1978). *J. Phys. C: Solid St. Phys.*, **11**, L405
21. White, A.M., Krebs, J.J. and Stauss, G.H. (1980). *J. Appl. Phys.*, **51**, 419
22. Martin, G.M., Verheijke, M.L., Jansen, J.A.J. and Poiblaud, G. (1979) *J. Appl. Phys.*, **50**, 467
23. Mircea, A., Mitonneau, A. and Vanimenus, J. (1977). *J. Phys. Lett. (France)*, **38**, L41
24. Mitonneau, A., Mircea, A., Martin, G.M. and Pons, D. (1979). *Rev. Phys. Appl.*, **14**, 853
25. Mircea, A., Mitonneau, A., Hollan, L. and Briere, A. (1976). *Appl. Phys.*, **11**, 153
26. Lang, D.V. and Logan, R.A. (1975). *J. Electron. Mat.*, **4**, 1053
27. Lindquist, P.F. (1977). *J. Appl. Phys.*, **48**, 1262
28. Zucca, R. (1977). *J. Appl. Phys.*, **48**, 1987
29. Look, D.C. (1977). *J. Appl. Phys.*, **48**, 5141
30. Brozel, M., Butler, J., Newman, R., Ritson, A., Stirland, D. and Whitehead, C. (1978). *J. Phys. C: Solid St. Phys.*, **11**, 1857
31. Visentin, N. (1979). Thesis, Toulouse
32. Martin, G.M., Farges, J.P., Jacob, G., Hallais, J.P. and Poiblaud, G. (1980). *J. Appl. Phys.*, in press
33. Goltzene, A., Schwab, C. and Martin, G.M. (1980). This volume
34. Hallais, J., Mircea-Roussel, A., Farges, J.P. and Poiblaud, G. (1977). *Inst. Phys. Conf. Ser.*, **33b**, 220
35. Zucca, R. (1977). *Inst. Phys. Conf. Ser.*, **33b**, 228
36. Nordquist, P., Klein, P.B. (1979). *EMC:* June, Boulder
37. Clegg, J.B. Private communication
38. Martin, G.M., Berth, M. and Venger, C. (1980). *Electron. Lett.*, in press

39. Huber, A.M., Morillot, G., Linh, W.T., Favennec, P.M., Deveaud, B. and Toulouse, B. (1979). *Appl. Phys. Lett.*, **34**, 858
40. Stolte, C.A. (1975). *Int. Electron Devices Meeting*, Tech. Digest, p. 585. New York; IEEE
41. Swiggard, E.W., Lee, S.H. and Von Batchelder, F.W. (1979). *Inst. Phys. Conf. Ser.*, **45**, 125
42. Partin, L., Chen, J.W., Milnes, A.G. and Vassamillet L.F. (1979). *J. Appl. Phys.*, **50**, 6845

MODELLING OF A MULTI-VALENT IMPURITY, SUCH AS GaAs:Cr

J.S. BLAKEMORE

Oregon Graduate Center, Beaverton, Oregon 97006, USA

Abstract

Chromium may participate as a deep lying impurity in GaAs in several ways: substitutional as Cr_{Ga} or Cr_{As}, or as a member of a complex. The usual Cr impurity studied in most work is believed to be Cr_{Ga}. Unusual conditions are pointed out that might create forms such as Cr_{As}. The remainder of the paper is specific to Cr_{Ga}, for which the lattice-neutral condition requires three electrons lost to bonding, Cr^{III}. The acceptor condition Cr^{II} is well-known, and there has been considerable evidence for the doubly-ionized acceptor Cr^{I}. Less evidence exists for Cr_{Ga} as an amphoteric impurity with a charge state Cr^{IV} (one trapped hole). However, that state is also taken into account in writing a set of coupled equations for transition processes among the charge conditions Cr^{I} through Cr^{IV}. Complicated splittings of the bound states are discussed, as arise from the anisotropic crystal field, spin-orbit coupling and Jahn-Teller distortions. Energy changes of the system (impurity + host) when an electron or hole is emitted or captured are contrasted with enthalpies of activation. An example from $Cr^{III} \rightleftharpoons Cr^{II}$ capture/emission data shows that internal consistency is possible in describing communication of Cr with the conduction and valence bands.

1 Introduction

Chromium may well be electrically and/or optically active as an impurity in the GaAs lattice in several ways, depending on its specific lattice location and on its neighbours. The paper by White [1] notes that Cr-related impurity complexes could account for some of the apparent anomalies and contradictions in phenomena seen with chromium doping. Optical transitions of one such type of complex might account for the well-known 'zero-phonon' luminescence line at 839 meV, a line which shows complicated and temperature-dependent fine structure [2], characteristic of levels split by amounts of ~ 1—3 meV. It would not be unreasonable to suppose that chromium could be stable in locations providing it an association with a vacancy, or with a foreign atom such as oxygen, nitrogen or carbon.

Vacancies are evidently important in affecting the motional characteristics of chromium at high temperatures. Diffusion studies [3] have shown that Cr diffuses rapidly as an interstitial. An encounter with a gallium vacancy:

$Cr_i + V_{Ga} \rightarrow Cr_{Ga}$, then immobilizes the chromium on a gallium site, with a nominally tetrahedral arrangement of arsenic ligands. That substitutional location on a gallium site, Cr_{Ga}, is considered to constitute the 'ordinary' chromium acceptor, and it is this site symmetry that will be presumed in most of what follows.

Bearing in mind these other possibilities for Cr participation as an electrically active defect, the possible states of Cr_{Ga} are now considered.

2   Charge states of substitutional chromium, Cr_{Ga}

Each gallium atom is required to contribute three outer electrons (the free atom states $4s^2 4p^1$) to bonding in GaAs. Thus Cr_{Ga} should be electrically neutral in the lattice with a charge state (oxidation state) of Cr^{3+}. As compared with the free atom configuration $3d^5 4s^1$ for chromium, the Cr^{3+} ion is apparently stripped down to $3d^3$, and that 'neutral' state of the Cr_{Ga} impurity is referred to simply as $3d^3$ in some of the literature. As with every other charge state of the centre, there are numerous possible total energies, with large crystal field splittings, and further splittings engendered by any displacement from the cubic (tetrahedral) symmetry of the perfect lattice. Clark [4] has recently discussed the terms of lowest energy to be expected for the $3d^3$, $3d^4$ and $3d^5$ configurations of Cr_{Ga}.

Clark [4] also commented on the opportunities that exist for confusion among terminologies as to what is a 'positive' or a 'negative' state of an impurity — and with respect to what? Thus, if Cr^{3+} is a neutral state of Cr_{Ga} then Cr^{2+} is a *negative* state; that is, one that has acquired an electron. Clark attempted to side-step these confusions by expressing the oxidation state of the 'neutral acceptor' Cr^{3+} condition in Roman numeral form, Cr^{III}. That terminology is used here.

Cronin and Haisty [5] suggested that GaAs:Cr could become semi-insulating most likely because Cr was providing a deep-level acceptor: $Cr^{III} + e^- \rightarrow Cr^{II}$. Many other studies have confirmed that hypothesis. The Cr^{III}:Cr^{II} balance in semi-insulating GaAs:Cr has been inferred from a diversity of experiments, including ambipolar conduction [6—8], photoelectronic response [9—11], capacitance spectroscopy [12—14] and low temperature ESR [15—18]. Martin *et al.* [14] make an interesting observation that no evidence for any *other* Cr-related deep lying state has been detected in the course of their extensive optical and capacitance studies of Cr-doped GaAs.

However, there has been evidence for other charge states of Cr_{Ga} from other kinds of experiment. An ESR response of the form to be expected for the 'doubly ionized acceptor' charge state Cr^I has been observed [16, 19, 20] at low temperatures and with appropriate optical pumping. Coulomb-repulsive electron capture in the reaction $Cr^{II} + e^- \rightarrow Cr^I$ probably accounts for a very slow photoconductive decay seen at liquid helium temperatures by Eaves and

Williams [21]. (A decay time constant of ~1 h, consistent with an electron capture cross-section of only $\sigma_n \sim 10^{-26}$ cm^2.)

How likely is it that an ionized donor condition Cr^{IV} would be stable in GaAs? The example of Au in Ge [22] comes to mind here, since this was the first multi-valent acceptor to be shown to be amphoteric as well as having a donor (hole trap) condition [23].

Diagrams of *electron* energy have universally been used for semiconductor discussions ever since the quantum theory of solids was founded. On a *single particle* electron energy diagram, a donor state (if it occurs) appears closer to the valence band than to the acceptor levels. Thus, a Cr^{IV} charge state of Cr$_{Ga}$ would be significant in the dark only for doping more p-type than is conducive to the presence of Cr^I and Cr^{II} in appreciable concentrations.

Since amphoteric impurity action was first shown with Ge:Au [22, 23], a number of other amphoteric systems have been identified, as reviewed, for example, by Milnes [24]. In any such case, the donor states become important for equilibrium when the electron Fermi energy is lower (closer to the valence band) than for acceptor conditions of the centre. However, when the various charge states differ essentially by the number of occupied 3d orbitals, the 'rules' concerning where successive 'levels' of an impurity centre should be shown on a single-particle electron energy diagram have to be interpreted with caution. Section 4 returns to this subject.

The spin-restricted cluster calculations of Hemstreet and Dimmock [25] indicate a ground state for Cr^{IV} within the intrinsic gap of GaAs. That by itself is not conclusive proof that Cr^{IV} corresponds to a localized charge state, since the authors point out several respects in which the calculation fails to account for seemingly well-known features of the ground states for Cr^{III} and Cr^{II}. However, evidence for the donor state Cr^{IV} has recently been reported from ESR experiments. Kaufmann and Schneider [26] find an ESR line for *non-illuminated* p-type Cr-doped GaP and GaAs with characteristics (isotropic g-factor, etc.) consistent with Cr^{IV} in those two semiconductor hosts. Observation of this line in p-type GaAs:Cr, and without illumination, gives us encouragement to model Cr$_{Ga}$ as an amphoteric centre, with the four states of charge shown in Table 1.

Table 1 **Possible charge states for chromium in GaAs, substitutional on a gallium site**

3d sub-shell configuration	Charge state, relative to free atom	Oxidation state	Status: acceptor, neutral, or donor
3d^5	Cr^{1+}	Cr^I	A^{2-}
3d^4	Cr^{2+}	Cr^{II}	A$^-$
3d^3	Cr^{3+}	Cr^{III}	N^0
3d^2	Cr^{4+}	Cr^{IV}	D$^+$

It has already been remarked that each state of Cr_{Ga} can correspond to many total system energies: a 'ground state', which itself may have fine and ultra-fine splittings and many states excited by larger amounts of energy. The states (orbitals) of a free Cr atom are complex enough in themselves, arising in part from spin-orbit interactions. Within the GaAs lattice, there are splittings of ~ 0.5—1 eV arising from the crystalline field. To a first approximation, these arise from the effects of the four As nearest neighbours; thus the site symmetry is tetrahedral in an unperturbed lattice.

However, orbital degeneracies of the Cr^{III} and Cr^{II} ground states with tetrahedral bonding render both of these Cr_{Ga} charge states liable to Jahn-Teller distortion. Low temperature ESR has provided rather convincing evidence of J-T distortions that are static for low enough temperatures. A distortion of orthorhombic symmetry has been reported [15] for Cr^{III}, with the $^4T_1$ state of spin S=3/2 lowest. A distortion of tetragonal symmetry is found by similar methods [16] for Cr^{II}, with the S=2 orbital singlet $^5B_2$ energetically the lowest member of the $^5T_2$ manifold.

Yu [27] has proposed a model to account for the well-known Cr-related GaAs 0.57 and 0.84 eV luminescence lines, parametrized by the crystal field and Jahn-Teller splitting coefficients for Cr^{II}. That model would require a 0.27 eV J-T split of the $Cr^{II}(^5T_2)$ manifold. Similarly, Krebs and Stauss [28] found evidence from stress effects on ESR of Cr^{II} that both the ^5E and $^5T_2$ states have splittings which exceeds 0.2 eV.

In a very different energy range, a ground state splitting of the Cr^{II} ground state of just 0.9 meV has been deduced by Wagner and White [29] from high resolution low temperature ESR using a far-IR laser source. Ballistic phonon transport experiments carried out in GaAs:Cr below 2 K have indicated [30] a tetragonal J-T distortion for Cr^{II} with a spin-orbit splitting of the S=2 ground state by some 1.2 meV.

It is therefore evident that each charge state of Cr_{Ga} has numerous energy options, with both large and small energy separations. Excited states several hundreds of meV above the ground state(s) can have striking consequences for optical experiments, in both absorption ($\uparrow$) and luminescence ($\downarrow$); but these highly excited states are occupied to a negligible degree under purely thermal conditions.

Very fine splittings of ground states, seen in high resolution types of low temperature experiment such as ESR [29], phonon transport [30] and luminescence [2], cease to be distinguishable for higher temperatures. Indeed, the static J-T distortions become unstable on warming, with disappearance of the characteristic ESR features. Tokumoto and Ishaguro [31] showed from ultra-sonic attenuation that the J-T distortion of Cr^{II} is subject to rapid reorientation on warming, with a relaxation time $\tau \sim 0.2\, T^{-7}$ s. Even for finely split states which maintain separate existences on warming, a multiplet with a total width $<kT$ will behave (from the standpoint of electron thermal statistics) as a single level of statistical weight (degeneracy factor) $g = \Sigma_i g_i$. Resolution of conflicts among the various experimental and theoretical results for Cr_{Ga} will have to proceed further before realistic

numbers can be proposed for g_1 through g_4 of the charge states Cr^I through Cr^{IV}.

Suppose a charge condition of Cr_{Ga} has some excited states a few tens of meV above its ground state(s). These would have a substantial — and temperature-dependent — effect on the apparent 'statistical weight' g of that charge condition. As it happens, there has been no spectroscopic evidence for excited states in that particularly troublesome energy range for any of Cr^I through Cr^{III}, and we should be grateful for the consequent simplification if they are, indeed, absent. The presence of such states would exacerbate all the other difficulties of interpreting thermal carrier data as a function of temperature, added to those arising from temperature dependence of ground state energy with respect to band energies, interference from other deep level species (such as oxygen), etc.

3 Transitions among charge states of Cr_{Ga}

Suppose GaAs to have a total concentration N_{Cr} of substitutional chromium, with concentrations N^I to N^{IV} of the four postulated localized charge conditions Cr^I to Cr^{IV}; then

$$N_{Cr} = (N^I + N^{II} + N^{III} + N^{IV}).$$

The total number of *electrons* on these states, over and above those required for all centres to be in the neutral N^{III} condition, is $(2N^I + N^{II} - N^{IV})$. Charge neutrality can thus be expressed as

$$n - p = N_{di} - N_{ai} - (2N^I + N^{II} - N^{IV}) \qquad (1)$$

supposing the semiconductor to contain N_d donors of various types of which N_{di} are ionized, and N_a assorted acceptors of which N_{ai} are ionized.

The individual populations N^I through N^{IV} depend on temperature, on externally applied illumination (if any), and on the supplies of electrons and holes to facilitate transitions among the various possible charge states. The four charge states, and the six sets of transition processes among them, are symbolized in Figure 1; thus R_{n1} is the net rate at which N^{II} is growing at the expense of N^I through an excess of conduction electron emission over conduction electron capture. Table 2 shows how each of these six sets of processes (three involving electron emission or capture, and three hole emission or capture) concerns an energy of transformation.

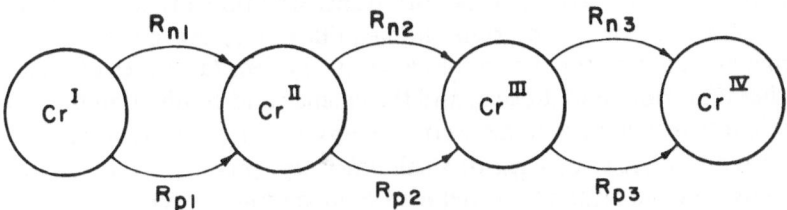

Figure 1   The four supposed localized charge states of Cr_{Ga} in GaAs, and the six sets of rate processes (three of electron capture/emission, and three of hole capture/emission) involved in the time derivatives (if non-zero) of the four state populations.

Table 2 **Transition processes among the four chromium states of charge shown in Figure 1**

Transitions involving electron emission or capture		Transitions involving hole emission or capture	
R_{n1}:	$Cr^I + \epsilon_{n1} \rightleftharpoons Cr^{II} + e^-$	R_{p1}:	$Cr^{II} + \epsilon_{p1} \rightleftharpoons Cr^I + h^+$
R_{n2}:	$Cr^{II} + \epsilon_{n2} \rightleftharpoons Cr^{III} + e^-$	R_{p2}:	$Cr^{III} + \epsilon_{p2} \rightleftharpoons Cr^{II} + h^+$
R_{n3}:	$Cr^{III} + \epsilon_{n3} \rightleftharpoons Cr^{IV} + e^-$	R_{p3}:	$Cr^{IV} + \epsilon_{p3} \rightleftharpoons Cr^{III} + h^+$

The directions of the arrows in Figure 1 are obviously arbitrary. These have been drawn, and expressions for the various rates written, so that the various R_n and R_p quantities are positive when their effects are to make a Cr centre gain a charge $+e$.

Let N^I through N^{IV} be denoted by N_0^I through N_0^{IV} for thermal equilibrium itself. Then a quantity such as (dN_0^{II}/dt) must be zero, apart from the instantaneous fluctuations of generation-recombination noise. When the conditions depart from equilibrium, it is still true that

$$dN_{Cr}/dt = d(N^I + N^{II} + N^{III} + N^{IV})/dt = 0. \tag{2}$$

However, the individual components of that derivative are, in general, non-zero:

$$\left.\begin{array}{l} (dN^I/dt) = - (R_{n1} + R_{p1}) \\[2mm] (dN^{II}/dt) = (R_{n1} + R_{p1}) - (R_{n2} + R_{p2}) \\[2mm] (dN^{III}/dt) = (R_{n2} + R_{p2}) - (R_{n3} + R_{p3}) \\[2mm] (dN^{IV}/dt) = (R_{n3} + R_{p3}) \end{array}\right\} \tag{3}$$

to an extent that depends on the extent to which equilibrium has been and is being perturbed, and on the *efficiency* of the various transformation processes.

The occupancies N_0^I to N_0^{IV} of thermal equilibrium itself are independent of the nature of, and efficiency of, the transformation processes. (However, that is not so for steady-state non-equilibrium.) At equilibrium itself, occupancy numbers for the entire electronic system (all states of Cr_{Ga} itself, all other impurities and defects, and the valence and conduction bands) must all be consistent with a single Fermi energy ϵ_F. Moreover, at equilibrium, each process, whether of large or small efficiency, proceeds at the same rate as its inverse. (That principle of detailed balance thus provides an invaluable link between capture and emission coefficients.) Since thermal equilibrium is a *dynamic* balance of generation-recombination, rate effects do manifest themselves instantaneously as g-r noise but this is not usually the simplest way to measure emission and capture coefficients.

When a semiconductor is driven away from equilibrium, either transiently or in steady state, the *most efficient* capture and emission processes are apt to dominate the redistribution among populations N^I through N^{IV} (and, of course, similarly for other defect species). An important exception is that sufficiently strong illumination can make the optical transition rates dominant even though non-radiative processes may be of higher inherent efficiency. Thus Kukimoto, Henry and Merritt [32] were able to discuss the kinetics of electron distributions among the levels of the divalent oxygen donor in GaP exclusively in terms of the optical excitation rates.

Each of the six rates R_{n1} to R_{p3} indicated in Figure 1, and for which the appropriate transformation energy is indicated in Table 2, can be written as a balance between carrier emission and carrier capture terms. Thus one may write R_{n1} as

$$R_{n1} = (e_{n1} + e_{n1}^0)N^I - nc_{n1} N^{II} \qquad (4)$$

Here, $c_{n1} = (V_n \sigma_{n1})$ is the capture coefficient for any Cr_{Ga} in condition Cr^{II} to capture any of the 'n' conduction band electrons; for an RMS thermal electron speed v_n, the capture efficiency can be equivalently quoted as the capture cross-section σ_{n1}. The quantity e_{n1}^0 indicates the probability (in units of s^{-1}) that external illumination will photo-ionize any centre in the condition Cr^I. Thus one may think of this as a product $e_{n1}^0 = \sigma_{n1}^0 F_{n1}$, where σ_{n1}^0 is the photo-ionization cross-section, and F_{n1} the flux of externally applied photons of suitable energy. (In many cases, the effective value for e_{n1}^0 and its five fellow coefficients has to be obtained from an integration over the applicable spectral range of light being used.)

The term involving e_{n1}^0 dealt only with photo-ionization response to *external* light. Any photo-ionization resulting from a radiative component of $Cr^I + \epsilon_{n1} \rightarrow Cr^{II} + e^-$ must appear as a portion of the quantity e_{n1}, the emission coefficient arising from all physical mechanisms that are natural consequences of the lattice temperature (phonon spectrum, black body photon spectrum, electronic velocity distribution, etc.). Thus e_{n1} and c_{n1} must lead to equal and opposite transition rates when the situation is of thermal equilibrium. Detailed balance thus requires that

$$e_{n1} = n_1 (v_n \sigma_{n1}) = n_1 c_{n1} \qquad (5)$$

where

$$n_1 = (N_c g_2 / g_1) \exp(-\epsilon_{n1}/kT). \qquad (6)$$

Energy ϵ_{n1} is the change in free energy of the entire system (impurity plus lattice), as thermalized for temperature T in the initial and final states, between having a chromium acceptor with two trapped electrons (Cr^I), and having a mobile electron plus a Cr^{II} with only one trapped electron. The minimum energy necessary to induce an optical transition may be substantially larger if either charge state of the impurity is prone to appreciable lattice relaxation — as appears most likely with Cr_{Ga}.

The 'apparent activation energy' of n_1 can also be quite different from ϵ_{n1} itself. Thus, suppose ϵ_{n1} has a temperature dependence that varies from quadratic at low temperatures to linear at high temperatures (in the manner suggested for intrinsic gaps by Varshni [33]):

$$\epsilon_{n1} = [\epsilon_{n10} - aT^2/(T+\beta)]. \tag{7}$$

Then data for n_1 over a moderate range of temperature can be fitted to the form

$$n_1 = [(N_c g_1/g_2) \exp(\Delta S_{n1}/k)] \exp(-\Delta H_{n1}/kT) \tag{8}$$

where the apparent activation energy is the enthalpy

$$\Delta H_{n1} = [\epsilon_{n10} + a\beta T^2/(T+\beta)^2]. \tag{9}$$

The pre-exponential part of n_1 is then modified by an entropy factor $\exp(\Delta S_{n1}/k)$, where

$$\Delta S_{n1} = aT(T+2\beta)/(T+\beta)^2. \tag{10}$$

Thus information about quantities such as n_1, derived from such techniques as combinations of measured e_n and c_n, is helpful about the location of an impurity state within the intrinsic gap only when intepreted carefully. Information of Mitonneau et al. [13] and Martin et al. [14] on the $Cr^{II} \rightleftharpoons Cr^{III}$ transitions, does have that necessary degree of completeness, which is discussed in detail elsewhere in this volume [34].

With terminology for the six sets of processes analogous to that introduced in connection with equations (4) to (6) for the R_{n1} process, one may write the set of all processes involving emission or capture of a conduction band electron as:

$$\left.\begin{array}{l} R_{n1} = (e_{n1} + e_{n1}^0)N^I - nc_{n1}N^{II} = (n_1 c_{n1} + e_{n1}^0)N^I - nc_{n1}N^{II} \\[2mm] R_{n2} = (e_{n2} + e_{n2}^0)N^{II} - nc_{n2}N^{III} = (n_2 c_{n2} + e_{n2}^0)N^{II} - nc_{n2}N^{III} \\[2mm] R_{n3} = (e_{n3} + e_{n3}^0)N^{III} - nc_{n3}N^{IV} = (n_3 c_{n3} + e_{n3}^0)N^{III} - nc_{n3}N^{IV} \end{array}\right\} \tag{11}$$

The corresponding set of equations for processes involving emission or capture of a valence band hole are:

$$\left.\begin{array}{l} R_{p1} = pc_{p1}N^I - (e_{p1} + e_{p1}^0)N^{II} = pc_{p1}N^I - (p_1 c_{p1} + e_{p1}^0)N^{II} \\[2mm] R_{p2} = pc_{p2}N^{II} - (e_{p2} + e_{p2}^0)N^{III} = pc_{p2}N^{II} - (p_2 c_{p2} + e_{p2}^0)N^{III} \\[2mm] R_{p3} = pc_{p3}N^{III} - (e_{p3} + e_{p3}^0)N^{IV} = pc_{p3}N^{III} - (p_3 c_{p3} + e_{p3}^0)N^{IV} \end{array}\right\} \tag{12}$$

Any equilibrium, static non-equilibrium, or dynamic situation is nominally capable of being modelled via equations (1), (3), (11) and (12). Even without light, this would have to be parametrized by the six quantities n_1 through n_3 and p_1 through p_3 (which say what the bound state energies are), and by the six capture coefficients c_{n1} through c_{n3} and c_{p1} through c_{p3}. The addition of external light, with the significance of the rate terms controlled by e_{n1}^0 through e_{n3}^0 and e_{p1}^0 through e_{p3}^0, makes a solution subject to 18 quantities (of which the last-mentioned six will vary with time if the illumination intensity or spectral form varies). Useful modelling can be attempted only under conditions whereby *most* of those 18 quantities can be rendered unimportant. Space does not permit the exposition of any modelled solutions here, apart from brief comments in Section 4 on possible energies of the various charge states, and their consequences.

4 Possible energies for Cr_{Ga} in its various charge states

A generally-agreed-on set of values for the 'ground state energies' of Cr_{Ga} in its various charge states does not exist, and there has been much controversy over this in the past decade. Any temperature dependence of the charge state transformation free energy can contribute to a misleading 'thermal activation energy', as discussed in Section 3; and activation energies derived from the temperature dependence of conductivity, etc., can also be affected by *other* deep lying states. This can result in an activation energy that looks like $(\epsilon_i/2)$, due to 'pseudo-intrinsic' conduction [35, 36].

Where should levels of Cr_{Ga} lie, within the gap and with respect to each other? The controlling factors are quite different from those for a shallow impurity, even for a shallow multi-valent impurity. Thus a shallow isocoric acceptor such as Zn in GaAs has properties that can be understood tolerably well in terms of effective mass properties of the upper valence bands. The isocoric divalent acceptor Zn in Ge appears to be a reasonable extension of those same ideas — from a 'hydrogenic' acceptor to a 'helium-like' acceptor — with successive ionization energies of 0.03 eV for the first electron accepted, and an additional 0.09 eV for the second [37]. In contrast, a satisfactory description of a truly deep impurity, dominated by short-range forces, requires the bound state wavefunction to be influenced by several of the one-electron lattice bands [25, 38]. Some much simpler models have been applied to transition element acceptors in GaAs with fair success [39, 40], although these deal only with the *first* stage of acceptor ionization.

One reason Figure 1 is drawn the way it is, rather than in terms of a conventional electron-energy diagram, is to get away from the idea that each stage of successive ionization is automatically higher than the preceding one. (One-electron energy diagrams are also confusing for a multi-valent impurity in any case, since each 'level' has to come into being as the 'level' for a different charge state disappears.) The literature on GaAs:Cr itself has produced a variety of ways to indicate the charge states and their relationships. It is hoped that the one of Figure 1 will be easier of interpretation than some.

Should one assume that the transition energy associated with placing a *second* electron on to a Cr_{Ga} site (i.e. the transition $Cr^{II} \rightarrow Cr^{I}$) *must* be larger than that necessary for placement of the *first* electron (i.e. the $Cr^{III} \rightarrow Cr^{II}$ transition)? To this observer, that may be so, but it does not automatically have to be. Whether ϵ_{p1} is larger than ϵ_{p2} depends not only on the spectrum of atomic 3d states, but on the extent of lattice relaxations in the two cases. As an example drawn from another solid, Baraff, Kane and Schlüter [41] made calculations of energies to be expected for states of a vacancy in silicon. They concluded that V^{0} and V^{2+} could be stable states, but that V^{1+} would not be, because of the energy necessary for associated lattice relaxation. Thus that type of calculation implies that V^{2+} would dominate for a low electronic Fermi energy, superceded by V^{0} (and then by the acceptor configurations V^{1-}, etc.) for increased Fermi energy.

Such a situation is not automatically to be dismissed for a transition element impurity in GaAs — and could play a role in some of the seemingly contradictory results reported with GaAs:Cr. However, the curve of Figure 2 is drawn on the supposition that $\epsilon_{p1} > \epsilon_{p2} > \epsilon_{p3}$ so that occupancy of the Cr centres would proceed, one step at a time, as an increased supply of shallow donors over shallow acceptors provided by the necessary electrons. This figure may be compared with that drawn by Brozel *et al.* [42] in indicating two 'subtypes' of semi-insulating GaAs:Cr, depending on whether the material is permitted to contain a Cr^{III}, Cr^{II} or a Cr^{II}, Cr^{I} mixture.

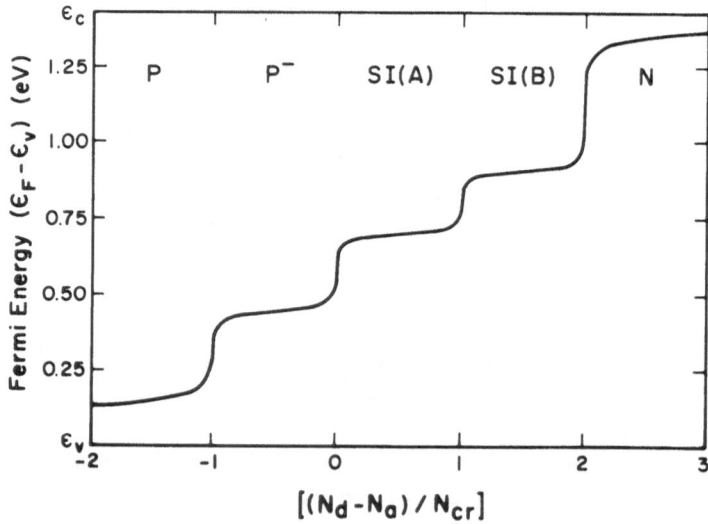

Figure 2   Possible variation of the non-illuminated Fermi energy ϵ_F as a function of shallow impurity doping, for GaAs:Cr. Effects of other deep lying states, such as oxygen, are not taken into account in drawing this curve. Drawn for T=300 K (ϵ_i=1.42 eV), supposing N_{Cr}=5×10^{16} cm^{-3}; and with $\epsilon_{p3} \simeq 0.45$ eV, $\epsilon_{p2} \simeq 0.7$ eV and $\epsilon_{p1} \simeq 0.9$ eV.

Additionally, it suggests that a slight excess of shallow acceptors over shallow donors should produce p-type material of moderately high resistivity, with the Fermi level controlled by a Cr^{III}, Cr^{IV} mixture. Kaufmann and Schneider [26] postulated $\epsilon_{P3} \sim 0.45$ eV, which would result in p-type material with a room temperature resistivity $\rho_{300} \sim 10^5$ ohm-cm. Of course, observation of that condition in GaAs:Cr could readily be impaired by the simultaneous presence of deep oxygen donors.

In the course of time, the quantities necessary for a total characterization of each localized charge state of GaAs:Cr will cease to be subject to speculation. If it does turn out that Cr^{IV}, as well as Cr^{III} through Cr^{I}, is a stable localized condition, then much work remains to be done. Interest in the underlying physics, as well as in the technical importance of this material, should help in furthering the necessary work.

ACKNOWLEDGEMENTS

I should like to thank G.M. Martin, U. Kaufmann, P.R. Jay and M.G. Clark, for communication of their results prior to publication; also to A.M. White for information concerning the scope of his discussion in this volume. Acknowledgement is made to the National Science Foundation for their support of this work.

References

1. White, A.M. (1980). This volume
2. Lightowlers, E.C., Henry, M.O. and Penchina, C.M. (1979). *Inst. Phys. Conf. Ser.*, **43**, 307
3. Tuck, B. and Adegboyega, G.A. (1979). *J. Phys. D: Appl. Phys.*, **12**, 1895
4. Clark, M.G. (1980). *J. Phys. C: Solid St. Phys.*, **13**, in press
5. Cronin, G.R. and Haisty, R.W. (1964). *J. Electrochem. Soc.*, **111**, 874
6. Balagurov, L.A., Omel'yanovskii, E.M. and Pervova, L.Y. (1975). *Sov. Phys. Semicond.*, **8**, 1051
7. Look, D.C. (1975). *J. Phys. Chem. Solids*, **36**, 1311
8. Zucca, R. (1977). *J. Appl. Phys.*, **48**, 1987
9. Lin, A.L. and Bube, R.H. (1976). *J. Appl. Phys.*, **47**, 1859
10. Omel'yanovskii, E.M., Pantyukhov, A.N., Pervova, L.Y., Fistul', V.I. and Vasil'ev, Y.A. (1976). *Sov. Phys. Semicond.*, **9**, 1267
11. Plesiewicz, W. (1977). *J. Phys. Chem. Solids*, **38**, 1079
12. Lang, D.V. and Logan, R.A. (1975). *J. Electron. Mater.*, **4**, 1053
13. Mitonneau, A., Mircea, A., Martin, G.M. and Pons, D. (1979). *Rev. Phys. Appl.*, **14**, 853
14. Martin, G.M., Mitonneau, A., Pons, D., Mircea, A. and Woodard, D.W. (1980). *J. Phys. C: Solid St. Phys.*, **13**, in press
15. Krebs, J.J. and Stauss, G.H. (1977). *Phys. Rev. B*, **15**, 17
16. Krebs, J.J. and Stauss, G.H. (1977). *Phys. Rev. B*, **16**, 971
17. Stauss, G.H., Krebs, J.J., Lee, S.H. and Swiggard, E.M. (1979). *J. Appl. Phys.*, **50**, 6251
18. White, A.M., Krebs, J.J. and Stauss, G.H. (1980). *J. Appl. Phys.*, **51**, 419
19. Kaufmann, U. and Schneider, J. (1976). *Solid St. Commun.*, **20**, 143
20. Goltzene, A., Poiblaud, G. and Schwab, C. (1979). *J. Appl. Phys.*, **50**, 5425
21. Eaves, L. and Williams, P.J. (1979). *J. Phys. C: Solid St. Phys.*, **12**, L725

22. Woodbury, H.H. and Tyler, W.W. (1957). *Phys. Rev.*, **105**, 84
23. Dunlap, W.C. (1955). *Phys. Rev.*, **100**, 1629
24. Milnes, A.G. (1973). *Deep Impurities in Semiconductors.* New York; Wiley-Interscience
25. Hemstreet, L.A. and Dimmock, J.O. (1979). *Phys. Rev. B*, **20**, 1527
26. Kaufmann, U. and Schneider, J. (1980). *Appl. Phys. Lett.*, in press
27. Yu, P.W. (1979). *Solid St. Commun.*, **32**, 1111
28. Krebs, J.J. and Stauss, G.H. (1979). *Phys. Rev. B*, **20**, 795
29. Wagner, R.J. and White, A.M. (1979). *Solid St. Commun.*, **32**, 399
30. Narayanamurti, V., Chin, M.A. and Logan, R.A. (1978). *Appl. Phys. Lett.*, **33**, 481
31. Tokumoto, H. and Ishaguro, T. (1979). *Inst. Phys. Conf. Ser.*, **43**, 299
32. Kukimoto, H., Henry, C.H. and Merritt, F.R. (1973). *Phys. Rev. B*, **7**, 2486; Prat, F. and Fortin, E. (1972). *Canad. J. Phys.*, **50**, 2551
33. Varshni, Y.P. (1967). *Physica*, **39**, 149
34. Martin, G.M. (1980). This volume
35. Roberts, G.G. (1971). *J. Phys. C: Solid St. Phys.*, **4**, 3167
36. Ashby, A., Roberts, G.G., Ashen, D.J. and Mullin, J.B. (1976). *Solid St. Commun.*, **20**, 61
37. Woodbury, H.H. and Tyler, W.W. (1956). *Phys. Rev.*, **102**, 647
38. Jaros, M. (1979). *Inst. Phys. Conf. Ser.*, **43**, 281
39. Bazhenov, V.K. and Solovev, N.N. (1972). *Sov. Phys. Semicond.*, **5**, 1589
40. Partin, D.L., Chen, J.W., Milnes, A.G. and Vassamillet, L.F. (1979). *Solid St. Electron.*, **22**, 455
41. Baraff, G.A., Kane, E.O. and Schlüter, M. (1979). *Phys. Rev. Lett.*, **43**, 956
42. Brozel, M.R., Butler, J., Newman, R.C., Ritson, A., Stirland, D.J. and Whitehead, C. (1978). *J. Phys. C: Solid St. Phys.*, **11**, 1857

REVIEW OF TECHNIQUES FOR EPITAXIAL GROWTH OF HIGH-RESISTIVITY GaAs — GROWTH SYSTEMS, PROBLEMS AND SUBSTRATE EFFECTS

H.M. COX and J.V. DILORENZO
Bell Laboratories, Murray Hill, New Jersey 07974, USA

Abstract

High-resistivity buffer layers have demonstrated important improvements in GaAs MESFET device performance and consistency over unbuffered structures. Higher active layer mobilities, especially near the interface, are obtained when buffer layers are used. The buffer layer also serves to isolate the active layer from problems generated at the substrate-epilayer interface and from some of the vagaries associated with semi-insulating substrates.

Several methods have now been developed for the epitaxial growth of high-resistivity GaAs. These methods have included the growth of GaAs that is undoped, oxygen-doped, iron-doped and chromium-doped, using a variety of epitaxial growth systems. The advantages and disadvantages of these various techniques will be reviewed along with the reported properties of the high-resistivity layers. Substrate effects on undoped buffer properties will also be discussed.

Introduction

Epitaxial buffer layers between the substrate and active layer of devices such as planar Gunn diodes and MESFETs are important for consistently good performance. Higher active layer mobilities, especially near the substrate interface, are obtained when buffer layers are used [1, 2]. The buffer layer also serves to isolate the active layer from interface states localized at the substrate-epilayer interface [3—5], which can give rise to back-biasing effects of the active layer when buffer layers are not used [3, 6]. One obvious effect of the introduction of a buffer layer is the elimination of light sensitivity and of loopiness in the I-V characteristics of MESFETs [1, 7, 8].

Early reports of improvements in device performance through the use of high-resistivity buffer layers included results for Gunn diodes [9] and for MESFETs [1, 10, 11]. By 1976, almost all GaAs MESFET results reported were for structures incorporating high-resistivity buffer layers. GaAs MESFET structures produced by ion implantation into buffer layers also show superior properties to those implanted directly into semi-insulating substrates [12].

Undoped buffers

Although MESFET structures incorporating undoped buffers have been grown by a variety of techniques, including the metal alkyl-hydride method [13] and molecular beam-epitaxy [14], the epitaxial growth system that has been most successful is the open flow tube $AsCl_3/Ga/H_2$ system.

SINGLE-BUBBLER $AsCl_3$ SYSTEMS

$AsCl_3$ was first used along with PCl_3 to grow $GaAs_xP_{1-x}$ in an $AsCl_3/Ga$ or $GaAs/PCl_3/H_2$ system [15]. Effer [16] later described the use of an $AsCl_3/Ga/H_2$ system for the growth of GaAs. High-purity material is readily grown with this system, which is a prerequisite for the growth of high-resistivity layers. The residual background level of undoped layers grown in this type of system was found to decrease with increasing $AsCl_3$ mole fraction [17].

Thermodynamic calculations of this type of transport system [18, 19] indicated that the most probable source of background contamination of undoped layers was Si from the vitreous quartz walls and hardware, transported as chlorosilanes. Consequences of such an Si contamination model are that Si contamination should decrease for:

1. Increase of $AsCl_3$ mole fraction.
2. Increase of H_2O pressure.
3. Decrease of source and/or substrate temperature.
4. Replacement of the H_2 carrier gas with an inert gas.

These approaches for the growth of low-background material have served as the starting point for several different methods of growth of high-resistivity buffer layers. Hollan [20] grew undoped buffer layers for MESFET applications by increasing the $AsCl_3$ mole fraction passing over the source of an $AsCl_3/Ga/H_2$ system. Komeno, Kitahara and Ohkawa [21] grew high-resistivity layers ($>10^5 \Omega$-cm) on Cr-doped substrates using nitrogen, rather than hydrogen, as the carrier gas in an $AsCl_3/Ga/N_2$ system [22]. Unusually low source and substrate temperatures have been used with an $AsCl_3/GaAs/H_2$ system [23] and with an $AsCl_3/Ga/H_2$ system [24].

DUAL-BUBBLER $AsCl_3$ SYSTEMS

Nozaki et al. [1] provided considerable versatility by the addition of a second $AsCl_3$ bubbler to an $AsCl_3/Ga/H_2$ system. A similar system has been used at our laboratory [25] for the growth of substrate/undoped buffer/n-active/n^+-contact structures for low noise [26] and high power [7] MESFETs. A schematic diagram of this system is shown in Figure 1. $AsCl_3/H_2$ from the main bubbler passes over the gallium source. The flow from the second bubbler is introduced into the reactor between the source and the substrate at a high flow during the *in situ* etch of the substrate prior to the growth, and at a lower flow during the growth of the undoped buffer. An H_2S/H_2 mixture is admitted to the reactor for doping of the active and n^+-contact layers. Complete details of system behaviour and epitaxial layer properties as a function of $AsCl_3$ mole fraction from the main bubbler and the bypass bubbler are given in the earlier reference [25] and will not be repeated here. One important factor that was discovered by the use of this system will be reviewed, however, because of its relevance to all methods of undoped high-resistivity buffer growth on semi-insulating substrates.

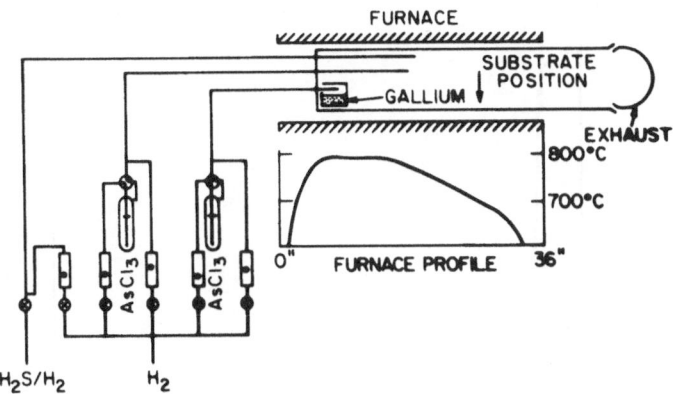

Figure 1　Two-bubbler $AsCl_3$/Ga/H_2 CVD system.

ACCEPTOR OUT-DIFFUSION

The properties of undoped layers of low background level were found to be dominated by substrate effects. Figure 2 shows a profile of an SI substrate/ undoped buffer/n-active structure grown under system conditions that give a background level of approximately 3×10^{14} cm^{-3} for a layer grown on an n^+ substrate. However, when a semi-insulating substrate is used, the first 10.5 μm of the buffer adjacent to the substrate are much lower in carrier concentration and, in fact, are high-resistivity. Only about 2μm of the buffer show the true background level. It was concluded from the variation of the thickness of the high-resistivity layer of several growth runs as a function of different growth rates, growth times and background level that the high-resistivity region was caused by acceptor out-diffusion from the substrate which compensates the donor population of the buffer in the vicinity of the substrate.

The acceptor out-diffusion can be used to advantage in the growth of undoped high-resistivity buffers. The growth conditions and time can be adjusted to obtain a good high-resistivity buffer, as shown in Figure 3. This requires a delicate balancing act, however. The acceptor level should be approximately equal to the uncompensated background level at the active-buffer interface. Variations in acceptor level found in different ingots or even from one end of the ingot to the other can cause problems. If the substrate level drops, an n-type conducting knee develops in the buffer, similar to that shown in Figure 2. If the acceptor level increases, a p-type conducting region is formed adjacent to the substrate-buffer interface. Variations in growth rates or background levels cause the same problems.

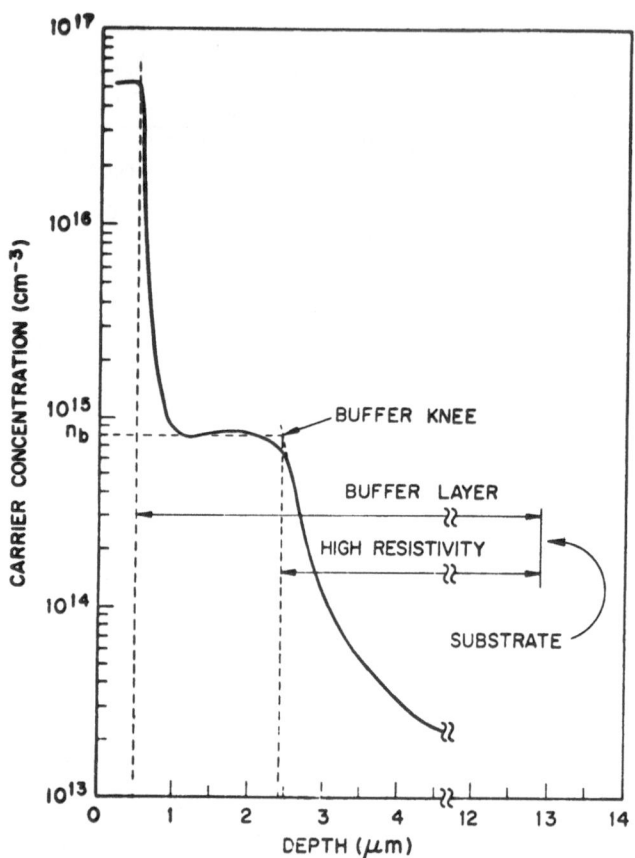

Figure 2 Undoped buffer showing knee.

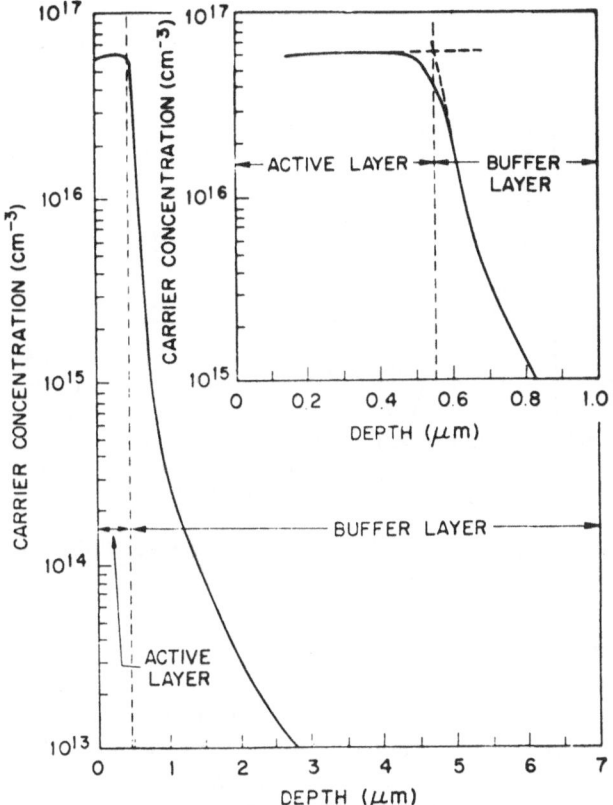

Figure 3 Undoped buffer and active layer.

SUBSTRATE ANNEALING

Bozler *et al.* [27] showed that the acceptors could be gettered out of the substrate by implanting the substrate with Si^+ or Ne^+ at 400 keV with a dose of 10^{16} cm^{-2} followed by Si_3N_4 and SiO_2 overcoats before a 16 hour 750°C thermal anneal. We repeated their experiment but found little difference between an annealed sample which had previously been implanted and one which had not been implanted before the anneal. Neither did the presence or absence of a Si_3N_4 encapsulation seem to affect the results significantly. Figure 4 shows profiles of two undoped buffers grown simultaneously on two samples of the same substrate material. One of the samples was previously annealed for 16 hours at 750°C in an $AsCl_3$/Ga/H_2 reactor. About 2μm were etched from the annealed sample prior to growth in order to remove any surface damage. The other was a control sample that did not receive an anneal. It can be seen that about 8 μm of the control sample have been compensated by out-diffusion, as expected, but there is little or no compensation of the previously annealed sample. The profile of a third sample, not shown, which was annealed but etched to a depth of 100 μm before growth, was identical to the profile of the lightly etched sample.

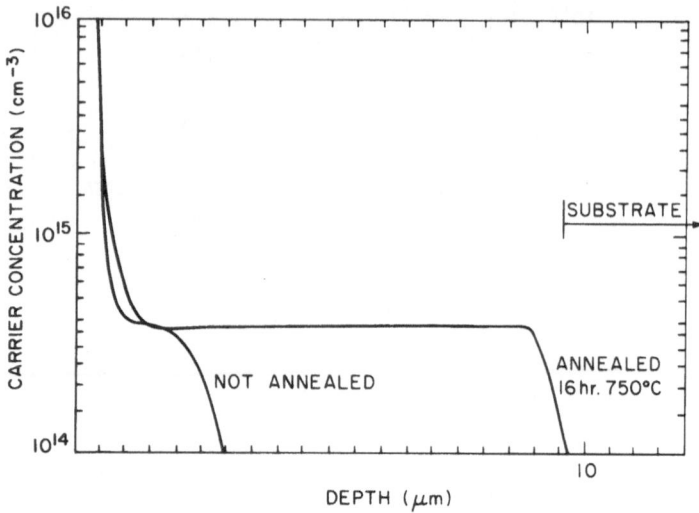

Figure 4 Effect of annealing on substrate out-diffusion.

ACCEPTOR IDENTITY

Despite the fact that acceptor out-diffusion is never observed with n+ substrates, the possibility that Cr might be the rapidly diffusing acceptor did not seem likely. The acceptor seemed to be virtually eliminated even deep within the substrate after the 16 hour 750°C anneal. In addition, we had seen out-diffusion behaviour from undoped SI substrates similar to that from Cr-doped substrates.

However, Tuck *et al.* [28] reported a series of radiotracer experiments with ^{51}Cr that showed diffusion behaviour consistent with our observations of the behaviour of the 'unknown acceptor'. ^{51}Cr was sealed along with 300 μm thick substrate samples into an evacuated ampoule and heated. Some of their findings are summarized here:

1. The ^{51}Cr would diffuse to a homogeneous level throughout the substrate thickness for all temperatures reported, 1100, 1000, 900 and 800°C within four hours except for about 10 μm near the surface where the concentration was much higher.

2. A second subsequent anneal of the homogeneously doped portion of such a sample for only one hour at 750°C showed a surface build-up of ^{51}Cr concentration and a reduction of the bulk concentration by more than a factor of 2.

3. Epigrowth on homogeneously ^{51}Cr-doped substrates showed rapid out-diffusion.

4. n-type GaAs, into which the ^{51}Cr was diffused, became high-resistivity (not semi-insulating).

More recent studies, using secondary-ion mass spectrometry, have also confirmed rapid diffusion of Cr under a variety of conditions [29, 30]. The evidence now is, admittedly, rather strong that Cr is the unknown acceptor. Puzzling questions remain, however. Acceptor out-diffusion from undoped substrates is similar to that from Cr-doped substrates and, apparently, epitaxial Cr-doped material does not show rapid Cr diffusion. It retains its semi-insulating properties after an 850°C anneal for four hours [31] and shows no movement of Cr after a one hour anneal at 750°C [32].

Regardless of the identity of the substrate acceptor, its presence will influence the properties of low-doped layers grown on semi-insulating substrates by any technique.

LIQUID PHASE EPITAXY

Both undoped and Cr-doped high-resistivity epitaxial layers have been grown by LPE. In either case, low-n material is a prerequisite. The most important factor for obtaining low carrier concentration layers, according to Morkoc and Eastman [33], is a pre-bake of the graphite boat in H_2 before growth. The higher the temperature the shorter the time required, but lower temperatures (down to 700°C) yield lower steady state background levels. The low doped layers grown by LPE have high resistivity regions near the substrate, presumably caused by acceptor compensation as in VPE growth. These high-resistivity regions have been successfully used for MESFET structures by Nanishi, Takahei and Kuroiwa [34] and Kim *et al.* [35], who found that the active layer mobility and device performance improved with buffered structures.

Doped buffers

Doped buffer growth techniques suffer from the disadvantage that the dopant that is added to the system during buffer growth is an undesirable impurity if incorporated into the active layer. The offsetting advantage is that buffers of higher resistivity may be grown, possibly independent of the variable substrate effects that plague undoped buffer growth [36].

O_2 and H_2O DOPING

The introduction of oxygen into a GaAs CVD system can possibly influence the resistivity of the grown layer in two different ways. The background doping level of undoped layers may be reduced because of the inverse relationship of Si incorporation to H_2O pressure discussed earlier. The oxygen may also be incorporated into the epitaxial layer as a dopant with uncertain results.

The growth of epitaxial layers of semi-insulating gallium arsenide by oxygen-doping was reported by Gudz, Maronchuk and Maronchuk [37]. They introduced water vapour into an $AsCl_3/Ga/H_2$ CVD system. Semi-insulating films with $>10^7\,\Omega$-cm were obtained for H_2O pressures greater than 20 Torr.

More recently, Palm et al. [38] have injected O_2 at a pressure of 10^{-3} bar into an $AsCl_3/Ga/H_2$ CVD system to produce high-resistivity buffer layers. Buffer layer growth was followed by the sequential deposition of an Se-doped active layer for MESFET structures. The mobility of Se-doped layers was found to be unaffected by the simultaneous introduction of O_2 into the system. Apparently the role of the O_2 is to reduce Si incorporation rather than to act as a dopant.

NH_3-DOPING

Stringfellow and Ham [39] introduced NH_3 into an $AsH_3/HCl/Ga/H_2$ hydride CVD system in order to produce high-resistivity buffer layers. The NH_3 reduced the background level of the grown layers by reacting with chlorosilanes to form Si_3N_4 analogous to the background reduction by oxygen doping.

Fe-DOPING

As early as 1966, Hoyt and Haisty [40] reported the growth of semi-insulating (ρ=3–9x$10^4\,\Omega$-cm) epitaxial GaAs using Fe-doping in an $AsCl_3/GaAs/H_2$ system. High purity Fe was simply placed in the reactor alongside the GaAs source during growth of the semi-insulating material. This technique did not allow multi-layer structures to be grown but clearly·demonstrated the feasibility of growing semi-insulating epi-material in a vapour phase reactor.

Fe-doping for the growth of multi-layer MESFET structures has more recently been reported by Dazai, Shibatomi and Kenya [41] and Nakai et al. [42] using an $AsCl_3/Ga/N_2$ system [22].

$FeCl_2$ was introduced between the Ga source and the substrate by passing HCl derived from $AsCl_3$ and H_2 over heated Fe powder. Resistivities of the Fe-doped layers were $10^4-10^5\,\Omega$-cm. When Fe-doped buffer layers were used in MESFET structures, 20% higher mobilities were reported for sequentially deposited S-doped active layers as well as higher power gains and·increased transconductance from devices with buffer layers.

CHROMIUM DOPING AT OTHER LABORATORIES

Chromium doping of epitaxial GaAs to form semi-insulating layers was first reported by Mizuno, Kikuchi and Seki [31]. Vapour from liquid chromyl chloride (CrO_2Cl_2) was introduced into an $AsCl_3/Ga/H_2$ system between the Ga source and the substrate. Resistivities in excess of $10^8\,\Omega$-cm were obtained. The same growth technique was used by Kato, Mori and Morizane [32] for MESFET application. They reported device results comparable to performance obtained using undoped buffers. There are some difficulties associated with this method of chromium doping which will be discussed in more detail later.

This same method was used by Kuroiwa, Aoki and Fujimoto [9] to grow epitaxial Cr-doped buffer layers for use in planar-type GaAs Gunn diodes. Diodes using buffer layers showed improved coherency of oscillation. A later paper by Mizutani *et al.* [43], about Gunn diode characteristics fabricated from this material, reported that diodes with 19 μm thick buffers were superior to ones with 5 μm thick buffers which were better than non-buffered devices. Hyder [44] used chromyl chloride for the doping of graded $In_xGa_{1-x}As$ buffers grown in an $AsCl_3/In/Ga/H_2$ system. The Cr doping provided layers with resistivities of about 2×10^8 Ω-cm while the In composition of the buffer was graded to provide lattice matching between the GaAs substrate and active layer of MESFET structure. Drukier *et al.* [11] reported improvements in gain and power out of MESFETs when Cr-doped buffers were used. Fairman [45] has recently produced semi-insulating layers by passing $AsCl_3$ over heated Cr before introduction into an $AsCl_3/Ga/H_2$ system.

Bass [46] has grown Cr-doped semi-insulating GaAs by introducing vapour from solid hexacarbonyl chromium into a cold wall $Ga(CH_3)_3/AsH_3/H_2$ system similar in design to that originally described by Manasevit and Simpson [47]. The substrate and susceptor are r.f. heated. He was able to compensate a donor concentration as high as 5×10^{16} cm^{-3} by introducing the hexacarbonyl chromium at a mole fraction of 4×10^{-9}.

Semi-insulating GaAs has been produced by Cr-doping of epitaxial layers grown by LPE, as reported by Otsubo and Miki [48]. Resistivities in excess of $10^7 \Omega$-cm were obtained by adding about one mole per cent Cr to the melt. Houng, Pearson and Mattes [49] grew material of greater than $10^6 \Omega$-cm resistivity by adding 0.5 mole per cent Cr to the melt. Woodard [50] has also grown by LPE semi-insulating GaAs buffer layers for MESFET structures by Cr doping.

Cr DOPING AT OUR LABORATORY

Work at our laboratory on the growth of Cr-doped epitaxial buffer material began with a system design which is basically similar to that described in [31]. Chromyl chloride was admitted into a standard $AsCl_3/Ga/H_2$ CVD system during the growth of the buffer. Resistivities in excess of $10^8 \Omega$-cm could be obtained with this method, but there were problems associated with it. Most of the chromyl chloride vapour that is introduced into the reactor thermally decomposes into mixed chromium oxides according to the reaction:

$$3CrO_2Cl_2(g) + 3H_2(g) \rightarrow CrO_3 \cdot Cr_2O_3(s) + 6HCl(g).$$

The deposits accumulate in the entry tube causing increased reactor contamination and resultant surface morphology problems. The deposits eventually clog the entry tube forcing replacement.

Because of the problems encountered, this system was abandoned in favour of the dual $AsCl_3$ bubbler system described earlier [25]. The discovery of the inconsistent effects that substrates could have on undoped buffers, however, forced us to reconsider Cr doping.

A new system was designed which consisted of the addition of a chromyl chloride bubbler to a dual $AsCl_3$ bubbler system [51], as shown in Figure 5. The chromyl chloride is not used during buffer growth as it is in the conventional chromyl chloride doping system. Instead, the chromyl chloride is injected only momentarily through the Cr-doping tube in order to form, *in situ*, at about $500°C$ the mixed chromium oxides described earlier. This deposit then serves as a high purity dopant source for subsequent Cr doping of the buffer layers [52].

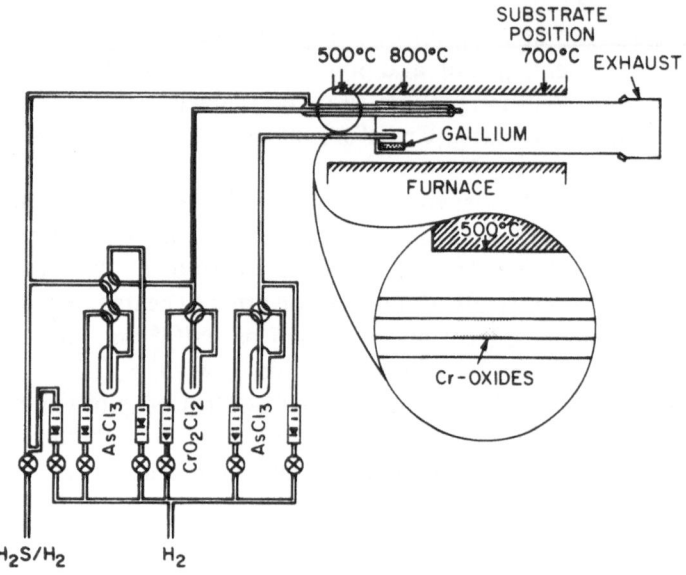

Figure 5   Three-bubbler $AsCl_3$/Ga/CrO_2Cl_2/H_2 CVD system. $AsCl_3$ bubbler temperatures=18.3°C. CrO_2Cl_2 bubbler temperature=0°C.

A four-way valve permits the output from the $AsCl_3$ bypass bubbler to be directed through either of the two entry tubes bypassing the source. When the $AsCl_3$ passes through the tube containing the chromium oxide, doping of the epitaxial layer is achieved. Transport of the Cr occurs most likely by the reversal of the deposition reaction because of excess HCl to form CrO_2-Cl_2. The generation of a volatile chromium chloride is also possible with perhaps the formation of $CrAs$ as an intermediate product.

Very little chromyl chloride is required with this technique. A one minute flow of 60 cc min^{-1} H_2 through the CrO_2Cl_2 bubbler provides sufficient source material for a month or more of Cr doping. The amount of compensation produced by the Cr doping is indicated in Figure 6, which shows the reduction in background level, n, as a function of $AsCl_3$ flow over the chromium oxide deposits. The data show a linear relationship.

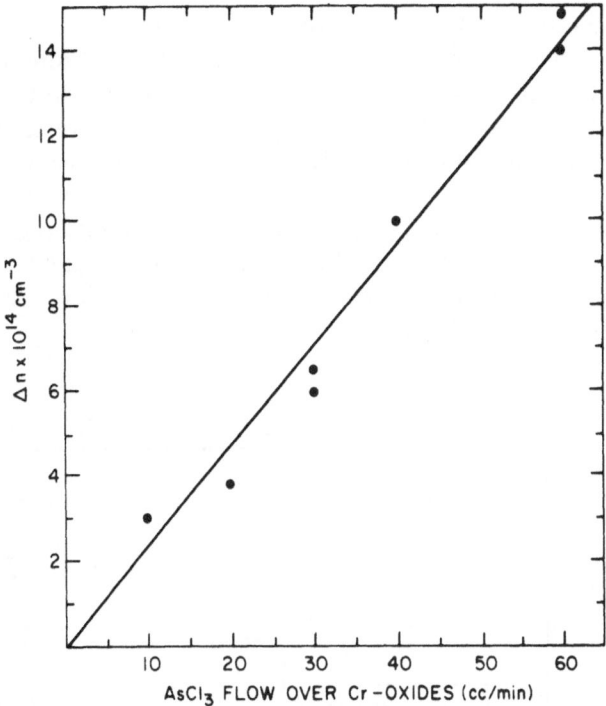

Figure 6 Reduction of background level (n) as a function of AsCl₃ flow over Cr-oxides.

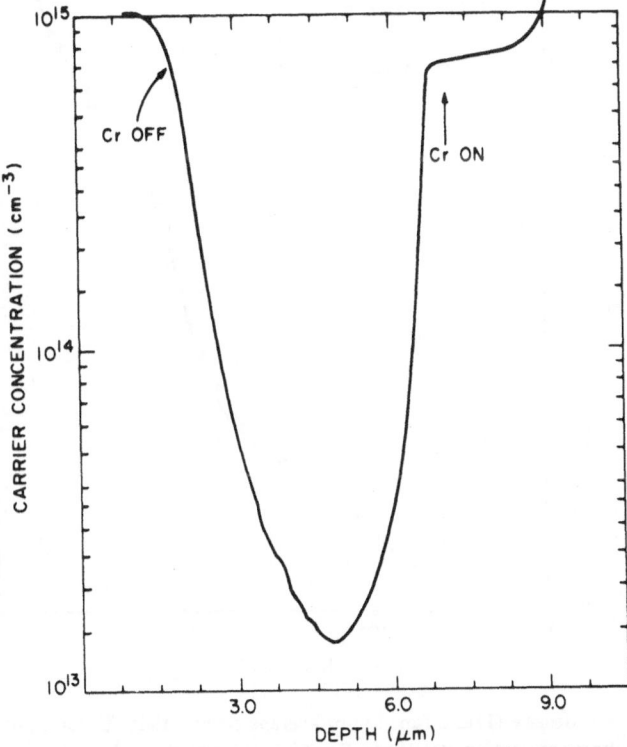

Figure 7 Cr-doped semi-insulating layer sandwiched between undoped layers grown on n⁺ substrate.

A flow rate is chosen for buffer growth which is sufficient to compensate fully the normal background level. This is shown in Figure 7, which is a profile of a layer grown on an n^+ substrate under normal buffer conditions. At the point marked 'Cr-on' the $AsCl_3$ bypass flow was diverted into the tube containing the chromium deposits. The background returns to a level higher than it was before the Cr doping because the $AsCl_3$ mole fraction was reduced by the termination of the $AsCl_3$ bypass flow. The Cr-doped region is, in fact, semi-insulating as indicated by the current density versus electric field characteristic of a similar structure shown in Figure 8. The resistivity of this sample was typical at $4 \times 10^8 \Omega$-cm.

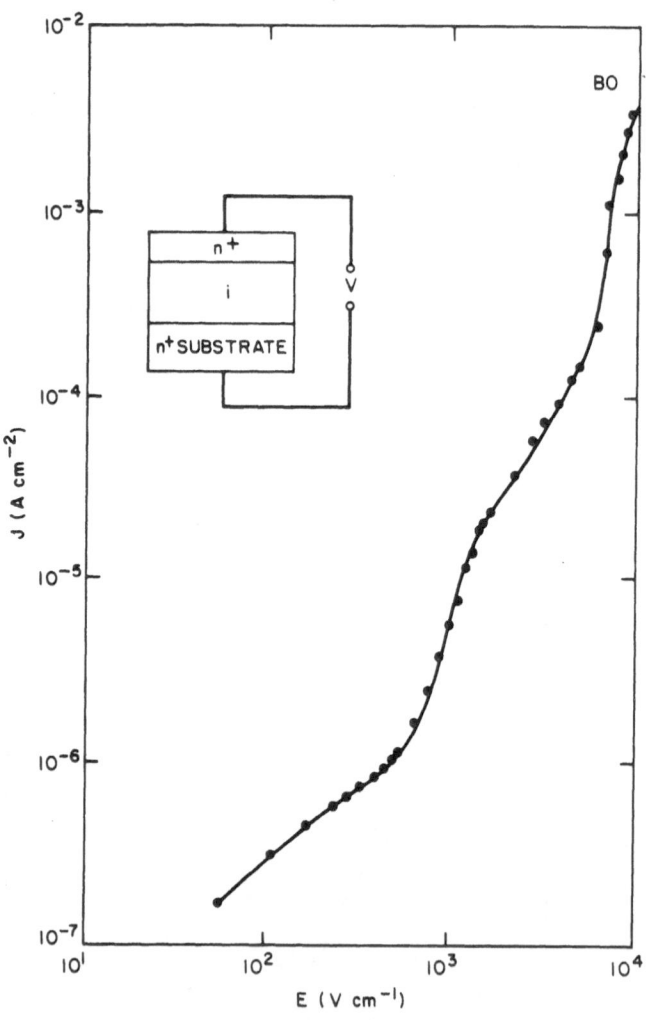

Figure 8   Current density (J) as a function of average electric field (E) for an n^+-i-n^+ structure. Thickness of i-region was 18μm. Sample was 2 mm cleaved square.

The threshold voltage at which the current changes from linear to superlinear with voltage is an important characteristic of the material. The threshold voltage varies with the square of the thickness of the semi-insulating layer, as shown in Figure 9. The lower solid line was established by the early work of Haisty and Hoyt [53] for bulk undoped SI samples as well as the bulk Cr-doped point. The upper solid line was determined by Otsubo and Miki [48] for LPE-Cr-doped material. The CVD-CrO_2Cl_2 data are from Mizuno, Kikuchi and Seki [31] for SI material grown with a continuous supply of CrO_2Cl_2 as the dopant. The solid circles are our data. The data indicate a threshold voltage for the epitaxial Cr-doped material about an order of magnitude higher than that of the bulk Cr-doped substrate material. Since structures of this type would generally burn out catastrophically at a point about an order of magnitude higher than the threshold voltage, the threshold voltage may be a measure of the burn-out propensity of the material. Although material is but one of many factors affecting device burn-out, devices using Cr-doped buffers grown by this technique have logged 1.8 million device hours under test at elevated temperatures with only one burn-out as reported by Fukui *et al.* [54].

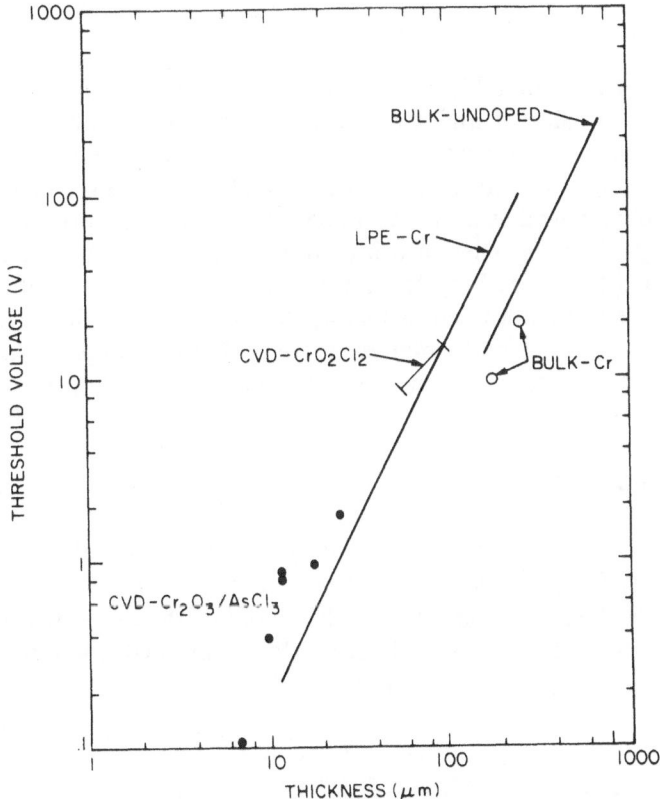

Figure 9 Threshold voltage as function of semi-insulating layer thickness. Bulk-undoped line and open circle after Haisty and Hoyt [53]. LPE-Cr line after Otsubo and Miki [48]. CVD-CrO_2Cl_2 line from Mizuno *et al.* [31]. The solid circles are our data.

The threshold voltage may also be important from the standpoint of buffer or substrate leakage current. The measured resistivity of the substrate material is meaningless at voltages much above the threshold voltage. Extrapolation of the bulk Cr-doped data from Figure 9 indicates that the threshold voltage is only 1 V at $60 \, \mu m$. This value is exceeded even for low noise devices.

HETEROSTRUCTURES

Perhaps the ultimate high-resistivity buffer for GaAs MESFETs will be $Ga_{1-x}Al_xAs$ which confines current to the active layer because of the heterojunction at the active-buffer interface. High-resistivity $Ga_{1-x}Al_xAs$ buffers of about $10^5 \, \Omega$-cm have been grown by Hallais *et al.* [55] in a $Ga(CH_3)_3/Al(CH_3)_3/AsH_3/H_2$ system. Heterostructure MESFET devices were made but, although buffer current was apparently eliminated as desired, device results were inferior to conventional buffered structures grown by the $AsCl_3$ process.

Summary

High-resistivity epitaxial GaAs has been grown that is undoped, H_2O-doped, NH_3-doped, Fe-doped and Cr-doped. The properties of these layers are strongly influenced by acceptor out-diffusion (probably Cr) when grown on semi-insulating substrates. The acceptor level present in the epilayer may be dramatically reduced by a thermal anneal of the substrate prior to growth.

Although the properties of a high-resistivity buffer, characteristic of the growth conditions, are best determined by growth on an n+ substrate where acceptor out-diffusion is not a factor, the properties of the buffer that are meaningful to device performance are those that it possesses when grown in the device structure, sandwiched between a semi-insulating substrate and an n-layer.

ACKNOWLEDGEMENTS

The authors would like to thank their co-workers at Bell Laboratories, especially A.R. Von Neida and L.J. Oster for the supply of substrate material, W. Urban for assistance with many of the experiments described, and L.J. Varnerin and J.E. Kunzler for continued encouragement and support.

References

1. Nozaki, T., Ogawa. M., Terao, H. and Watanabe, H. (1975). *Inst. Phys. Conf. Ser.*, **24**, 46
2. Butlin, R.S., Parker, D., Crossley, I. and Turner, J. (1977). *Inst. Phys. Conf. Ser.*, **33a**, 237
3. Yokayama, N., Shibatomi, A., Ohkawa, S., Fukuta, M. and Ishikawa, H. (1975). *Inst. Phys. Conf. Ser.*, **33b**, 201
4. Houng, Y.M. and Pearson, G.L. (1978). *J. Appl. Phys.*, **49**, 3348
5. Meignant, D., Boccon-Gibod, D. and Bourgeois, J.M. (1979). *Electron. Lett.*, **15**, 779
6. Hower, P.L., Hooper, W.W., Tremere, D.A., Lehrer, W. and Bittmann, C.A. (1969). *Inst. Phys. Conf. Ser.*, **7**, 187
7. Niehaus, W.C., Cox, H.M., Hewitt, B.S., Wemple, S.H., DiLorenzo, J.V., Schlosser, W.O. and Magalhaes, F.M. (1977). *Inst. Phys. Conf. Ser.*, **33b**, 271
8. Crossley, I., Goodridge, I.H., Cardwell, M.J. and Butlin, R.S. (1977). *Inst. Phys. Conf. Ser.*, **33b**, 289
9. Kuroiwa, K., Aoki, T. and Fujimoto, M. (1974). *J. Cryst. Growth*, **25**, 229
10. Slaymaker, N.A. and Turner, J.A. (1973). *Proc. 1st European Microwave Conf.*, A.5.1
11. Drukier, I., Camisa, R.L., Jolly, S.T., Huang, H.C. and Narayan, S.Y. (1975). *Electron. Lett.*, **2**, 104
12. Stolte, C.A. (1975). *Int. Electron. Dev. Meeting*, Tech. Digest, p. 585
13. Bass, S.J. (1975). *J. Cryst. Growth*, **31**, 172
14. Cho, A.Y., DiLorenzo, J.V., Hewitt, B.S., Niehaus, W.C., Schlosser, W.O. and Radice, C. (1977). *J. Appl. Phys.* **48**, 346
15. Finch, W.F. and Mehal, E.W. (1964). *J. Electrochem. Soc.*, **111**, 814
16. Effer, D. (1965). *J. Electrochem. Soc.*, **112**, 1020
17. Cairns, B. and Fairman, R.D. (1968). *J. Electrochem. Soc.* **117**, 197c
18. DiLorenzo, J.V. and Moore, G.E. (1971). *J. Electrochem. Soc.*, **118**, 1824
19. Weiner, M.E. (1972). *J. Electrochem. Soc*, **119**, 496
20. Hollan, L. (1975). *Inst. Phys. Conf. Ser.*, **24**, 22
21. Komeno, J., Kitahara, K. and Ohkawa, S. (1979). *J. Cryst. Growth*, **47**, 601
22. Ihara, M., Dazai, K. and Ryuzan, O. (1974). *J. Appl. Phys.*, **45**, 528
23. Hallais, J., Boccon-Gibod, D., Chane, J.P., Durand, J.M. and Hollan, L. (1977). *J. Electrochem. Soc.*, **124**, 1290
24. Shaw, D.W. (1977). *152nd Electrochem. Soc. Meeting*, Ext. Abst., No. 77, p. 863
25. Cox, H.M. and DiLorenzo, J.V. (1977). *Inst. Phys. Conf. Ser.*, **33b**, 11
26. Hewitt, B.S., Cox, H.M., Fukui, H., DiLorenzo, J.V., Schlosser, W.O. and Iglesias, D.E. (1977). *Inst. Phys. Conf. Ser.*, **33a**, 246
27. Bozler, C.O., Donnelly, J.P., Lindley, W.T. and Reynolds, R.A. (1976). *Appl. Phys. Lett.*, **29**, 698
28. Tuck, B., Adegboyega, G.A., Jay, P.R. and Cardwell, M.J. (1979). *Inst. Phys. Conf. Ser.*, **45**, 114
29. Evans, C.A., Deline, V.R., Sigmon, T.W. and Lidow, A. (1979). *Appl. Phys. Lett.*, **35**, 291
30. Huber, A.M., Morillot, G., Merenda, P. and Linh, N.Y. (1979). *2nd Int. Conf. on Sec. Ion Mass Spectr.*, Palo Alto
31. Mizuno, O., Kikuchi, S. and Seki, Y. (1971). *Jap. J. Appl. Phys.*, **10**, 208
32. Kato, Y., Mori, Y. and Morizane, K. (1979). *J. Cryst. Growth*, **47**, 12
33. Morkoc, H. and Eastman, L.F. (1976). *J. Cryst. Growth*, **36**, 109
34. Nanishi, Y., Takahei, K. and Kuroiwa, K. (1978). *J. Cryst. Growth*, **45**, 272
35. Kim, C.K., Malbon, R.M., Omori, M. and Park, Y.S. (1979). *Inst. Phys. Conf. Ser.*, **45**, 305
36. Fairman, R.D., Morin, F.J. and Oliver, J.R. (1979). *Inst. Phys. Conf. Ser.*, **45**, 134
37. Gudz, E.S., Maronchuk, I.E. and Maronchuk, Y.E. (1971). *Izv. Akad. Nauk. SSSR, Neorg. Mater.*, **7**, 1616
38. Palm, L., Bruch, H., Bachem, K.H. and Balk, P. (1979). *J. Electron. Mat.*, **8**, 555
39. Stringfellow, G.B. and Hom, G. (1977). *J. Electrochem. Soc.*, **124**, 1809
40. Hoyt, P.L. and Haisty, R.W. (1966). *J. Electrochem. Soc.*, **113**, 296

41. Dazai, K., Shibatomi, A. and Kenya, N. (1976). *Fujitsu Sci. Tech. J.*, **June**, 179
42. Nakai, K., Kitahara, K., Shibatomi, A. and Ohkawa, S. (1977). *J. Electrochem. Soc.*, **124**, 1635
43. Mizutani, T., Kurumada, K., Kuroiwa, K., Aoki, T. and Fujimoto, M. (1975). *J. Phys. D: Appl. Phys.*, **8**, L6
44. Hyder, S.B. (1976). *J. Electrochem. Soc.*, **123**, 1503
45. Fairman, R.D. (1979). *Proc. 7th Biennial Cornell Elec. Eng. Conf.*
46. Bass, S.J. (1978). *J. Cryst. Growth*, **44**, 29
47. Manasevit, H.M. and Simpson, W.I. (1969). *J. Electrochem. Soc.*, **116**, 1725
48. Otsubo, M. and Miki, H. (1977). *J. Electrochem. Soc.*, **124**, 441
49. Houng, Y.M., Pearson, G.L. and Mattes, B.L. (1978). *J. Electrochem. Soc.*, **125**, 2058
50. Woodard, D.W. (1979). PhD Thesis, Cornell University
51. Cox, H.M., DiLorenzo, J.V. and D'Asaro, L.A. (1979). *1st Ann. IC Symp.* (spons. by IEEE Elec. Dev. Soc), Lake Tahoe, Nevada
52. Cox, H.M. and DiLorenzo, J.V. (1979). *Wkshp. Cmpnd. Semicond. Mat. and Dev. (WOCSEMMAD) Conf.*, Atlanta, Georgia
53. Haisty, R.W. and Hoyt, P.L. (1967). *Solid St. Electron.*, **10**, 795
54. Fukui, H., Wemple, S.H., Irvin, J.C., Niehaus, W.C., Hwang, J.C.M., Cox, H.M., Schlosser, W.O. and DiLorenzo, J.V. (1980). *IEEE Int. Reliab. Phys. Symp.*, Las Vegas
55. Hallais, J., André, J.P., Baudet, P. and Boccon-Gibod, D. (1979). *Inst. Phys. Conf. Ser.*, **45**, 361

Section 2:
Preparation and growth of semi-insulating GaAs and InP

THE GROWTH AND PROPERTIES OF LARGE SEMI-INSULATING CRYSTALS OF INDIUM PHOSPHIDE

D. RUMSBY, R.M. WARE and M. WHITTAKER
Metals Research, Melbourn, Royston, Herts, UK

Abstract

Semi-insulating crystals of indium phosphide have been grown with diameters up to 80 mm and weight up to 3 kg using the liquid encapsulated Czochralski (LEC) process with boric oxide as encapsulant. Polycrystalline indium phosphide was first synthesized by direct reaction between the elements in graphite or pyrolytic boron nitride tubes, enclosed in sealed quartz ampoules. This gives dense polycrystalline ingots in high yield with only a slight excess of free indium. This material is used in the LEC process.

Undoped crystals are n-type with carrier concentrations of $5 \times 10^{15} - 1 \times 10^{16}$ cm^{-3}. The dominant donor impurities are sulphur and silicon. Provided that the donor concentration is less than 5×10^{16} cm^{-3} the iron doping yields semi-insulating material with a resistivity greater than 10^7 ohm-cm, which is thermally stable under epitaxial growth conditions. Etch pit densities for large iron-doped crystals fall in the range $2 - 5 \times 10^4$ cm^{-2}, increasing along the crystal from seed to tail.

Introduction

Indium phosphide is a material of increasing importance for microwave device and other applications, for many of which it is important to have uniform large area substrates for epitaxial deposition of the active device layer.

The liquid encapsulated Czochralski technique, first reported by Metz, Miller and Mazelsky [1], was applied to indium phosphide by Mullin *et al.* [2] and the growth of semi-insulating indium phosphide using chromium doping was reported by Straughan *et al.* [3]. The solubility of chromium in indium phosphide is very low and it is necessary to achieve very low background impurity levels before chromium doping is successful. However, iron doping presents a less demanding alternative [4] and the present work describes the establishment of a process for the regular production of large semi-insulating indium phosphide crystals of either <100> or <111> orientation.

Experimental

Polycrystalline indium phosphide was synthesized from the elements using 6N indium and 6N phosphorus from various sources. The reaction was carried out in a standard Metals Research PGR 2100 reactor, shown schematically in Figure 1. In this process the indium metal is contained in a graphite tube closed with graphite end caps (Schunk and Ebe Ltd, Grade FE 15 G1 or Ringsdorf Grade 5061) and supported by an inner quartz tube. After machining, the Ringsdorf graphite is ultra-sonically cleaned to remove dust and treated at very high temperature ($>2500^{\circ}C$) in a halogen atmosphere. Red phosphorus chunks are loaded directly into the outer quartz envelope and separated from the graphite tube by a quartz wool plug. The quartz envelope is evacuated to a pressure of approximately 10^{-5} Torr and sealed. The sealed tube is loaded into the steel vessel which is pressurized to 300 psi to balance the phosphorus pressure generated inside the quartz tube during the reaction. The equipment is essentially a three-zone furnace. Zone 1 controls the phosphorus pressure in the tube. Zone 3 is set at a higher temperature than Zone 1 to prevent distillation of phosphorus from one end of the tube to the other. Zone 2 consists of an r.f. induction heating coil coupled into the graphite tube which contains the indium.

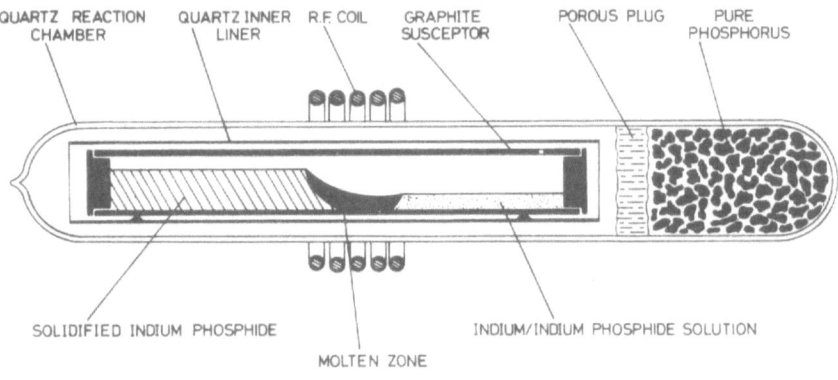

Figure 1   Synthesis of polycrystalline InP.

To carry out the reaction, the quartz tube is traversed through the induction coil at a rate of 6 cm h^{-1} yielding a polycrystalline bar of 1.3 kg of InP in six hours.

Some material was synthesized in pyrolytic boron nitride crucibles (ex Fulmer Research Institute). In this case the induction heating coupled directly into the indium metal.

The polycrystalline material prepared by this means is used as the starting charge for LEC growth.

Crystals were grown from two types of high pressure crystal puller: the Metals Research 'Malvern' MSR6R and the 'Melbourn'. Both pullers are resistance heated.

All the results reported here were obtained using silica crucibles and boric oxide (either Johnson Matthey 'Puratronic' grade or BDH 'Optran'). The boric oxide encapsulation remains clear throughout the run so that visibility on the television monitor is excellent, with the meniscus clearly visible. Moreover, in contrast to gallium phosphide, the growth of indium phosphide seems to be a stable process and the manual growth of crystals of good diameter control is possible, as evidenced in Figure 2 for a <111> crystal of over 800 g weight. A <100> crystal of similar size is shown in Figure 3. It will be noted that while the <111> crystal has a tapered top cone, the <100>

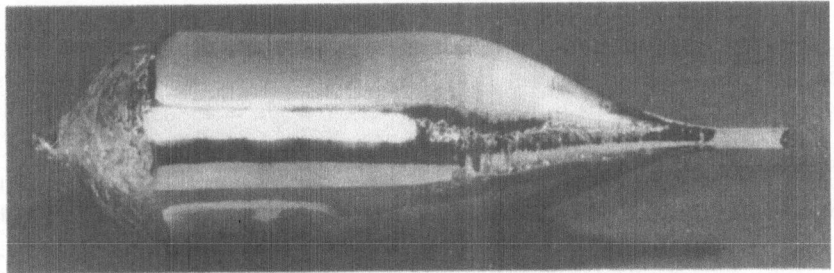

Figure 2   InP crystal, 800 g, <111> orientation.

Figure 3   InP crystal, 850 g, <100> orientation.

crystal has a flat top. It has been found empirically, in agreement with the results of Bachmann *et al.* [5], that the incidence of twinning in <100> crystals is much reduced in flat top crystals compared with tapered ones. Both <100> and <111> crystals were grown at a pull rate of 9 mm h^{-1}.

Results

Crystals of the types shown in Figures 2 and 3 are repeatably obtainable and a large number of crystals of both <100> and <111> orientations have been grown.

Crystals were assessed by cutting slices from each end and measuring conductivity type, resistivity, carrier concentration, mobility by the van der Pauw technique and etch pit density.

In order to obtain high resistivity semi-insulating material it is first necessary to produce indium phosphide with a low background impurity level.

'Undoped' indium phosphide is usually n-type with carrier concentration in the range 10^{15}—5×10^{16} cm^{-3}. For instance, Mullin *et al.* [2] reported carrier concentrations of 2—5×10^{16} cm^{-3} for growth from silica crucibles while Henry and Swiggard [6] obtained 2×10^{15} cm^{-3} by synthesis and growth in pyrolytic boron nitride. The lowest value reported is that of Antypas [7], measured on selected grains of polycrystalline material. He obtained 6.8×10^{14} carriers cm^{-3} at the start of the ingot and 4×10^{15} carriers cm^{-3} at the end. LEC crystals grown from the high purity material had a carrier concentration of 2×10^{15} cm^{-3} suggesting that some impurities are segregated at the grain boundaries of the polycrystalline material.

It has been reported that the solid solubility of iron in InP [4] is considerably higher than that of chromium [3] and thus InP with background carrier concentrations of less than 5×10^{16} carriers cm^{-3} can be made semi-insulating by doping with iron.

Results of measurements on some undoped crystals are shown in Table 1, with mass spectrometric analysis in Table 2. These crystals were grown from a number of different source materials and the second column shows the source of the raw materials (all nominally 6N purity) and the crucible material in which they were synthesized. As expected, doping of material of this purity with about 1 ppmA of iron, produces semi-insulating material. Measured resistivities are greater than 10^7 ohm-cm.

Table 3 shows the resistivity, etch pit density and weight of a number of iron-doped crystals grown in the Malvern crystal puller. The diameters of these crystals were in the range of 45—60 mm. These crystals were grown with the addition of 0.1 wt % iron to the melt, and it can be seen that with one exception (302) all values of resistivity are greater than 10^7 ohm-cm and many are greater than 10^8 ohm-cm. The etch pit densities fall very consistently in the range 2—5×10^4 cm^{-2}. For 80 mm diameter 3 kg crystals grown in the Melbourn puller, etch pit densities are slightly higher at 3—8×10^4 cm^{-2}.

Table 1 Properties of 'undoped' InP

Crystal No.	Source material		Synthesized in	Carrier concentration (cm^{-3})		Mobility (cm^2 V^{-1} s^{-1})		Etch pit* density (cm^{-2})	
	In	P		Seed	Tail	Seed	Tail	Seed	Tail
294	Kawecki	Alusuisse	Graphite	3.8×10^{16}	1.6×10^{17}			3×10^4	4.8×10^4
260	Kawecki	Alusuisse	Graphite	7.5×10^{16}	7×10^{16}			2.8×10^4	4.1×10^4
251	Kawecki	Alusuisse	Boron nitride	8.7×10^{15}	–	–		–	–
263	Johnson Matthey (JM) AlA	Alusuisse	Graphite	6×10^{15}	1.3×10^{16}	3917	3745	6.4×10^4	2.5×10^5
265	JM AlA	Alusuisse	Graphite	1×10^{16}	2.1×10^{16}	3738	3850	5.4×10^4	8.17×10^4
305	JM AlA	MCP	Graphite	7.8×10^{15}	8.7×10^{15}	3771	3297	7.6×10^4	8.9×10^4
344	JM AlA	MCP	Boron nitride	6.8×10^{15}	1.3×10^{16}	4089	3439	5.3×10^4	9.7×10^4
272	MCP polycrystalline	InP	–	6×10^{15}	8.5×10^{15}	3738	3850	4.06×10^4	6.32×10^4

*Average of five points, 2 mm from edge, half radius at centre.
(100) by Haber-Linh etch [9].
(111) by HNO_3/$AgNO_3$/HF/H_2O.

Table 2   **Properties of iron-doped InP**

Crystal No.	Resistivity (ohm–cmx10^7)		Etch pit density (10^4 cm^{-2})		Crystal weight (g)
	Seed	Tail	Seed	Tail	
302	0.38	11	2.8	4.1	543
317	11	11	1.8	2.7	453
318	6.9	15	2.7	3.2	431
327	9.0	8.7	2.9	4.2	627
338	12	20	2.1	4.5	619
346	3.2	4.4	3.1	3.6	563
354	2.7	16	2.0	3.7	669
364	31	41	2.2	2.9	490
369	8.5	21	2.9	3.8	588
379	9.5	9.1	3.7	4.8	569
383	11	1.1	2.0	5.1	586
407	12	18	2.7	3.3	650
409	8.7	–	3.3	–	630
411	9	27	3.0	3.9	650

Table 4 shows the mass spectrographic analysis of a semi-insulating iron-doped crystal. It can be seen that the major electrically active impurity is sulphur at 0.75—1.05 ppmA. This crystal was grown from starting material of 4—7×10^{16} carriers cm^{-3}. Since the distribution coefficient of sulphur in InP is close to unity we would expect the sulphur content in the crystal to be about 1 ppmA, in agreement with the mass spectrometric results.

For a number of similar crystals it has been found that with a concentration of iron in the melt of 0.2 wt%, i.e. 10^{20} cm^{-3}, the concentration of iron in the seed end of the crystal is 0.7—1.0 ppm. Mass spectrographic analysis is normally quoted subject to a factor of three in either direction, i.e. a figure quoted as 1.0 ppmA lies between 0.33 and 3.0 ppmA. However, since the sulphur content of the crystal is close to 1 ppm and the crystal is semi-insulating, there must be at least 1 ppmA iron in the crystal giving a lower limit for the effective segregation coefficient of 4×10^{-4} and an upper limit of 1.2×10^{-3}, compared with values of 1.6×10^{-3} reported by Lee et al. [10] and 2×10^{-4} by Iseler [11]; the latter was for very low concentrations of iron in the melt.

Using 0.1 wt% of iron, inclusions are found in the lower end of the boule. The inclusions are crystalline in nature and are of two types: one needle shaped, the other rhombohedral shaped platelets. They are randomly orientated within the boule and although of relatively large size, there does not appear to be any significant strain field associated with them. From these facts it seems likely that the crystallites are grown from solution in the melt and later incorporated into the growing crystal.

X-ray microprobe analysis showed that these inclusions consist of iron and phosphorus only and contain approximately 43 wt% iron, suggesting the composition FeP_2 (47.4% iron).

Some of the crystallites were extracted by dissolving away the InP matrix with aqua regia and X-ray diffraction carried out. For both needles and platelets, all the X-ray diffractions were accounted for on the assumption that the crystal was of the FeP_2 structure, C^{18} type. The cell dimensions were measured and found to be the same as those given in the literature [8], namely a=2.725 Å, b=4.975, c=5.675 (orthorhombic).

Table 3 **Mass spectrometric analysis of 'undoped' InP**

Element	Crystal No.				Polycrystalline		
					MR	MR	MCP
	263	272	305	344	2776	2842	1210
C	0.6	0.3	0.6	<3	0.6	0.6	1.0
N	0.03	0.03	0.1	0.3	0.1	0.04	0.1
O	0.1	0.3	0.3	0.06	0.03	0.3	1.0
Na	0.007	0.03	0.03	0.06	0.03	0.008	0.02
Al	<0.006	<0.02	0.02	<0.006	<0.02	0.01	0.02
Si	0.2	0.8	0.3	0.3	0.15	0.2	0.3
P	Matrix						
S	<0.1	<0.2	<0.2	<0.2	<0.2	<0.2	<0.2
K	0.003	0.008	0.003	0.007	<0.002	0.007	0.02
Ca	0.01	<0.02	<0.008	<0.02	0.01	0.03	0.1
Fe	<0.02	0.045	0.065	0.045	<0.007	0.02	0.1
Cu	0.2	0.2	0.4	0.07	0.065	0.1	0.065
Zn	<0.04	<0.03	<0.01	<0.04	<0.02	<0.03	0.04
Ga	0.5	0.1	0.3	0.2	<0.015	<0.015	<0.02
As	0.3	<0.03	<0.03	0.02	<0.006	0.3	<0.009

Notes:

1. Figures expressed as ppmA.
2. All other elements undetected; limits range from 0.01 to 0.006 ppmA.
3. Poly 2842 was synthesized in PBN and used for crystal No. 344.
4. MCP 1210 was used for crystal No. 272; it is usual to find that silicon appears to have a segregation coefficient >1.

Table 4   Mass spectrographic analysis of iron-doped InP

Element	Impurity levels (ppmA)		
	Seed	Tail	
Li	0.02	0.025	
B	0.2	0.45	
C	<0.45	2.4	
N	<0.2	0.35	
O	<1.0	7.0	
Na	0.007	0.035	
Mg	0.015	0.045	
Al	0.007	0.01	
Si	0.03	0.1	
S	0.75	1.05	
Cl	0.1	0.2	Etchant
K	0.015	0.1	
Ca	0.07	0.035	
Fe	1	4	
Zn	0.05	0.03	
Ga	7.4	0.25	
As	0.007	0.01	
Br	0.04	0.01	Etchant
Nb	0.7	0.7	Masked by P_3
Ta	0.14	0.7	Spectrometer slits
Bi	0.07	0.1	

Notes:

1. All other elements undetected; limits vary from 0.01 to 0.003 ppmA.
2. This sample was analysed in a mass spectrometer without cryopumping, hence values for C, N, O are upper limits only.

Table 5   Thermal stability of InP at 650°C

	Crystal 1		Crystal 2	
	As grown	Annealed	As grown	Annealed
Orientation	<100>		<100>	
Type	N	N	N	N
Resistivity (ohm−cm)	8.1×10^7	6.7×10^7	1.3×10^8	7.6×10^7
Carriers (cm^{-3})	1.2×10^6	4.9×10^8	2.6×10^7	4.1×10^7
Mobility (cm^2 V^{-1} s^{-1})	2720	1910	1795	2030

THERMAL STABILITY

Since the most likely use of these crystals is as substrates for epitaxial growth, it is important that the semi-insulating properties survive the thermal treatment involved in the epitaxy process. Samples have therefore been treated in flowing hydrogen at 650°C for $1\frac{1}{2}$ hours and the results are shown in Table 5.

It can be seen that, although there is a slight drop in resistivity, the values are still above 10^7 ohm-cm and there is no type conversion.

Conclusions

Thermally stable semi-insulating indium phosphide can be produced by the LEC process using iron as dopant. Large crystals of both <100> and <111> orientation can be obtained in good yield, at economic rates, provided the top cone is grown in the appropriate shape.

The effective distribution coefficient of iron in indium phosphide is between 4×10^{-4} and 1.2×10^{-3}.

References

1. Metz, E.P.A., Miller R.C. and Mazelsky R. (1962). *J. Appl. Phys.*, **33**, 782
2. Mullin, J.B., Heritage, R.J., Holliday, C.H. and Straughan, B.W. (1968). *J. Cryst. Growth*, **3**, 281
3. Straughan, B.W., Hurle, D.T.J. Lloyd, K. and Mullin, J.B. (1974). *J. Cryst Growth*, **21**, 117
4. Muzuno, O., and Watanabe, H. (1975). *Electron. Lett.*, **11**, 10
5. Bachmann, K.J., Buehler, E., Shay, J.L. and Strand, A.R. (1975). *J. Electron. Mat.*, **4**, 389
6. Henry R.L. and Swiggard, E.M. (1976). *Inst. Phys. Conf. Ser.*, **33b**, 28
7. Antypas, G.A. (1977). *Inst. Phys. Conf. Ser*, **33b**, 55
8. *Strukturberichte*, **3**, 310
9. Huber, A. and Linh, N.T. (1975). *J. Cryst. Growth*, **29**, 80
10. Lee, R.N., Norr, M.K., Henry, R.L. and Swiggard, E.M. (1977). *Mat. Res. Bull.*, **12**, 651
11. Iseler, G.W. (1978). *Inst. Phys. Conf. Ser*, **45**, 144

LOW PRESSURE ORGANOMETALLIC GROWTH OF CHROMIUM-DOPED GaAs BUFFER LAYERS

M. BONNET, J.P. DUCHEMIN, A.M. HUBER and G. MORILLOT
Thomson-CSF, Domaine de Corbeville, 91401 Orsay, France

Abstract

A strong memory effect takes place during chromium incorporation in low pressure organometallic growth of GaAs layers. SIMS analysis is well adapted to measure the chromium profile in the layer and its incorporation rate in GaAs. The results agree with a two-step adsorption-desorption mechanism on the reactor wall. A decrease of 15% of the Hall mobility for a free carrier concentration of 10^{17} cm^{-3}, is observed when there is 2×10^{16} at. cm^{-3} of chromium in the active layer of a FET structure.

Introduction

The organometallic growth of GaAs is now a convenient method for producing epitaxial layers for FET applications on a large scale. One of the most important remaining problems is the purity of the starting compounds such as AsH$_3$ and trimethyl gallium (TMG) which does not allow us to obtain a low residual doping concentration [1] (about 5×10^{15} cm^{-3} and sometimes 1×10^{16} cm^{-3}). In this situation, it is not possible to grow a thick semi-insulating buffer layer required to avoid 'substrate effect' or a low mobility near the FET pinch-off voltage. Chromium doping of the buffer layer can be easily achieved by using CrO$_2$Cl$_2$ [2] in the halide process or hexacarbonyl chromium (HCC) as a volatile source of chromium as was previously shown by Bass [3] for organometallic growth at atmospheric pressure. Secondary ion mass spectrometry (SIMS), because of its low detection limit for chromium (5×10^{13} at. cm^{-3} [4]), is the best way of determining exactly the actual chromium profile in the epitaxial layer.

Experimental

The horizontal large reactor used has already been described [5] and was able to process 30 cm^2 of GaAs per run with good homogeneity. The growth temperature was fixed at 600°C and the TMG flow-rate was adjusted to obtain a growth rate of 0.1 μm min^{-1}. The total flow rate in the reactor was

20 litres min^{-1} measured at atmospheric pressure. The total pressure in the reactor was about 76 Torr during the epitaxial process. The hexacarbonyl chromium source was in a diffusion cell as shown in Figure 1. The cell's temperature varied between 0 and 18°C. We used several capillary diameters ranging from 1 to 2.5 mm.

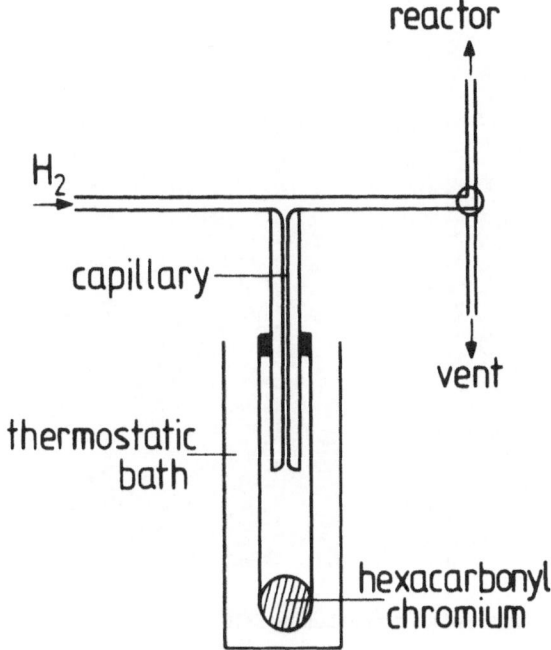

Figure 1   Hexacarbonyl chromium diffusion cell.

Results

Semi-insulating buffer layers were obtained when the chromium concentration was larger than $5-10\times10^{15}$ cm^{-3} measured by SIMS analysis. It seems that the chromium solubility limit in GaAs is around 5×10^{16} cm^{-3} [3]. The layers grown with more than 10^{17} at. cm^{-3} of chromium exhibit precipitates and an important surface roughness (Figure 2).

The semi-insulating threshold with increasing HCC concentration in the gas phase is not reliable, as we have observed by SIMS analysis. The reason is that the chromium profile in the layer is not flat. A typical chromium profile obtained by SIMS analysis in a FET structure with a chromium-doped buffer layer is shown in Figure 3. The chromium incorporation is clearly a very slow process, even at low pressure, and the maximum chromium concentration is reached in the active layer. Nevertheless, the buffer layer is semi-insulating. When we shut the chromium valve 25 minutes before injecting the dopant gas (SiH$_4$), to grow the active layer, we can see the strong memory effect of the chromium shown in Figure 4. The typical chromium concentration versus capillary diameter is difficult to evaluate because of the long transition times due to the memory effect. A rough comparison can be made by measuring chromium concentrations 30 minutes (3 μm from the interface) after the beginning of the growth, as shown in Table 1 where the temperature and the H_2 flow rate in the HCC cell are respectively 18°C and 45 cm³ min⁻¹.

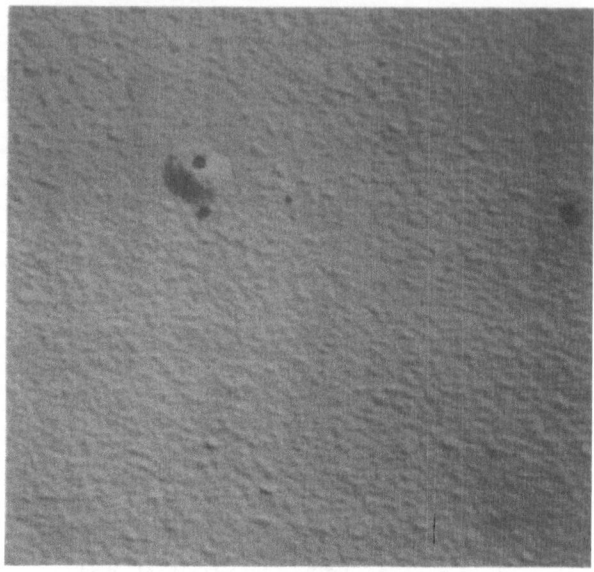

Figure 2   Surface of a layer containing 5×10^{17} cm⁻³ chromium. (×500)

The effect of chromium in the $1-2 \times 10^{16}$ cm⁻³ range in the active layer of a FET structure is to decrease the Hall mobility (measured at 300 K) at 1×10^{17} cm⁻³ by about 15% (Figure 5) compared with similar active layers grown without buffer layers. To measure the doping concentration profile of the active layer, we use the C(V) method with a mercury probe Schottky contact. The mercury contact is reverse biased with respect to a grounded large area mercury ring (Figure 6). The substrate is reverse biased with respect to the mercury ring. If the carrier concentration profile measured at −25 V substrate bias is different from the one obtained at 0 V, we call this the 'substrate effect' [6]. The reason is a developing space charge from the interface in the active layer due to charged states at the interface.

The main advantage of a chromium-doped buffer layer is to suppress the substrate effect if the buffer thickness is over 3 μm (Figure 7). If there is a 0.5 μm thick undoped buffer layer or if the chromium-doped buffer layer thickness is less than 3 μm, we always observe the substrate effect (Figure 8). The 0.5 μm undoped buffer layer is highly compensated by impurities from the semi-insulating substrate.

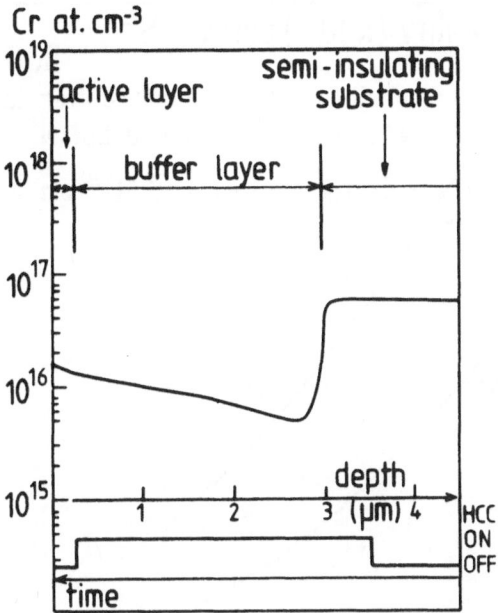

Figure 3 Typical chromium profile obtained in a FET structure by SIMS analysis.

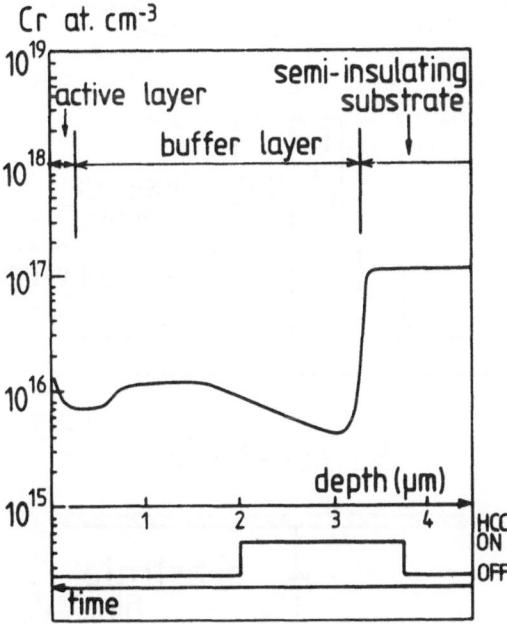

Figure 4 Chromium profile in a FET structure by SIMS analysis. The chromium valve was shut 25 minutes before the active layer growth.

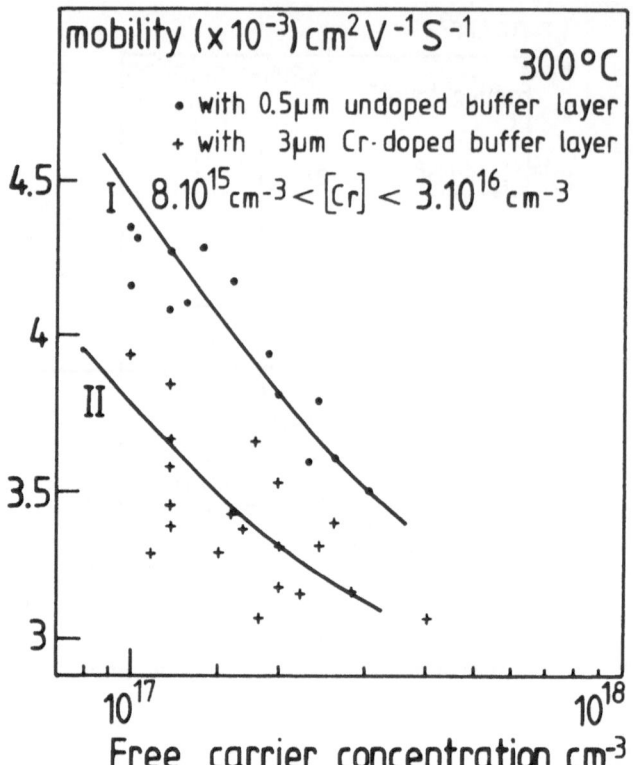

Figure 5 Average Hall mobility measured at room temperature on a FET structure: I — without; II — with a chromium-doped buffer layer.

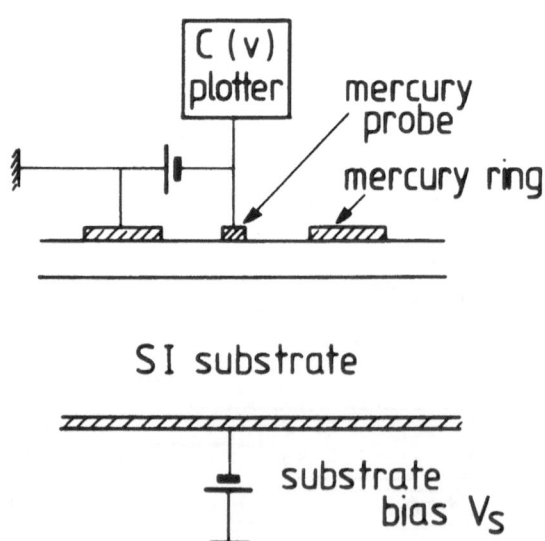

Figure 6 Reverse biased substrate C(V) measurement.

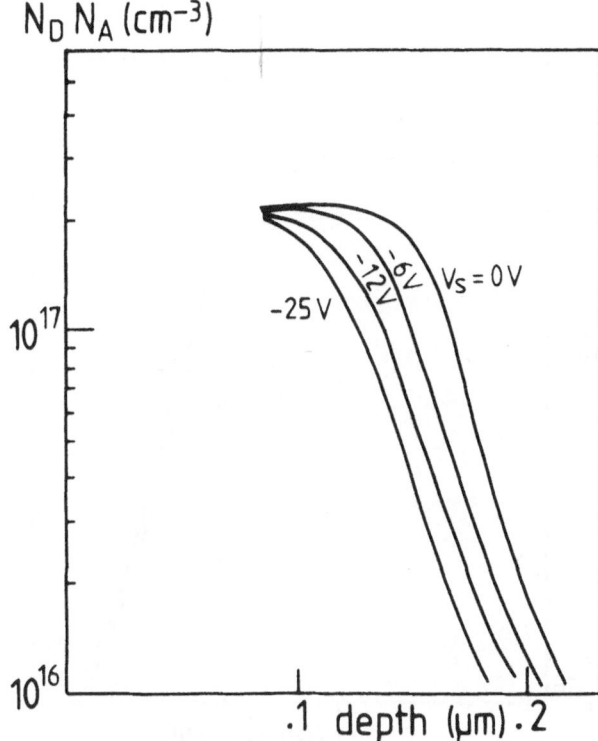

Figure 7   Carrier concentration versus depth on a FET structure with a 0.5 μm undoped buffer
layer at different substrate biases.

Table 1 **Chromium concentrations measured by SIMS at 3 μm from the
interface versus capillary diameter**

Capillary diameter (mm)	Chromium concentration measured by SIMS (cm^{-3})
1.0	6×10^{15}
1.5	9×10^{15}
2.0	1.3×10^{16}
2.5	2.3×10^{16}

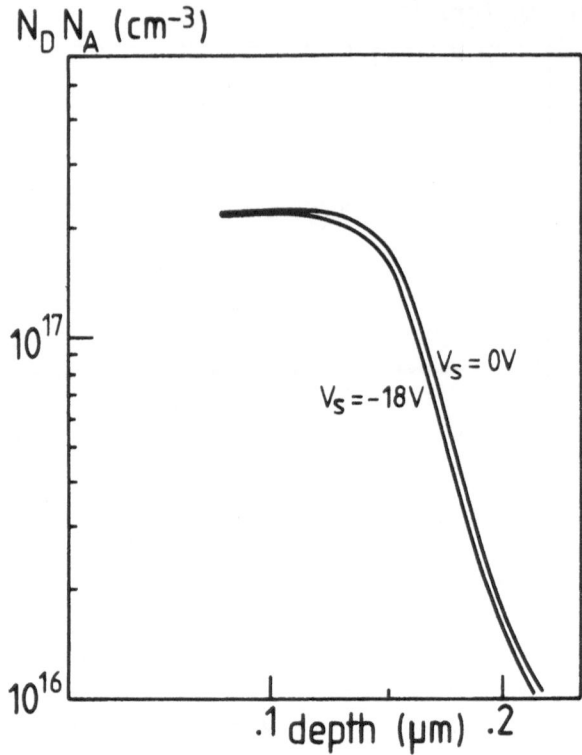

Figure 8   Carrier concentration versus depth on a FET structure with a 4 μm chromium-doped buffer layer at different substrate biases.

Discussion

All the SIMS analysis results agree with a two-step mechanism when the chromium is introduced during the growth in the form of HCC:

1. The hexacarbonyl chromium is probably quickly and strongly adsorbed on the reactor wall which is covered by solid arsenic and GaAs.
2. The desorption process is very slow and a steady state of the chromium concentration in the solid is obtained after a long time.

This effect can be described in terms of a gas chromatography process in which the reactor system is the column. The hexacarbonyl chromium retention time would then be very large (about 20 minutes), and it would probably be a consequence of the large area of the wall in contact with the gas flow. This effect could be reduced by changing the input line geometry of hexacarbonyl chromium.

Conclusion

FET structures are easily grown with a chromium-doped semi-insulating buffer layer. The problem is the memory effect of chromium which is incorporated in the active layer. Chromium decreases the Hall mobility of the layers. SIMS analysis is a convenient tool for monitoring the chromium profile in the layers in order to study the chromium incorporation rate versus reactor geometry.

ACKNOWLEDGEMENTS

The authors would like to thank D. Huyghe and M. Hug for their excellent technical assistance and maintenance of the experimental apparatus.

References

1. Hallais, J.H. (1978). *Acta Electron.,* **21,** 129
2. Kato, Y., Mori, Y. and Morizane, K. (1979). *J. Cryst. Growth,* **47,** 12
3. Bass, S.J. (1978). *J. Cryst. Growth,* **44,** 29
4. Huber, A.M. Morillot, G., Merenda, P. and Linh, N.T. (1979). *SIMS II Conference,* Stanford, 27—31 August, p. 91. New York; Springer Verlag
5. Duchemin, J.P., Bonnet, M., Koelsch, F. and Huyghe, D. (1979). *J. Electrochem. Soc.,* **126,** 1134
6. Rossel, P., Tranduc, H., Graffeuil, J., Azizi, C., Nuzillat, G. and Bert, G. (1978). *Rev. Phys. Appl.,* **13,** 503

LARGE DIAMETER, UNDOPED SEMI-INSULATING GaAs FOR HIGH MOBILITY DIRECT ION IMPLANTED FET TECHNOLOGY

R.N. THOMAS, H.M. HOBGOOD, D.L. BARRETT
and G.W. ELDRIDGE
Westinghouse Research and Development Center, 1310 Beulah Road,
Churchill Boro, Pittsburgh, Pennsylvania 15235, USA

Abstract

The growth of 2 and 3 inch diameter, <100> oriented semi-insulating GaAs crystals of improved purity by liquid encapsulated Czochralski (LEC) growth from silicon-free, pyrolytic boron nitride (PBN) crucibles in a high pressure Melbourn crystal puller, is described. Undoped and Cr-doped LEC GaAs crystals pulled from PBN crucibles exhibit bulk resistivities in the 10^7-10^9 and 10^8-10^9 ohm-cm ranges, respectively. High sensitivity SIMS demonstrates that GaAs crystals grown from PBN crucibles contain residual silicon concentrations in the low 10^{15} cm^{-3} range, compared to concentrations up to the 10^{16} cm^{-3} range for growths in silica containers. The residual chromium content in LEC/PBN-grown crystals is below the 10^{15} cm^{-3} range.

The achievement of direct ion implanted channel layers of near-theoretical mobilities is further evidence of the improved purity of undoped, semi-insulating GaAs prepared by LEC/PBN crucible techniques. Direct implant FET channels with $1-1.5 \times 10^{17}$ cm^{-3} peak donor concentrations exhibit channel mobilities of 4800 to 5000 cm^2 V^{-1} s^{-1} in undoped, semi-insulating GaAs substrates, compared with mobilities ranging from 3700 to 4500 cm^2 V^{-1} s^{-1} for various Cr-doped GaAs substrates. Discrete power FETs, which exhibit 0.7 W mm^{-1} output and 6 dB associated gain at 8 GHz, have been fabricated using these implanted high mobility, semi-insulating GaAs substrates.

Introduction

High frequency GaAs MESFETs have received increasing attention over the past decade. Monolithically integrated power and low noise/high gain amplifiers operating at X-band frequencies, as well as high speed GaAs digital logic ICs, are now being developed at several laboratories throughout the world. Fabrication of these monolithic circuits is often based on selective, direct ion implantation of semi-insulating GaAs substrates, because of its potential as a reliable, low-cost manufacturing technology.

It is widely recognized that the variable and often poor quality of semi-insulating substrates is one of the major limitations at present. To address this problem of unpredictable substrate properties, a crystal growth facility which utilizes a Melbourn liquid encapsulated Czochralski (LEC) puller has recently been established at our laboratories. LEC growth was selected over other growth technologies because of its current capability for producing round (instead of D-shaped), 2 and 3 inch diameter, <100> and <111> crystals and its adaptability to silicon-free, pyrolytic boron nitride crucible techniques [1, 2], offering the potential of semi-insulating GaAs crystals of significantly improved purity.

Crystal growth

Liquid encapsulated Czochralski (LEC) growth was first demonstrated experimentally in 1962 for the growth of volatile PbTe crystals [3] and has since been developed by Mullin *et al.* [4] for several III—V crystals. In this Czochralski technique, the dissociation of the volatile As from the GaAs melt, which is contained in a crucible, is avoided by encapsulating the melt in an inert molten layer of boric oxide and pressurizing the chamber with a non-reactive gas, such as nitrogen or argon, to counterbalance the As dissociation pressure.

Using a Melbourn high pressure LEC puller (manufactured by Metals Research Ltd, England), which enables *in situ* compound synthesis to be carried out from the elemental Ga and As components since the boric oxide melts before significant sublimation starts to take place, a number of 2 and 3 inch diameter <100> GaAs crystals free of major structural defects, such as twin planes, inclusions and precipitates, have been grown successfully. Several important modifications to the growth procedure have been instituted to achieve reproducible growth of high quality crystals, including the gradual increase to the desired crystal diameter during growth — to reduce dislocation generation — and use of vacuum baking of the boric oxide encapsulant to remove residual moisture — an important factor in maintaining high visibility of the melt-crystal interface during growth.

Substrate quality

CRYSTALLINE PERFECTION

Large diameter (≲3 inch) LEC GaAs crystals are usually characterized by high dislocation densities (10^4 to 10^5 cm^{-2}). Our initial investigations on improving quality of large diameter GaAs crystals have concentrated on the effects of seeding and crystal cone angle formation. Figure 1 shows an X-ray reflection topograph of a longitudinal section of a 2 inch diameter <100> GaAs crystal grown with a 27° cone angle. Dislocation-free growth is initiated by the Dash-type seeding used but dislocations are generated as the diameter increases. Dislocation multiplication is the result of the high internal stresses which develop as the crystal cone emerges from the B_2O_3 encapsulant [5] and our results indicate that the severe dislocation generation

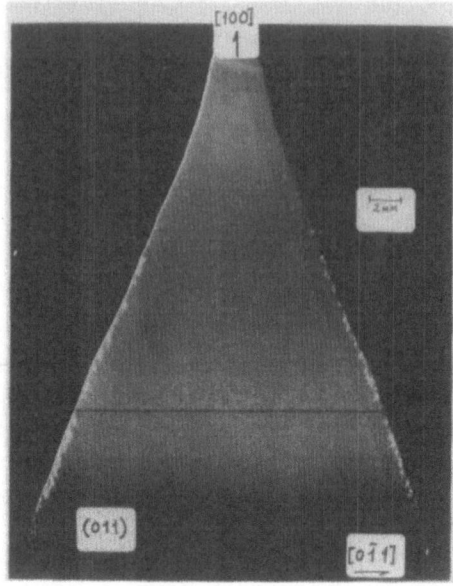

Figure 1 Reflection X-ray topograph (g=<260>) of (011) longitudinal section for a cone angle of 27°. Crystal pulled from pyrolytic boron nitride crucible.

and activation of glide planes in flat-topped crystals are significantly reduced by steep cone formation during growth.

RESIDUAL IMPURITY CONTENT

Secondary ion mass spectrometry bulk analysis of LEC GaAs material pulled from both quartz and pyrolytic boron nitride crucibles at our laboratories, as well as large-area boat-grown substrates purchased from an outside supplier have been carried out. The analyses were performed at Charles Evans and Associates, San Mateo, CA, using a Cameca 1MS—3f ion microanalyser. A wide range of impurity species was examined and the data generally show that the lowest impurity content is achieved in LEC growths from PBN crucibles. Quantitative estimates of impurity concentrations were obtained from GaAs samples which had been ion implanted with known doses of specific impurities. The results are illustrated in Figure 2, where the horizontal markers indicate different crystal samples. Residual silicon concentrations in the low 10^{15} cm^{-3} range are observed in GaAs/PBN samples compared to levels which range up to 10^{16} cm^{-3} in crystals grown in quartz containers. The residual chromium content in undoped GaAs is below the low 10^{15} cm^{-3} range (the Cr detection limit of the SIMS instrument), with Cr dopant concentrations typically approaching 1×10^{17} cm^{-3} in Cr-doped GaAs. Additional SIMS studies [6] do, however, indicate that LEC growths from PBN crucibles generally result in high boron concentrations (10^{17} cm^{-3} range) in the GaAs material versus low 10^{15} cm^{-3} concentrations in quartz crucible growth. No significant differences in carbon, 10^{17} cm^{-3}, oxygen, 10^{16} cm^{-3}, selenium, $<10^{15}$ cm^{-3} and tellurium, $<10^{15}$ cm^{-3} contents of different GaAs samples are revealed by these investigations.

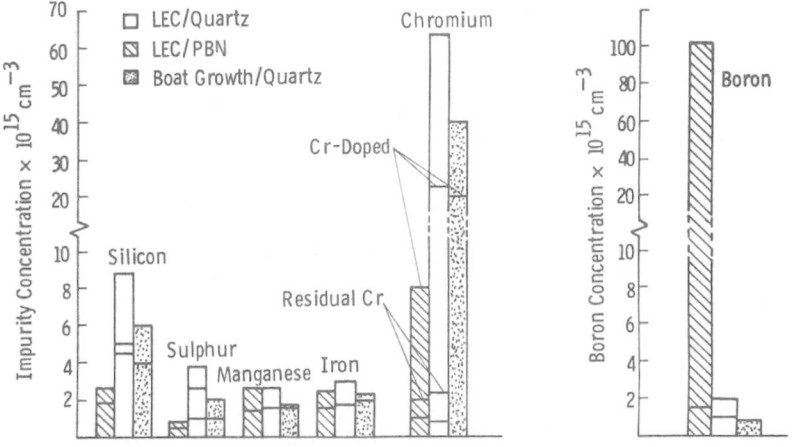

Figure 2   Bulk SIMS analysis of semi-insulating GaAs prepared by LEC and horizontal Bridgman growth. Horizontal markers indicate different samples. (Residual Cr is below the $1-2 \times 10^{15}$ cm^{-3} SIMS detection limit.)

ELECTRICAL CHARACTERIZATION

The axial resistivity variation along undoped and Cr-doped GaAs crystals is shown in Figure 3. Substrate resistivities in the $10^8 - 10^9$ ohm-cm range are observed in Cr-doped substrates compared to resistivities of $10^7 - 10^8$ ohm-cm in undoped GaAs/PBN substrates. Lightly Cr-doped GaAs/PBN crystals exhibit resistivities greater than 3×10^8 ohm-cm. Undoped GaAs crystals pulled from fused silica crucibles show lower resistivities and a greater variation along the crystal. Variable results were obtained from crystal to crystal varying from 10^6 to 10^7 ohm-cm in one to resistivities in the 10^3 ohm-cm range in another crystal. High impedance Hall measurements performed at ambient temperature indicate high apparent electron mobilities of between 5000 and 7200 cm^2 V^{-1} s^{-1} in undoped GaAs pulled from quartz or PBN crucibles and is a qualitative indication of the high purity of these samples. In contrast, the measured electron mobilities in all semi-insulating Cr-doped samples are in the 1000 to 2700 cm^2 V^{-1} s^{-1} range.

Thermal stability of substrates for ion implantation was assessed by means of resistivity measurements following an encapsulated anneal of the semi-insulating slice to determine whether any conducting surface layers had formed as a result of thermal treatment. The encapsulation consisted of a 900 Å plasma enhanced Si$_3$N$_4$ deposition overlaid with 1500 Å of 420°C CVD phosphorus-doped glass (PSG) applied to both surfaces. The wafers are annealed for 15 minutes at 860°C in flowing forming gas. Typically, sheet resistances of $\geq 10^8$ ohms/□ were observed for semi-insulating undoped GaAs/PBN and $\geq 10^9$ ohms/□ for Cr-doped substrates after implantation anneal. Surface sheet resistances exceeding 10^6 ohms/□ are desired for low leakage FETs and passive elements in monolithic circuits.

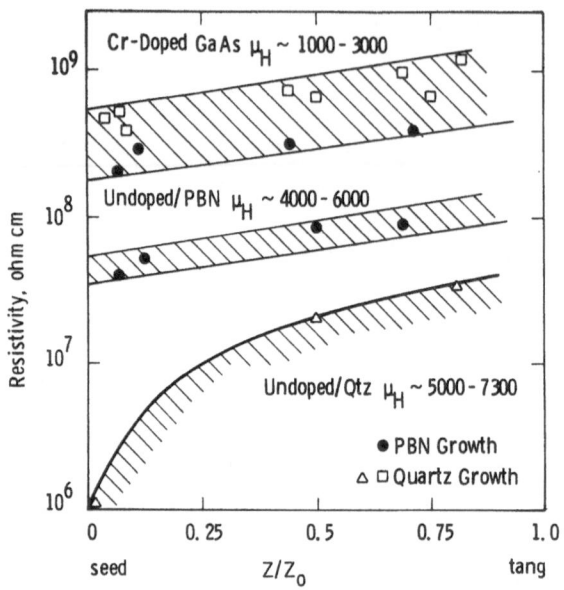

Figure 3   Axial resistivity variation along LEC grown GaAs crystals pulled from quartz and pyrolytic boron nitride crucibles.

Ion implantation

Direct ion implantation of undoped and Cr-doped GaAs substrates was performed at ambient temperature using the ^{29}Si ion beam of a 400 kV Varian/Extrion ion implanter. Implants were carried out on bare and encapsulated substrates at typically 350 kV. Measurements of channel mobility, implant activation efficiency and profile integrity were made using van der Pauw and capacitance-voltage techniques. Low-field Hall mobilities as a function of the peak donor concentration of direct implanted channels in undoped and Cr-doped GaAs substrates are shown in Figure 4. ^{29}Si implants into semi-insulating GaAs/PBN substrates show the highest mobilities: at peak donor concentrations of about 1×10^{17} cm^{-3}, mobilities of 4800 to 5100 cm^2 V^{-1}s^{-1} are measured compared with mobilities in the 3700 to 4500 cm^2 V^{-1}s^{-1} range for Cr-doped substrates prepared by LEC or boat growth. In lightly doped channel layers (2×10^{16} cm^{-3}) the observed differences are considerably larger — an electron mobility of 5600 cm^2 V^{-1} s^{-1} in undoped GaAs/PBN substrates versus 3000 cm^2 V^{-1} s^{-1} in Cr-doped substrates. The measured mobilities in directly implanted channel layers are compared to theoretical bulk mobilities [7] in Figure 4. The higher mobilities observed in undoped GaAs/PBN substrates are consistent with an excess compensating acceptor density of about 1×10^{16} cm^{-3} (and of unknown origin) in this higher purity substrate material. In addition to these high direct implant channel mobilities, implants into undoped GaAs/PBN substrates exhibit consistent and reproducibly high activation efficiencies of about 75—80%, low threshold dosages for the onset of activation (corresponding to an excess acceptor compensation of 1×10^{16} cm^{-3}) and predictable implant profiles which exhibit high uniformity over large substrate areas and from slice-to-slice ($\pm 5\%$).

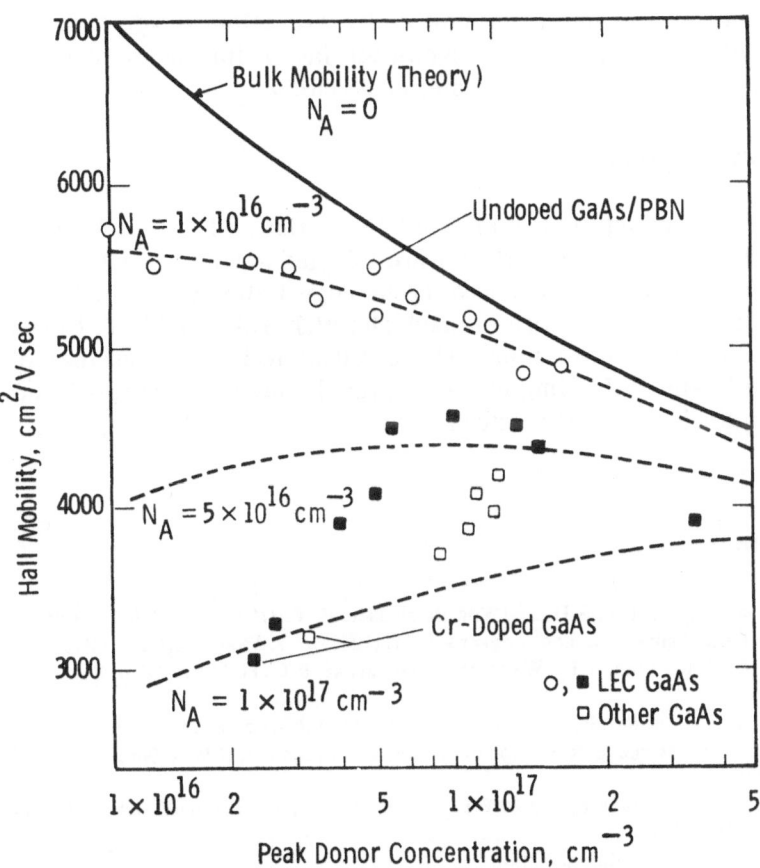

Figure 4 Measured direct ^{29}Si implant mobility in semi-insulating undoped and Cr-doped GaAs compared with theoretical bulk mobility.

Discussion and summary

The progress made towards developing higher purity, high resistivity GaAs crystals through the use of LEC pulling from pyrolytic boron nitride crucibles is reported in this paper. The intent is to achieve stable, semi-insulating substrate properties without resorting to intentional doping with chromium (or at least, to reduce the Cr content of GaAs significantly) in order to avoid the serious redistribution problems associated with this impurity at quite low temperatures and their potentially harmful effects on FET device performance and reliability.

Excellent power FET performance has been demonstrated from 1 μm gate length devices fabricated on semi-insulating GaAs/PBN substrates in our laboratories [8]. Directly implanted devices exhibit g_m values of about 100 mmhos mm^{-1}, small signal gains at 8 GHz of over 12 dB and an rf output power of 0.7 W mm^{-1}. Preliminary evaluations also indicate that device characteristics of improved uniformity are attained on semi-insulating GaAs/PBN substrates — a significant factor for the design of circuit parameters in monolithic integrated GaAs circuits.

ACKNOWLEDGEMENTS

The authors gratefully acknowledge the advice and encouragement of Dr H.C. Nathanson in this work. We are indebted to Dr J.G. Oakes and Dr M. Cohn (Westinghouse Advanced Technology Laboratories, Baltimore) for their support. We also wish to thank Drs W.J. Takei and T.T. Braggins for their technical contributions. The excellent technical assistance of L.L. Wesoloski and W.E. Bing in crystal growth and T.A. Brandis in substrate preparation is gratefully acknowledged.

References

1. Swiggard, E.M., Lee, S.H. and Von Batchelder, F.W. (1977). *Inst. Phys. Conf. Ser.,* **336,** 23; see also Henry, R.L. and Swiggard, E.M. (1977). *Inst. Phys. Conf. Ser.,* **336,** 28
2. AuCoin, T.R., Ross, R.L., Wade, M.J. and Savage, R.O. (1979). *Solid St. Technol.,* **22,** (1), 59
3. Metz, E.P.A., Miller, R.C. and Mazelsky, R. (1962). *J. Appl. Phys.,* **33,** 2016
4. Mullin, J.B., Heritage, R.J., Holliday, C.H. and Straughan, B.W. (1968). *J. Cryst. Growth,* **34,** 281
5. Roksnoer, P.J., Huybregts, J.M.P.L., van de Wiggert, W.M. and deKock, A.J.R. (1977). *J. Cryst. Growth,* **40,** 6
6. Evans, C. Private communication
7. Walukiewicz, W., Lagowski, L., Jastrzebski, L., Lichtensteiger, M. and Gatos, H.C. (1979). *J. Appl. Phys.,* **50** (2), 899
8. Oakes, J.G., Driver, M.C., Wickstrom, R.A., Eldridge, G.W., Wang, S.W. and Watkins, E.T. (1979). *IEEE GaAs IC Symposium,* Lake Tahoe, Nevada, Sept. 1979

GROWTH AND CHARACTERIZATION OF SEMI-INSULATING GaAs FOR USE IN ION IMPLANTATION

R.D. FAIRMAN and J.R. OLIVER
Rockwell International/Electronics Research Center,
Thousand Oaks, California 91360, USA

Abstract

A review of semi-insulating GaAs produced by three current bulk growth technologies has been made in which the electrical properties of the resultant semi-insulating GaAs have been studied in detail. High resistivity GaAs grown by liquid encapsulated Czochralski techniques is contrasted with horizontal Bridgman and gradient freeze materials. Both undoped and chromium-compensated materials have been studied. Materials qualification for ion implantation as determined by measurement of resistivity, thermal stability and active ion implant distributions will be described. Degree of impurity activation and relationship to LSS range statistics are discussed for selenium ion implants. Assessment of deep centres with concentration $<10^{14}$ cm^{-3} have been made on high resistivity samples using photo-induced current transient and dark conductivity versus temperature measurements. Impurity redistribution effects resulting from thermal annealing with Si_3N_4 capping layers are assessed by analysis with secondary ion mass spectroscopy (SIMS). Conclusions drawn from qualification tests, electrical transport and mass spectrographic measurements are correlated with the specific materials growth method used.

Introduction

Semi-insulating GaAs is currently being used in the development and production of a number of high performance electronic devices. These include high speed digital integrated circuits, low and high power field effect transistors, high speed charged coupled devices and monolithic microwave integrated circuits. Most of these device applications utilize ion implantation as a major technology for developing active layers and planar device structures.

Development of horizontal Bridgman and gradient freeze technology for LED applications has not been beneficial for advancement of semi-insulating GaAs. Resulting electrical yields for ion implantation by Bridgman grown crystals is generally low. GaAs boules grown in the <100> direction by liquid encapsulated Czochralski methods have shown considerable promise for use in high purity applications. Workers as early as 1965 [1] had shown the virtues of the B_2O_3 encapsulation technique. Weiner, Lassota and Schwartz [2] were the first to show the relationship between silicon contamination and the thermal stability of undoped semi-insulating GaAs. Recent work by Swiggard and Henry [3] and Rumsby [4] have demonstrated the merits of the LEC method. Recent high purity results reported by Aucoin *et al.* [5] for *in situ* synthesis of undoped and Cr-doped semi-insulating GaAs have confirmed the use of the liquid encapsulated Czochralski method for reliable growth of semi-insulating GaAs.

Liquid encapsulated Czochralski crystal growth

Recent results by Ware [6] have attracted much interest in the growth of large, highly uniform ingots using a new coracle method. *In situ* synthesis techniques developed over the past years by Metals Research have offered great promise in improving the electrical properties of semi-insulating GaAs. Problems associated with the electrical yield of GaAs for direct ion implantation involve:

1. Thermal conversion, resistivity and type degradation.
2. Low activation of implanted species.
3. Control of carrier profiles.
4. Variable implant properties over length of ingot.

The attainment of lower background doping levels is essential in maintaining control over high resistivity materials. In this work 'state of the art' 52 and 75 mm diameter crystals have been grown with the expressed intent to apply this technology to direct ion implantation.

Qualification criteria for ion implantation

Successful ion implantation has been demonstrated in semi-insulating GaAs showing the following properties:

1. High purity substrate with $< 5 \times 10^{15}$ cm^{-3} donor and acceptor impurities.
2. Freedom from harmful crystalline defects such as inclusions, precipitates and segregates.
3. Resistivity $> 10^7 \Omega$-cm or sheet resistance $> 10^7 \Omega/\square$.
4. Electrical compensation by deep donors, acceptors, or both.
5. Thermal stability during an 850°C annealing process with Si_3N_4 caps.
6. Achievement of high activation efficiency of implanted species.

Qualification of semi-insulating GaAs requires two basic tests:

1. *Thermal anneal:* semi-insulating substrates must demonstrate sheet resistance $\geqslant 10^7 |\Omega/\square$ following an anneal at 850°C, 30 min in H_2 with 1100 Å Si_3N_4 cap.
2. *Active ion implantation:* following a Se ion implant with 300 keV, 3×10^{12} cm^{-2} dose and annealing as in (1) the implanted layer must show

the following: (*a*) high activation of donor impurity, ⩾80% of dose; (*b*) correlation with LSS range statistics for above conditions; (*c*) uniformity in above properties over length of ingot tested.

Two slices, one from the front and the tail of a boule, are examined to provide the necessary data for acceptance of an ingot. Samples which meet the above criteria have been suitable in defining properties of the entire boule under study.

Experimental results

Both undoped and chromium-doped Czochralski ingots were sampled at front and tail locations and the following tests were performed:

1. Thermal anneal (sheet resistance measurements).
2. Selenium implanted (carrier profiles).
3. Secondary ion mass spectroscopy (elemental analyses).
4. Secondary ion mass profiling (elemental depth profiles for Cr).
5. Photo-induced transient spectroscopy (deep trap analysis).

Results of the thermal anneal test are shown in Table 1 along with the dopant used and crucible involved in the crystal growth work. Samples R1 to R5 were all grown from SiO_2 crucibles and showed excellent thermal stability, in excess of our specification. Sample R7 was grown from high purity pyrolytic boron nitride. It is interesting to note that all of these crystals have sheet resistance properties in the $10^9 \, \Omega/\square$ range, regardless of whether they are grown in PBN, SiO_2, undoped or chromium-compensated materials. The success indicated here is attributable to the high purity *in situ* synthesis technique employed in the growth of all of these ingots.

Table 1 Thermal stability — LEC GaAs

| Sample | Sheet resistance ($\Omega/\square$) | | Crucible material | Dopant |
	Front	Tail		
R1/C	4.2×10^9	4.2×10^9	SiO_2	None
R2/C	3.6×10^9	5.0×10^9	SiO_2	None
R3/C	2.0×10^9	4.2×10^9	SiO_2	Cr
R4/C	2.0×10^9	1.5×10^9	SiO_2	Cr
R5/M	6.7×10^9	5.0×10^9	SiO_2	Cr
R7/C	2.9×10^9	1.0×10^9	PBN	None

Selenium implantation of these materials, as shown by carrier profiles in Figure 1, indicates the high activation of Se in these substrates and the excellent correlation with LSS range statistics over the entire carrier profile. No indication of straggling is apparent for either the undoped or the chromium-compensated crystals. Implantation profiles from both front and tail samples show excellent agreement in profile depth at 10^{16} cm^{-3}. Hall mobility measurements at 300 K have shown carrier mobilities of 4500—4800 cm^2 V^{-1} s^{-1} with peak doping of 1.5×10^{17} cm^{-3} for selenium implanted samples of chromium-doped and undoped LEC GaAs.

ERC 80-7506

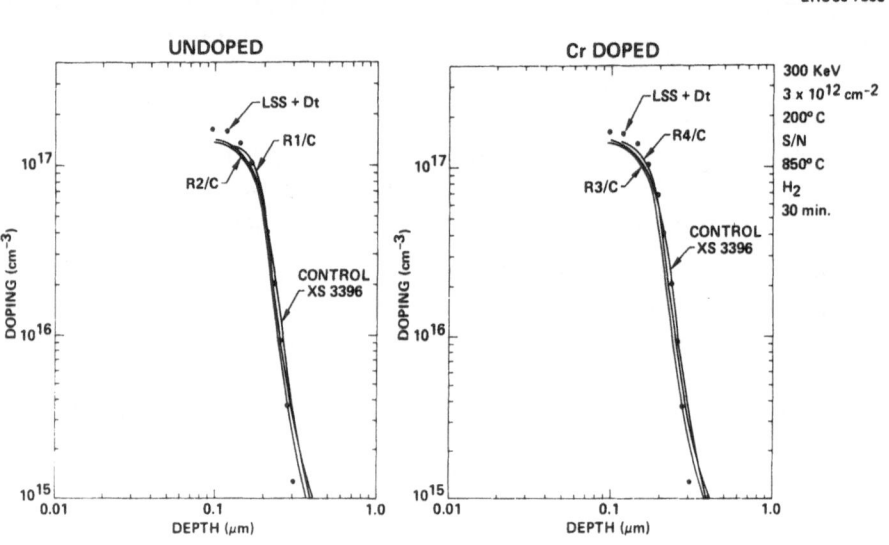

Figure 1   Selenium implanted LEC GaAs.

Secondary ion mass spectroscopy was carried out on Czochralski as well as Bridgman grown crystals in order to understand the major differences for their application to ion implantation. Table 2 shows the common donor, acceptor and deep level impurities studied in our materials.

It is important to note the low silicon content of all LEC crystals produced by *in situ* synthesis as compared to the Bridgman sample. Sulphur is a significant impurity in all GaAs samples as a common trace impurity in high purity arsenic. Magnesium in the LEC materials is due to a contaminant in the Si_3N_4 coracle. Low chromium levels are observed in the undoped crystals grown from SiO_2 or PBN crucibles. The concentration of iron is of some concern but does not limit the use of these materials in present device applications.

Table 2 Chemical analysis of LEC and Bridgman semi-insulating GaAs by secondary ion mass spectrometry (at.cm^{-3})

Sample	Si	S	Mg	Cr	Mn	Fe	B	Growth crucible
R2F	8.1×10^{14}	2.1×10^{15}	3.1×10^{15}	3.9×10^{14}	8.6×10^{14}	$<10^{15}$	4.2×10^{14}	SiO$_2$
R2T	1.0×10^{15}	2.9×10^{15}	5.6×10^{15}	1.2×10^{15}	1.7×10^{15}	4.0×10^{15}	7.0×10^{14}	SiO$_2$
R7F	1.3×10^{15}	3.2×10^{15}	1.3×10^{15}	1.7×10^{14}	$<5 \times 10^{14}$	2.1×10^{15}	6.2×10^{14}	PBN
R7T	2.4×10^{15}	6.3×10^{15}	3.2×10^{15}	1.9×10^{14}	8.2×10^{14}	2.3×10^{15}	3.4×10^{15}	PBN
R4F	6.3×10^{14}	1.4×10^{15}	3.6×10^{15}	1.1×10^{16}	1.5×10^{15}	1.9×10^{15}	1.2×10^{15}	SiO$_2$
R4T	1.2×10^{15}	3.9×10^{15}	2.8×10^{15}	4.8×10^{16}	2.4×10^{15}	1.1×10^{15}	3.8×10^{14}	SiO$_2$
3787F	6.3×10^{15}	4.2×10^{15}	4.0×10^{14}	5.8×10^{15}	5.4×10^{14}	3.6×10^{15}	$<10^{14}$	SiO$_2$
3787T	1.3×10^{15}	2.6×10^{15}	3.2×10^{14}	1.3×10^{16}	4.3×10^{14}	3.8×10^{15}	$<10^{14}$	SiO$_2$
4033F	1.2×10^{16}	1.2×10^{15}	3.9×10^{14}	4.2×10^{15}	3.5×10^{14}	2.5×10^{15}	$<10^{14}$	SiO$_2$
4033T	4.9×10^{15}	4.2×10^{15}	7.2×10^{14}	2.0×10^{16}	4.2×10^{14}	3.0×10^{15}	$<10^{14}$	SiO$_2$
GI–9HF	2.0×10^{16}	2.5×10^{15}	4.5×10^{14}	$17. \times 10^{16}$	5.9×10^{14}	3.7×10^{15}	$<10^{14}$	SiO$_2$
GI–9HT	4.6×10^{15}	3.4×10^{15}	4.6×10^{14}	8.7×10^{15}	4.2×10^{14}	2.9×10^{15}	$<10^{14}$	SiO$_2$
GI02–011GF	1.6×10^{16}	1.6×10^{15}	1.7×10^{14}	5.6×10^{15}	2.2×10^{14}	3.2×10^{15}	$<10^{14}$	SiO$_2$
GI02–011GT	8.9×10^{15}	4.2×10^{15}	6.8×10^{14}	2.3×10^{16}	8.1×10^{14}	2.2×10^{15}	$<10^{14}$	SiO$_2$
Secondary ion mass detection limits	(5×10^{14})	(1×10^{14})	(1×10^{14})	(1×10^{14})	(5×10^{14})	(1×10^{15})	(1×10^{14})	

The high Si concentration for all the boat grown crystals should also be noted. The effect of Si segregation here is significant where in LEC growth a normal (k<1) dependence is observed, but in boat grown materials a retrograde segregation of Si is shown. Sample GI02-011GF failed the thermal anneal test due to the high Si and low Cr concentrations present. The mechanism for thermal conversion in Cr-doped GaAs was shown by Asbeck et al. [7] and other workers to be the result of Cr out-diffusion as first described by Tuck et al. [8]. Figure 2 indicates how the high donor background becomes uncompensated following the thermal redistribution of chromium. A typical redistribution profile also shown in Figure 2 illustrates effects similar to those described by Asbeck et al., but in our case the background concentration of Si was much lower than the Cr intercept and consequently the crystal remained semi-insulating.

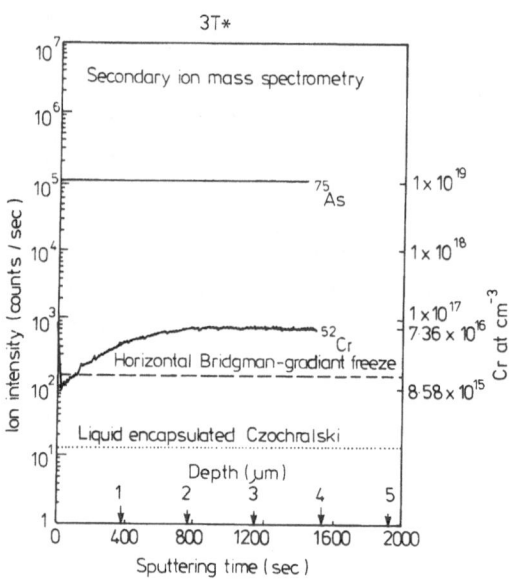

Figure 2   Chromium distribution by SIMS.

Transport measurements were performed on several epilayer and substrate samples using photo/dark conductivity and photo-induced transient spectroscopy (PITS), first described by Hurtes et al. [9]. PITS involves the detection of the current decay due to the emission of trapped carriers after illumination by chopped bandgap light. The current decay is sampled at two points in time, with the difference logarithmically recorded as a function of temperature. Peaks will occur when the trap emission rate is equal to the sampling rate in the case of discrete levels and hence, by varying the sampling time constant and observing the peak shift in temperature, the trap energy and cross-section can be computed. In the case of the measurements performed here, slightly greater than bandgap light (1.55 eV) and 5 μm separation ohmic surface contacts were used so that the measurements would be confined to the epilayer or near-surface regions.

PITS spectra for two epilayers and two substrates are shown in Figures 3 and 4, with the data plotted as the initial value of the current decay ΔI, normalized by the photocurrent, I_0 for a 3 ms sampling time constant. The VPE layers 1043 and 1157A in Figure 3 are Cr-doped grown on Cr-doped substrates. The substrates 3F and 3921F in Figure 4 are Cr-doped samples from the front of LEC and Bridgman ingots, respectively. A prominent feature of these spectra is the broad 0.90 eV peak associated with chromium acting as a hole trap. A second major peak occurs at $T_m = 330$ K as an inversion for the substrate samples and as a normal peak for sample 1043, with an activation energy of 0.65 eV. This has been associated with chromium acting as an electron trap. The hole trap at 0.83 eV, $T_m = 370$ K for 1157A is assigned to an (unidentified) deep recombination centre. Another prominent feature of these spectra is the peak at 170—190 K due to two 0.34 eV levels. A summary of the major levels measured is given in Table 3. An interesting interaction between Cr and the recombination centre is observed in the suppression of the 0.83 eV level (HL10) upon increasing the Cr concentration in VPE materials.

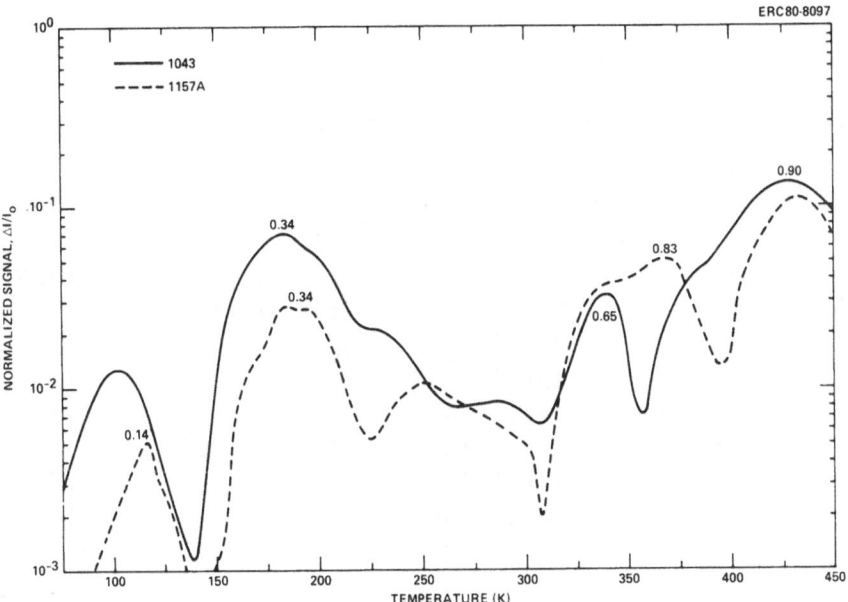

Figure 3 Normalized photo-induced transient spectroscopy, VPE Cr-doped GaAs.

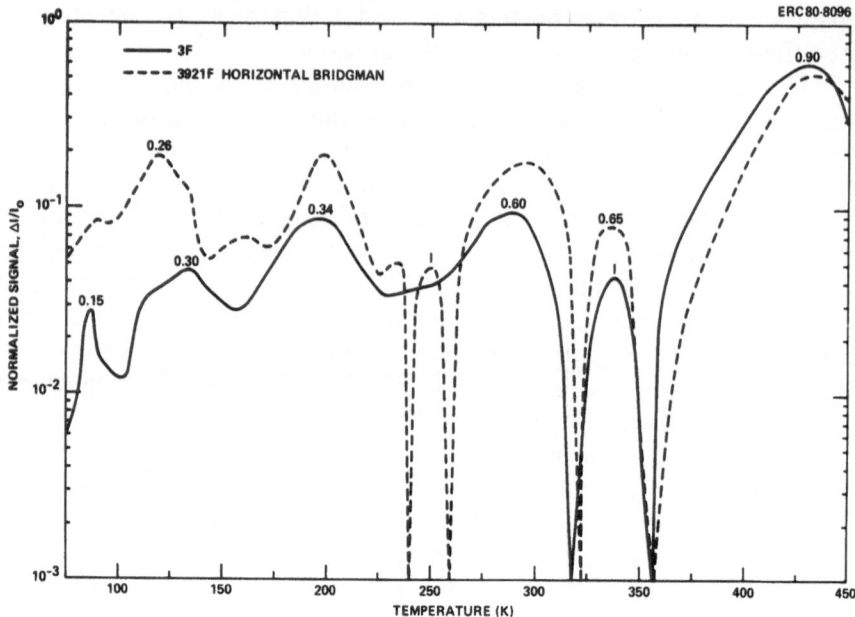

Figure 4 Normalized photo-induced transient spectroscopy, bulk Cr-doped GaAs.

Table 3 Trap summary from PITS measurements

T_M (K)	E_t (eV)	σ (cm^2)	Sample	Identity
88	0.15	8×10^{-14} (n)	3F	—
117	0.14	1×10^{-16} (n)	1157A	—
126	0.26	2×10^{-12} (n)	3921F	—
149	0.30	7×10^{-14} (p)	3F, 3921F	HL12
170	0.34	6×10^{-13} (n)	1043, 1157A	—
190	0.34	4×10^{-14} (n)	A11	EL6
245	0.51 (I)	1×10^{-12} (n)	3921F	EL4
280	0.60	1×10^{-12} (n)	3F	EL3
330	0.65 (I)	1×10^{-13} (n)	3F, 3921F, 1043	—
373	0.83	2×10^{-13} (p)	1157A	HL10
430	0.90	2×10^{-14} (p)	A11	HL1

The 300 K dark conductivity varied by no more than a factor of 4 for the measured samples, with the LEC substrate 3F showing the lowest conductivity at $\sigma = 5\times10^{-9}$ ohm^{-1} cm^{-1}. A summary of these data is shown in Table 4.

Table 4 **Dark conductivity data**

Sample	σ (ohm^{-1}-cm^{-1}) at 300 K	E_F (eV) at 300 K	E_A (eV)
1043 VPE	1.3×10^{-8}	0.69	0.75
1157 A VPE	1.2×10^{-8}	0.69	0.60
3F	5.0×10^{-9}	0.72	0.72
3921F	2.0×10^{-8}	0.68	0.70

E_F and E_A referenced to 0 K bandgap.

Conclusion

Conclusions drawn from thermal stability tests, ion implantation and SIMS analyses indicate silicon as the major impurity responsible for conversion in semi-insulating GaAs. The fact that chromium redistributes upon annealing at high temperatures influences the upper permissible donor concentration for Cr-compensated GaAs. In LEC crystal growth, a low silicon content may be achieved by *in situ* synthesis which permits undoped semi-insulating crystals to be obtained. Moderately high sulphur concentrations observed in this work may be offset by nearly equivalent compensation by iron also detected by SIMS analysis. Electrical activity of the impurities detected by SIMS was shown by photo-induced transient spectroscopy measurements. PITS data for LEC samples show up to a factor of 5 reduction in shallow states in comparison to Bridgman samples, with a similar reduction in dark conductivity. Eleven discrete trap levels have been identified in VPE and bulk crystal samples, including those due to chromium and the mid-gap recombination centre.

ACKNOWLEDGEMENTS

The authors wish to thank Drs Daniel Ch'en, Peter Asbeck and Ricardo Zucca for helpful discussion of this work, and Messrs Rolle Ware, Roger Waldock of Metals Research and Jeff Dreon for their assistance in crystal growth and characterization.

References

1. Mullins, J.B., Heritage, R.J., Holliday, C.H. and Str..ughan, B.W. (1965). *J. Phys. Chem. Solids,* **26**, 782
2. Weiner, M.E., Lassota, D.T. and Schwartz, B. (1971). *J. Electrochem. Soc.,* **118**, 301
3. Henry, R.L. and Swiggard, E.M. (1977). *Inst. Phys. Conf. Ser.,* **33b**, 28
4. Rumsby, D. (1979). *IEEE Workshop on Compound Semiconductors for Microwave Materials and Devices,* Atlanta

5. Aucoin, T.R., Ross, R.L., Wade, M.J. and Savage, R.O. (1979). *Sol. St. Tech.,* **22,** 59
6. Ware, R.M. (1979). *Int. Conf. Crystal Growth,* ICCG-5, Boston
7. Asbeck, P., Tandon, J., Babcock, E., Welch, B., Evans, C. and Deline, V. (1979). *Trans. Electron Dev. Ed-26,* **11,** 1853
8. Tuck, B., Adegboyega, G.A., Jay, P.R. and Cardwell, M.J. (1979). *Inst. Phys. Conf. Ser.,* **45,** 114
9. Hurtes, Ch., Boulou, M. Mintonneau, A. and Bois, D. (1978). *Appl. Phys. Lett.,* **32,** 821

THE INFLUENCE OF SUBSTRATE PROPERTIES ON THE ELECTRICAL CHARACTERISTICS OF ION-IMPLANTED GaAs

C.A. STOLTE
Solid State Laboratory, Hewlett-Packard Laboratories,
Palo Alto, CA 94304, USA

Abstract

The type of substrate used for the formation of n-type layers in GaAs by ion implantation has a pronounced effect on the electrical properties of the implanted and annealed layers. Certain Cr-doped semi-insulating substrates exhibit a marked decrease in sheet resistance during the anneal cycle. We find that the carrier concentration of the converted material increases with the background Si concentration in the starting material. The carrier concentration profile and mobility of implanted and annealed GaAs are dependent on the substrate. The most reproducible and the highest quality layers with regard to mobility and carrier profile are obtained in high purity LPE or high purity bulk starting material. Results from 13 high purity ingots, grown during a one-year period at HP, demonstrate the quality and reproducibility of the layers produced by direct implantation into these substrates.

Introduction

The use of ion implantation to produce n-type layers in GaAs suitable for device fabrication is well established. The techniques of implant and anneal are understood and many organizations achieve satisfactory results with their particular procedures. One area of major concern is the influence of the substrate material on the properties of the implanted and annealed layers. There are many procedures used to minimize the substrate problem, including the screening of bulk material [1], the use of semi-insulating epitaxial layers employing Cr doping [2] and the use of high purity, undoped epitaxial layers [3]. The goal of most investigators has been to develop materials and procedures which employ implantation into bulk material to produce satisfactory results without the necessity of epitaxial layers.

Experimental procedures

The results reported here were obtained using several different substrate types. Standard implant and anneal conditions were used to allow valid comparison of results. The ions were implanted into bare (100) GaAs substrates held at 350°C and oriented to minimize channelling.

The post-implant anneals were performed at 850 or 900°C in a flowing Ar atmosphere with a 1500 Å, Si_3N_4 anneal cap. The nitride layers were deposited at 650°C by a silane-ammonia reaction at a rate of 100 Å min^{-1}. He-backscattering analysis of these films indicates less than 2% oxygen concentration.

The bulk substrates were obtained from commercial sources, from other laboratories and from our internal sources. The properties of these substrates will be specified when the results are presented. The high purity LPE layers were grown on (100) semi-insulating substrates in a horizontal slider reactor at 700°C. The layers used for implantation were 2—3 μm thick. This epi material, characterized using Hall measurements on 15—20 μm thick samples, is n-type with mobilities measured at room temperature of approximately 8000 cm^2 V^{-1} s^{-1} and mobilities greater than 120 000 cm^2 V^{-1} s^{-1} measured at 77 K. The measured free carrier concentration at room temperature is less than 2×10^{14} cm^{-3}.

The properties of the implanted and annealed layers were measured using Hall measurements employing the van der Pauw geometry, differential Hall measurements [4] with chemical stripping to obtain carrier concentration and mobility profiles, and by automated C-V profile measurements.

Substrate conversion studies

The phenomenon of substrate conversion during the post-implant anneal and during epitaxial growth cycles is well known. Recent work using high sensitivity SIMS [5] and radioactive tracer techniques [6] has verified the Cr out-diffusion model proposed to explain this conversion. The data plotted in Figure 1 illustrate the dependence of the degree of conversion, N_s determined by Hall measurements, on the background Si concentration, determined by emission spectroscopy, in Cr-doped substrate materials which have not been implanted. Measurements of the carrier concentration profiles of these samples verify the dependence of the degree of conversion on the background Si concentration. The depth of the conversion layer is approximately 1 μm under these anneal conditions, 900°C for 15 min, which indicates a diffusion coefficient for Cr of about 1×10^{-11} cm^2 s^{-1} at 900°C, in agreement with the value derived by Asbeck [7].

Conversion experiments with low impurity, $<10^{15}$ cm^{-3}, bulk material or with high purity LPE layers produce no detectable decrease in sheet resistance. The Cr added during the growth to render the material semi-insulating by carrier compensation has deleterious effects on the carrier concentration profile and mobility of the implanted layers, as discussed below.

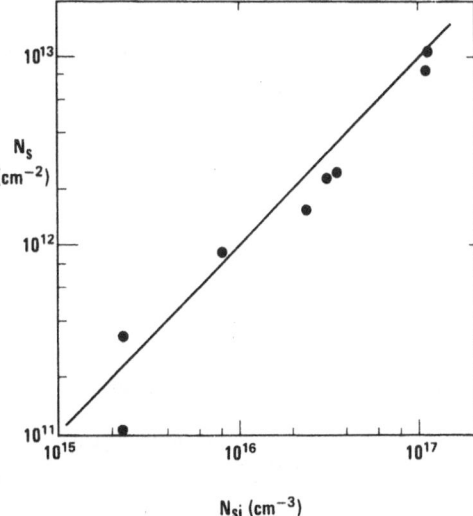

Figure 1   The dependence of the change in electron concentration, N_s, on the background Si concentration, N_{Si}, for unimplanted Cr-doped GaAs bulk material annealed at 900° C for 15 minutes with a Si_3N_4 cap.

Profile and mobility studies

The carrier concentration profile of implanted and annealed GaAs depends on the substrate used. This dependence can be very dramatic in cases where a high background impurity concentration of donors is revealed by the out-diffusion of Cr. The dependence is more subtle in cases where the background concentration is low and an excess Cr concentration is present to compensate the implanted species. In addition, it is believed that the defect density in the substrate can produce enhanced diffusion in some bulk material during the anneal process as reported for S implants [3].

The carrier concentration profiles of Figure 2 for Se implanted into different substrates illustrate some of the more subtle effects observed. The high purity bulk substrates were grown at Hewlett-Packard [8]. This material has very low impurity concentration and no Cr added to produce the semi-insulating properties. As seen in Figure 2, the profiles in the high purity LPE or bulk substrates are essentially the same and are as predicted for the implant and anneal conditions used. The profiles obtained in the high purity bulk material with small amounts of Cr added during the growth exhibit a carrier compensation effect. This effect is greater than predicted by the level of Cr concentration and is believed to be due to the Cr pile-up in the implant damage region which has been observed using the SIMS technique [5].

Differential Hall mobility data for S-implanted substrates are presented in Figure 3 where the effective mobility is plotted as a function of carrier concentration in the sample. These data were taken at room temperature and each data point represents the mobility and carrier concentration measured in a 400 Å layer. The dashed curves on the figure represent the theory of Rode and Knight [9] for the degree of compensation indicated. The influence of the background Cr concentration is most apparent in the low carrier concentration region and the agreement with theory is demonstrated.

The influence of the substrate material on the mobility measured at liquid nitrogen temperature will be presented in the next section.

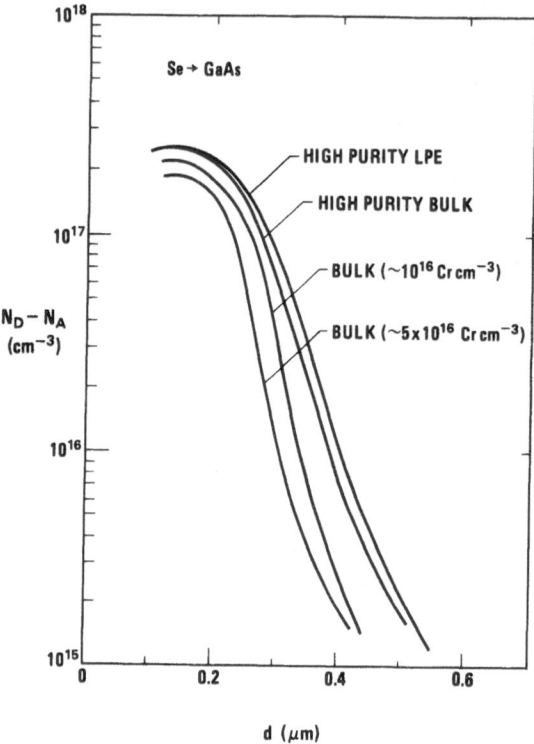

Figure 2   Carrier concentration profiles for Se implanted at 500 keV with a dose of 6×10^{12} cm^{-2} into the GaAs substrates indicated. Samples annealed at 850°C for 15 minutes with a Si$_3$N$_4$ cap.

High purity substrate results

The use of standard Se implant and anneal conditions over a period of three years has allowed the comparison of results from many different substrate types including high purity semi-insulating material grown at Hewlett-Packard [8]. The results of these long-term investigations are summarized in Figure 4 where we have plotted the mobility measured at room temperature and at 77 K for standard Se implant and anneal conditions for different substrate types. There was no conversion detected in the unimplanted anneal tests on these substrates.

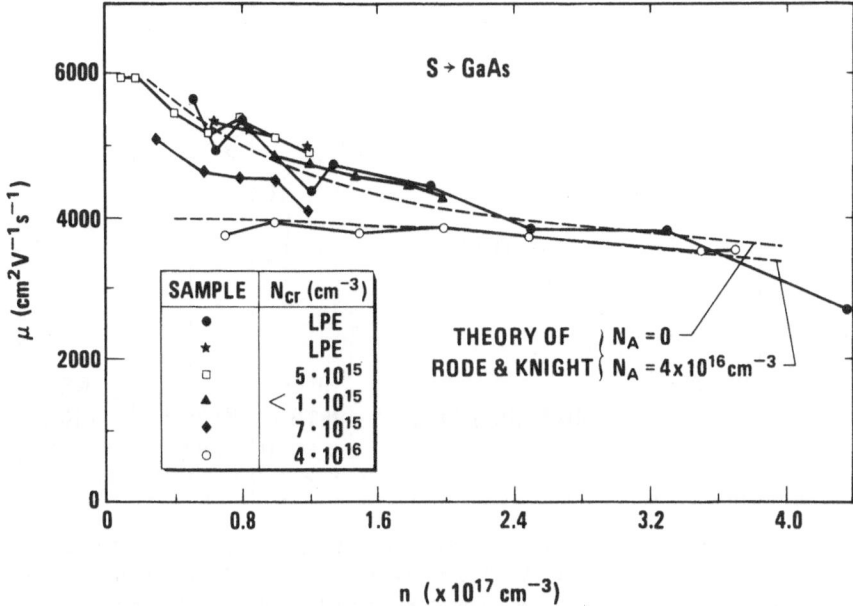

Figure 3 Hall mobility versus carrier concentration for GaAs samples implanted with S at 250 keV and annealed at 900°C for 15 minutes. The data were determined from differential Hall measurements on the indicated substrate types.

The data for the high purity LPE samples illustrate the desired mobility behaviour, greater than 4000 cm² V^{-1} s^{-1} at room temperature and a significant increase when the sample is cooled to 77 K, indicating a low amount of impurity compensation. The data for the HP high purity LEC substrates represent a total of 13 ingots grown over a one-year period. There are two sets of data points for each ingot representing the head and the tail of the ingot. As seen in Figure 4, in all cases except the one sample from the head of one ingot, the samples have the desired mobilities.

The data representing the HP LEC material with Cr added illustrate the decrease of mobility which results from a low Cr concentration, less than 3×10^{16} cm^{-3}, in the high purity material. This same behaviour is seen in the commercial material, from two different sources, represented in Figure 4. The data for the material supplied by the Naval Research Laboratory (NRL) [10] is comparable to the HP high purity material and indicates further promise that high quality bulk material can be produced for direct implantation.

The profile data of Figure 5 further illustrate the reproducibility of results using the standard implant and anneal conditions in the high purity material. The wafer-to-wafer reproducibility of the sheet resistance is ±5%, one sigma value, and the uniformity measured over a single wafer is less than ±3%. These uniformity values are verified by measurement of the saturated source-drain current on MESFET devices. The reproducibility and uniformity of the implanted and annealed properties is important for device fabrication, particularly for increased degrees of circuit integration.

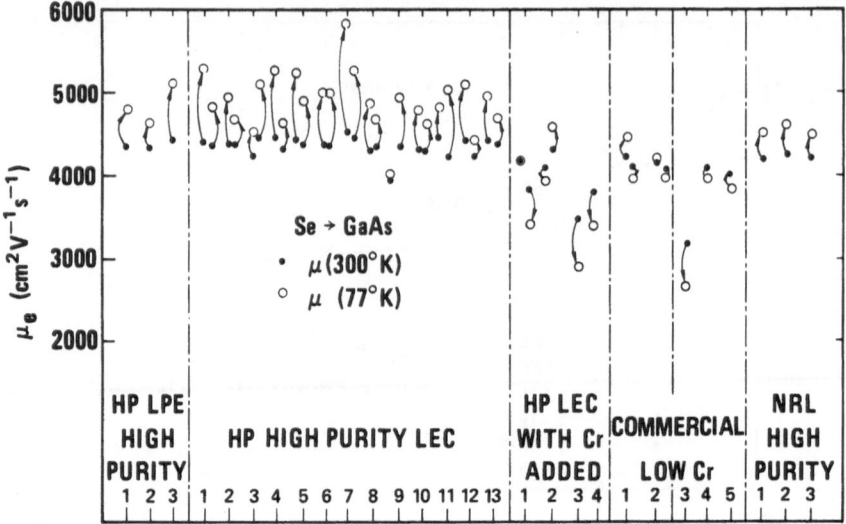

Figure 4   Hall mobility measured at 300 K (filled circles) and at 77 K (open circles) for GaAs implanted with Se at 500 keV with a dose of 6×10^{12} cm^{-2} and annealed at 850°C for 15 minutes with a Si$_3$N$_4$ cap. Each number along the horizontal axis represents one ingot, the double sets of data for some ingots represent samples taken from the top and bottom of the ingot.

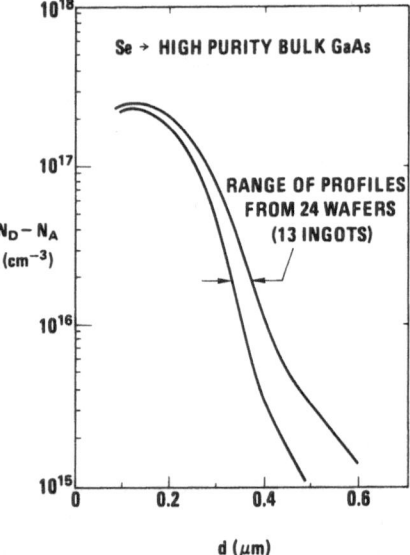

Figure 5   Carrier concentration profile uniformity for high purity bulk GaAs grown at Hewlett-Packard implanted with Se at 500 keV with a dose of 6×10^{12} cm^{-2} and annealed at 850°C for 15 minutes with a Si$_3$N$_4$ cap.

Summary and conclusions

The data presented summarize the effects of the substrate material on the conversion behaviour, the carrier concentration profiles, and the mobility of implanted and annealed layers. The results demonstrate the necessity for high purity starting material to obtain high quality, reproducible implanted and annealed layers. The use of Cr to compensate background impurities leads to inconsistent and inferior results as measured by mobility and profile analysis. The correlation of these material evaluation techniques and conclusions with microwave device performance is under investigation. The initial results indicate that high purity material is superior to Cr-doped material as an implant substrate.

References

1. Higgins, J.A., Kuvas, R.L., Eisen, F.H. and Chen, D.R. (1978). *IEEE Trans. Elec. Dev.,* **ED-25,** 587
2. Cox, H.M., DiLorenzo, J.V. and D'Asaro, L.A. (Sept. 1979). *GaAs IC Symposium,* Lake Tahoe, Nev., paper 12
3. Stolte, C.A. (1975). *IEDM Dig. Tech. Pap.,* p. 585. New York; IEEE
4. Mayer, J.W., Erickson, L. and Davies, J.A. (1970). *Ion Implantation in Semiconductors,* p. 186. New York; Academic Press
5. Evans, C.A. Jr, Deline, V.R., Sigmon, T.W. and Lidow, A. (1979). *Appl. Phys. Lett.,* **35,** 291
6. Tuck, B., Adegboyega, G.A., Jay, P.R. and Cardwell, M.J. (1978). *Inst. Phys. Conf. Ser.,* **48,** 114
7. Asbeck, P., Tandon, J., Siu, D., Fairman, R., Welch, B., Evans, C.A. Jr and Deline, V.R. (Sept. 1979). *GaAs IC Symposium,* Lake Tahoe, Nev., paper 14
8. Ford, W.M. To be published
9. Rode, D.L. and Knight, S. (1971). *Phys. Rev. B,* **3,** 2534
10. Swiggard, E.M., Lee, S.H. and Von Batchelder, F.W. (1978). *Inst. Phys. Conf. Ser.,* **48,** 125

HEAT TREATMENT BEHAVIOUR OF Cr IMPLANTED IN GaAs SI MATERIAL

F.SIMONDET, C. VENGER, G.M. MARTIN and J.CHAUMONT*
Laboratoire d'Electronique et de Physique Appliquée,
3, avenue Descartes, 94450 Limeil-Brévannes, France

Abstract

Cr has been implanted at different doses in GaAs (10^{13}—10^{15} at. cm^{-2}, 190 keV). Following implantation, the samples were annealed in two ways: with a silicon nitride cap or face to face in hydrogen flow. SIMS has been used to study the Cr distributions. Redistribution of chromium atoms occurs during heat treatment for the two types of annealing. In the case of the silicon nitride capped samples, the Cr migration involves two processes with very different rates. The faster one, which is observed as soon as the silicon nitride has been deposited, corresponds to a chromium motion towards the surface. The presence of a deeper structure for both types of heat treatment shows how important is the influence of the defects on the Cr diffusion mechanism. In the case of face to face annealing, the tail of the Cr distribution demonstrates that in-diffusion takes place.

Introduction

This study aims at a better understanding of the Cr behaviour (migration, solubility, etc.) during thermal treatments similar to those taking place in device processing (epitaxial growth, post implantation annealing, impurity diffusion, etc.). Chromium has been implanted in GaAs, and its distribution has been examined after different types of heat treatment.

Chromium-doped semi-insulating substrates supplied by RTC Caen have been used. The ingots were grown by the horizontal Bridgman technique and 0.4 mm thick slices were cut from the central part of the ingot with a (001) orientation. Their initial preparation involved a conventional lapping and a sodium hyperchlorite chemomechanical polishing procedure followed by a rinse in de-ionized water.

Chromium ion implantations were performed at room temperature. The incident beam was set at 6° off any low-index axis to minimize channelling effects. The ion doses ranged from 10^{13} to 10^{15} cm^{-2} and the ion energy was 190 keV.

*IN2 P3, Laboratoire R. Bernas, BP 1, 91406 Orsay, France.

SIMS procedure

Secondary ion mass spectrometry (SIMS) was used to measure the chromium profiles. The Cr^+ secondary ion signal in GaAs was converted into concentration by means of two calibration methods:

1. Use of standard samples previously analysed by spark source mass spectrometry, nuclear activation method and infra-red measurements [1].
2. Comparison of the implanted dose to the integral of the as-implanted chromium profile [2].

The agreement between these two methods is within 20%. In Figure 1, the chromium depth distributions measured by SIMS for doses of 10^{13}, 10^{14}, 10^{15} at. cm^{-2} at 190 keV are given. The peak concentrations are found to be in the same ratio as the doses, within the experimental precision. From these data it can be deduced that implantation damage due to 190 keV Cr^+ has little effect on $^{52}Cr^+$ secondary ion emission, and thus the concentration scale is valid.

Since the maxima of the curves can also be deduced from one another by translation, this means that the sputtering rate does not seem to be affected by the matrix damage due to different chromium implant doses. This justifies the validity of the depth scale.

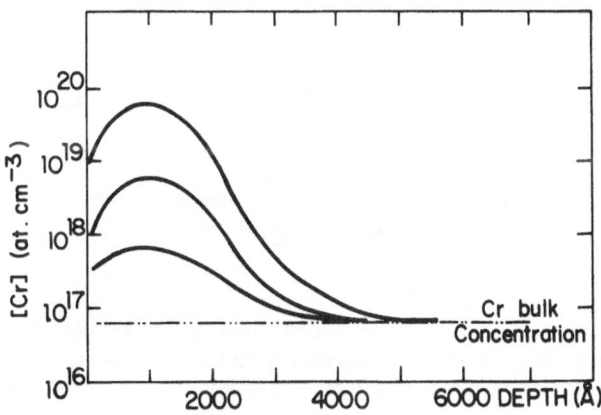

Figure 1 Chromium distribution, measured by SIMS, of as-implanted Cr in GaAs: energy 190 keV, doses 10^{13}, 10^{14} and 10^{15} Cr at. cm^{-2} .

Heat treatments of the implanted samples

Two types of heat treatment were performed on the implanted samples: in flowing argon with a silicon nitride cap and in flowing hydrogen with samples face to face.

(a) EFFECT OF SILICON NITRIDE DEPOSITION ON CHROMIUM PROFILE

A good quality silicon nitride film was grown by CVD on the surface of an implanted sample by a process described elsewhere (780°C, deposition time ~30 s)[3]. In Figure 2, two chromium profiles are shown: one (dashed line) is the as-implanted chromium distribution (10^{15} at. cm^{-2}, 190 keV), the other (continuous line) the chromium profile of the same sample just after silicon nitride deposition and without further annealing. The comparison of these data shows that during the growth of the silicon nitride film:

1. Most chromium atoms already move towards the first 100Å of the substrate[4].
2. The distribution of the chromium atoms which have not migrated shows a maximum at a depth around $R_p + \Delta R_p$ (projected range, projected standard deviation).

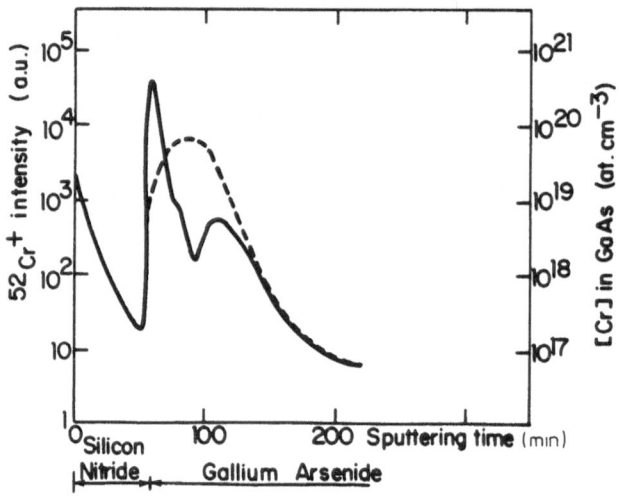

Figure 2 $^{52}Cr^+$ depth profile of an as-implanted sample: solid line=190 keV, 10^{15} Cr at. cm^{-2} and the same sample after silicon nitride CVD deposition; dashed lines = 780°C, deposition time~30 s. The concentration scale is only valid in the GaAs.

(b) ANNEALING WITH A SILICON NITRIDE CAP (IN ARGON FLOW)

Following the silicon nitride deposition, the samples were annealed at 870°C for 15—60 min in an argon flow. The chromium distribution of samples previously implanted at a dose of 10^{15} at. cm^{-2} (190 keV) are given in Figure 3. It appears that:

1. The thickness of the surface chromium film is fairly independent of the annealing time.

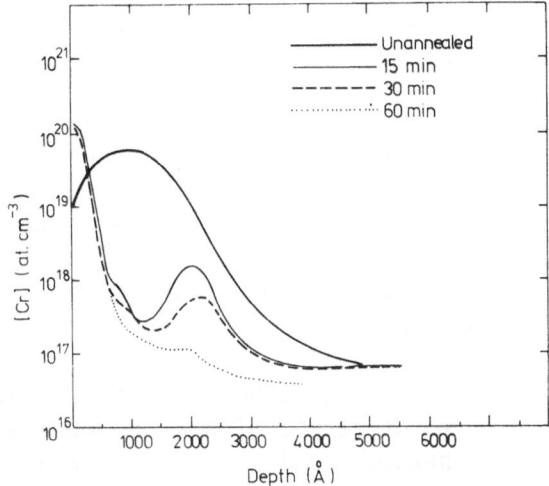

Figure 3 Chromium SIMS profile of Cr implanted GaAs samples (10^{15} Cr at. cm^{-2}, 190 keV): as-implanted, after silicon nitride deposition and annealing at 870°C for 15, 30 and 60 min.

2. The distribution of the chromium atoms that lie below the chromium surface film tends to flatten with annealing and there is no evidence for in-diffusion.

*(c)*EFFECT OF THE IMPLANTED DOSE

The heat treatment described (*b*) has also been performed for 15 min at 870°C on samples implanted at 10^{13}, 10^{14} and 10^{15} at. cm^{-2} (190 keV). Their chromium distributions are depicted in Figure 4.

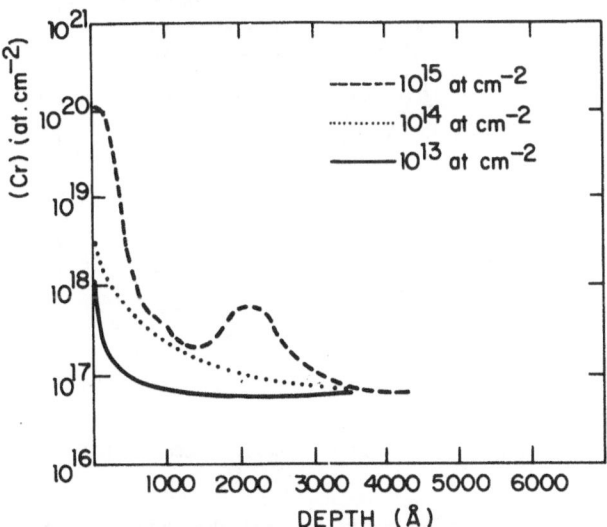

Figure 4 Comparison of chromium depth profile for Cr implanted GaAs samples ($10^{13}-10^{14}$ and 10^{15} at. cm^{-2}, 190 keV) after heat treatment for 15 min at 870°C.

Such different doses correspond to quite different implant damages: it has been shown that a dose of 10^{15} cm^{-2} practically amorphizes the material, while doses of 10^{13} cm^{-2} do not produce such an effect, as evidenced by spectroscopic ellipsometry [5]. The comparison with Figure 1 clearly demonstrates that in all cases the Cr distribution vanishes, and this seems to be explained by the same fast exo-diffusion in the three cases. This suggests that such a behaviour cannot be explained only by implantation damage.

(*d*)CAPLESS ANNEALING WITH FACE TO FACE SAMPLES, IN HYDROGEN FLUX AS A FUNCTION OF TEMPERATURE

Using this type of annealing, it is possible to investigate heat treatment at low temperature (here, from 670 to 870°C). This was not possible in the previous case where some annealing had already taken place at 780°C leading to the Cr redistribution reported in (*a*) and to a substantial recrystallization of amorphized material as evidenced by spectrometric ellipsometry [6].

Figure 5 shows Cr profiles of face to face annealed samples as a function of the annealing temperature for Cr implanted bulk samples (3×10^{14} at. cm^{-2}, 190 keV).

As in the case of silicon nitride capping there is an accumulation of Cr at the surface and a bump deeper in the sample, located around $R_p+\Delta R_p$. It is worth noticing that these effects already take place at a temperature as low as 673°C.

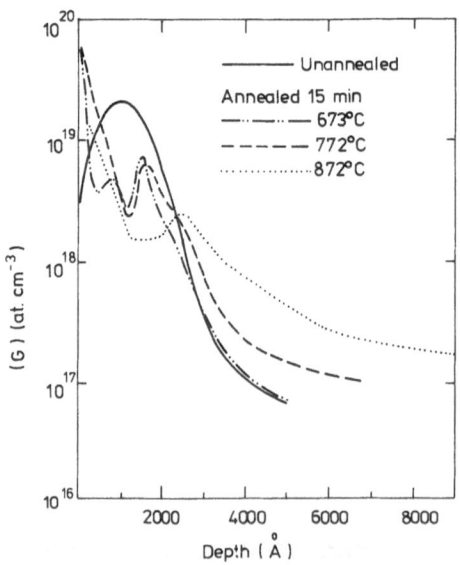

Figure 5   Chromium concentration profile (Cr implanted samples: 3×10^{14} Cr at. cm^{-2}, 190 keV): as implanted samples, face to face annealed samples 673, 772 and 872°C (in flowing H_2).

Furthermore, contrary to capped samples there is evidence for a Cr in-diffusion in the deeper part of the profile. From these data, a crude calculation of the diffusion coefficient D leads to the following values:

$$D \sim 2.0 \times 10^{-14} \, cm^2 \, s^{-1} \qquad \text{for } T=772^\circ C$$
$$D \sim 1.5 \times 10^{-13} \, cm^2 \, s^{-1} \qquad \text{for } T=872^\circ C.$$

These values are in qualitative agreement with previously published data [7].

(*e*)COMPARISON BETWEEN CAPPED AND CAPLESS SAMPLES

Figure 6 gives two profiles of implanted samples which have been annealed for 15 min at 870°C. Keeping in mind that the dotted curve refers to the samples implanted at 3×10^{14} at. cm^{-2} and the full curve to the one implanted at 10^{15} at. cm^{-2}, two comments can be made:

1. Exo-diffusion occurs in both samples, but is apparently larger when silicon nitride is used.
2. A tail, which is never present in capped samples, is now clearly observed in the uncapped material.

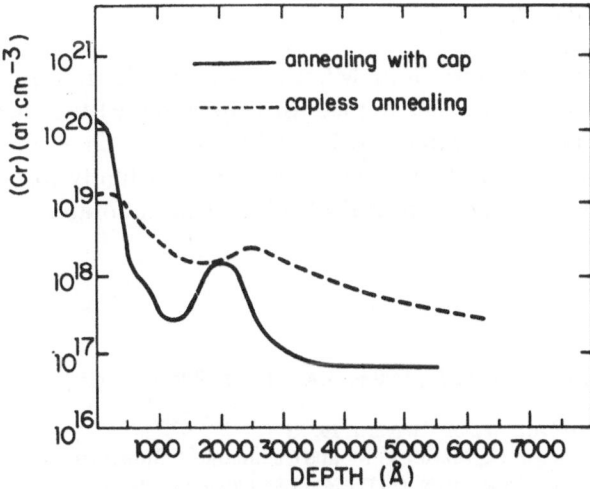

Figure 6 Comparison of Cr distribution of silicon nitride cap annealed sample 870°C, 15 min (Cr implanted 10^{15} at. cm^{-2}, 190 keV) and face to face annealed sample 872°C, 15 min (Cr implanted 3×10^{14} cm^{-2}, 190 keV).

Discussion and conclusions

We have demonstrated a complex redistribution of chromium, the main characteristics of which are the following:

1. Cr segregates at the GaAs surface. The migration towards the surface is very fast: it takes place within less than 30 s at 780°C in the case of silicon nitride deposition. It is not induced by implantation damage since it also occurs for a dose as low as 10^{13} Cr at. cm^{-2}. Such a segregation has also been observed in Cr-doped samples [8—10].
2. The diffusion mechanisms are influenced by the presence of the silicon nitride cap.
3. The exodiffusion predominates when a silicon nitride cap is used. This could be related to a possible gettering effect, already suggested by other authors [8]. This effect seems to be so strong that it prevents in-diffusion, which is only observed when a silicon nitride cap is not used.
4. Cr diffusion processes are also affected by the presence of defects in the bulk. This influence appears through the analysis of deeper structures in the profiles of capped or capless samples. The data also demonstrate that Cr prefers to migrate rather than to precipitate even when it has been introduced at very high concentrations. The formation of Cr precipitates close to the surface is still questionable. Thus the question related to the limit of solubility of Cr in GaAs remains an open one.

The position of chromium in the lattice as well as the existence of possible phases near the surface remain to be studied.

ACKNOWLEDGEMENTS

The authors wish to thank Dr A. Mircea and Dr S. Makram-Ebeid for many fruitful discussions, and J.Y. Aupied for help with the SIMS measurements. Further, they are very grateful to Dr H.W. Werner, Dr A.E. Morgan and H.A.M. de Grefte, who have given them the opportunity to make some of the measurements on the Cameca IMS 300 of their Laboratory.

References

1. Martin, G.M., Verheijke, M.L., Jansen, J.A.J. and Poiblaud, G. (1979). *J. Appl. Phys.*, **50**, 467
2. Werner, H.W. (1976). *Acta Electronica*, **19**, 53
3. Simondet, F., Venger, C., Martin, G.M. and Chaumont, J. Submitted to *Appl. Phys. Lett.*
4. Deveaud, B. and Favennec, P.N. (1977). *Solid St. Commun.*, **24**, 473
5. Theeten, J.B. Private communication
6. Martin, G.M., Steers, M., Venger, C., Simondet, F. and Rigo, S. (1979). *Laser and Electron Beam Processing of Materials*, November, Cambridge, Mass.
7. Casey, H.C. (1973). *Atomic Diffusion in Semiconductors* (Ed. D. Show) p. 417. New York; Plenum Press

8. Evans, C.A. Jr, Deline, V.R., Sigmon, T.W. and Lidow, A. (1979). *Appl. Phys. Lett.,* **35,** 291

9. Huber, A.M., Morillot, G., Linh, N.T., Favennec, P.M., Deveaud, B. and Toulouse, B. (1979). *Appl. Phys. Lett.,* **34,** 858

10. Tuck, B., Adegboyega, G.A., Jay, P.R. and Cardwell, M.S. (1979). *Inst. Phys. Conf. Ser.,* **45,** 114; Tuck, B. and Adegboyega, G.A. (1979). *J. Phys. D.,* **12,** 7090

REDISTRIBUTION AND VAPORIZATION OF Cr IMPURITIES IN SEMI-INSULATING GaAs

T. UDAGAWA, M. HIGASHIURA AND T. NAKANISI
Electronics Equipment Division, Toshiba Corporation,
1 Komukai Toshiba-Cho, Saiwai-ku, Kawasaki 210, Japan

Abstract

Cr impurity redistribution and its origin in Cr-doped semi-insulating GaAs substrates annealed without encapsulation have been investigated using secondary ion mass spectrometry and flameless atomic absorption spectrometry. It is concluded that, as a result of the Cr vaporization from the substrate surface, Cr diffuses rapidly towards the surface at 800°C and above. Also, the influence of surface thermal dissociation on the Cr diffusion is discussed.

Introduction

Cr-doped semi-insulating GaAs has been extensively used as a substrate for GaAs microwave FETs and high speed ICs. These devices require active layers formed either epitaxially on the substrates or directly by ion implantation into the substrates. During the course of layer formation, the substrates are heat treated to relatively high temperatures. Such annealing frequently gives rise to device degradation [1, 2].

It has been reported that, during annealing of the substrates with encapsulation, Cr diffuses rapidly towards the substrate surface and accumulates, leaving a Cr-depleted region underneath [3]. Favennec and L'Haridon [4] stated that thermal degradation of the substrate is caused by such Cr depletion. From these results, it seems that a detailed thermal degradation mechanism has not been established.

This paper shows that, during annealing without encapsulation, Cr vaporization from the surface occurs, which gives rise to rapid Cr diffusion towards the surface. It is also shown that surface thermal dissociation plays an important role in Cr diffusion.

Experimental

Cr-doped semi-insulating GaAs substrates were purchased from Mitsubishi Monsanto Chemical Co. The substrate orientation was <100> tilted 2° towards <110>. One side of the substrates was chemically, mechanically polished until a mirror-like surface was attained. The substrates were boiled in trichloroethylene, methanol and concentrated HCl, and then rinsed in deionized water.

Annealing was carried out in an apparatus essentially similar to that used by Kasahara, Arai and Watanabe [5]. The substrate was loaded on a quartz pedestal, where the mirror-like surface of the substrate was exposed to the gas stream. The substrates were annealed in pure H_2 or H_2 ambient containing a small amount of AsH_3 gas. The AsH_3 was introduced to reduce thermal dissociation of the substrates. The partial pressure of AsH_3, P_{AsH_3}, was defined as the ratio of AsH_3 flow rate to total flow rate.

After annealing, the substrates were removed from the furnace and mounted on a secondary ion mass spectrometer (CAMECA IMS-300). In-depth profiles of Cr concentration were measured.

In an attempt to determine whether Cr vaporizes from the surface, the substrates were annealed in pure N_2 ambient in a graphite atomizer (Perkin Elmer HGA 2100) mounted in a flameless atomic absorption spectrometer (Perkin Elmer Model 360). After removal of the substrate from the atomizer, adsorbents on the inside wall of the atomizer were analysed.

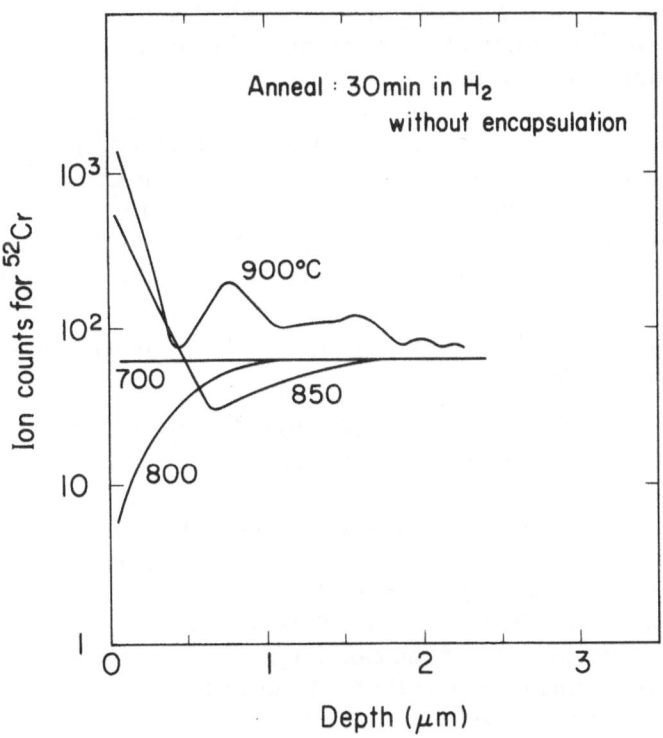

Figure 1 Cr concentration profiles for 30 minutes annealing in H_2 ambient at 700, 800, 850 and 900°C.

Results and discussion

Figure 1 shows the Cr concentration profiles for substrates annealed in H_2 ambient. After annealing at 700°C, no Cr redistribution is observed. At 800°C, a region is found where Cr is depleted. In this case, no Cr accumulation is observed near the surface region.

At 850°C and above, the profiles were not reproducible. They differed according to the measured position of the surface, although the general trend for the profiles was maintained. It seems that Cr tends to accumulate near the surface. The accumulation became intense when the temperature was increased. It was found that the higher the annealing temperature, the rougher the surface became. An optical micrograph of the surface showed the presence of well-known Ga droplets [6]. The situation is analogous to that of Cr doping into an LPE layer [7], where GaAs solution is in contact with the substrates. Because of the large Cr solubility in the solution and small segregation coefficient [7, 8], it is probable that Cr diffuses into Ga droplets. This may be the origin of the apparent accumulation and poor reproducibility of the profiles.

At 900°C, a very irregular profile is seen below the near surface region. Moreover, it seems that the Cr concentration is higher than that in the initial sample. However, it is not clear at present whether the profile shows the accurate depth distribution of Cr, since the surface becomes very rough and large Ga droplets grow at 900°C.

Figure 2 shows Cr concentration profiles for substrates at 900°C under $P_{AsH_3} = 1$ Torr as a function of annealing time. The AsH_3 was introduced in an attempt to reduce thermal dissociation. Actually, the Cr-depleted region extends towards the inner part of the substrate as the annealing time is increased. In all cases, no surface accumulation of Cr is observed. This may suggest that Cr diffuses rapidly towards the surface and vaporizes, leaving a Cr-depleted region underneath.

Based on a simple out-diffusion model, Cr concentration profiles are given by the following equation [9]:

$$N_{Cr}(x,t)/N^0_{Cr} = erf(x/2\sqrt{Dt}) + exp(ht)/\sqrt{Dt})^2 \cdot exp(ht/\sqrt{Dt})(x/2\sqrt{Dt}) \cdot erfc(ht/\sqrt{Dt} + x/2\sqrt{Dt}), \tag{1}$$

where N^0_{Cr} denotes the original Cr concentration, and h is the gas-phase mass-transfer coefficient. Other symbols have conventional meanings. Figure 2 shows calculated profiles (dashed lines) matching experimental data by choosing appropriate D and h values. Relatively good agreement is found between the experiments and the calculations for anneals shorter than 30 minutes. In the fitting procedure the Cr diffusion constant, D, at 900°C was estimated to be 6×10^{-12} cm^2 s^{-1}. This value is half an order of magnitude smaller than that reported by Sato [10]. When annealing for 60 minutes, however, agreement is poor. The actual Cr diffusion is much more rapid than the calculation shows. In accordance with this, the initial mirror-like surface appeared to be slightly matt indicating the initiation of thermal dissociation. Contrary to the annealing in pure H_2 ambient at temperatures higher than 850°C, Ga droplets could hardly be observed at the surface, except for the substrate edge region. It is reasonable, then, to suppose that such slight surface thermal dissociation tends to enhance Cr diffusion.

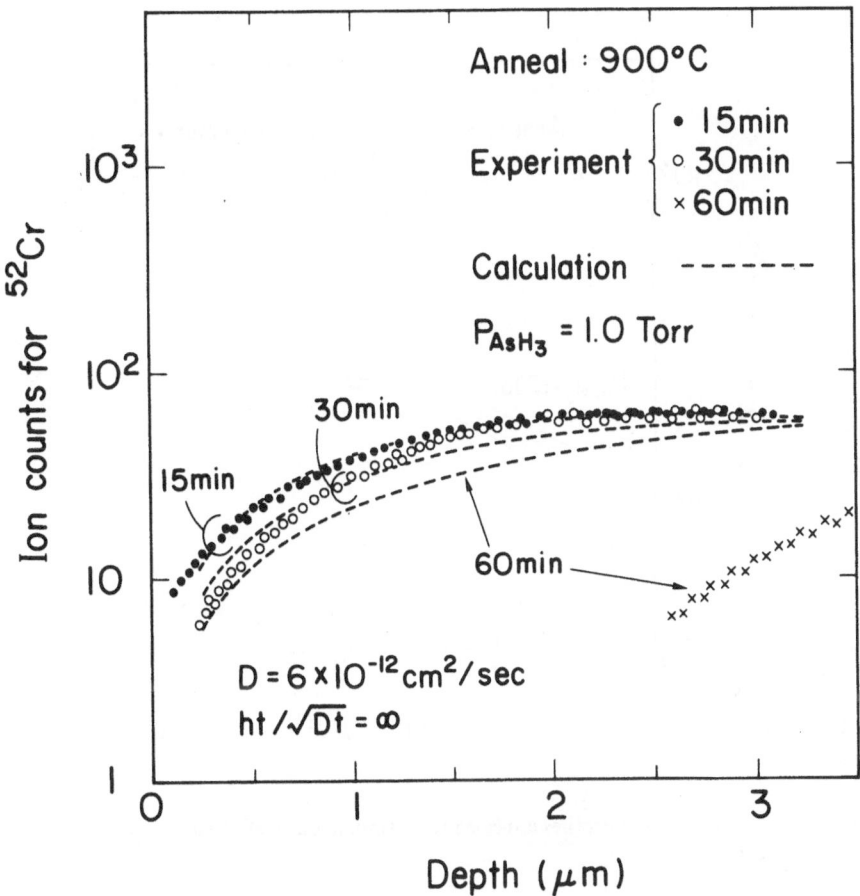

Figure 2 Cr concentration profiles annealed at 900°C under P_{AsH_3}=1 Torr for 15, 30 and 60 minutes. Data points represent experimental results, while dashed lines are calculated from equation (1).

Figure 3 shows the influence of AsH_3 partial pressure, P_{AsH_3}, on Cr diffusion. When P_{AsH_3}=10 Torr, the profile is well explained by equation (1). However, at lower P_{AsH_3}, the Cr diffusion is enhanced. Similar phenomena were observed by Kasahara [11], with substrates annealed by the face-to-face method [5]. The present results cannot be explained by a model on the basis of a simple Ga vacancy in-diffusion from the substrate surface [12]. In this model, arsenic overpressure tends to increase the Ga vacancy surface concentration and hence to enhance the Cr diffusion. However, this trend is contrary to the present results. It is found, from Figures 1 to 3, that surface thermal dissociation plays an important role in the Cr diffusion, but the detailed diffusion mechanism is still to be solved.

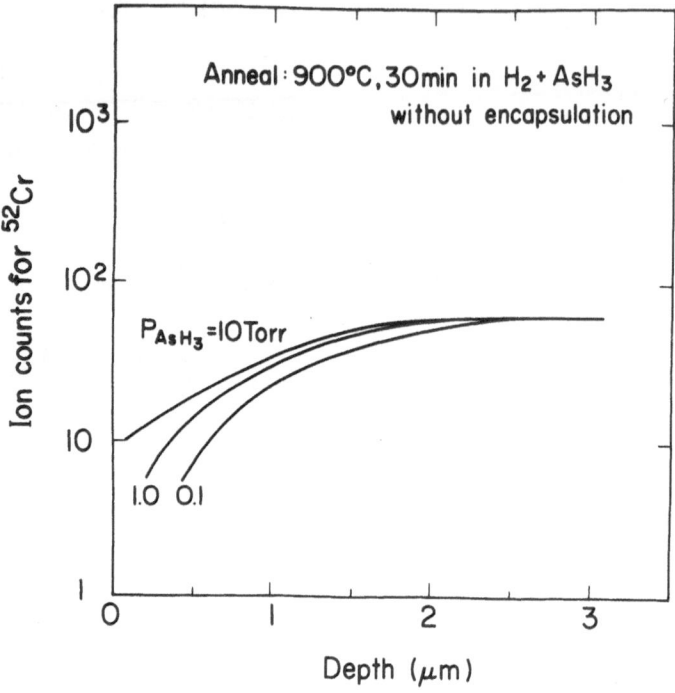

Figure 3   Cr concentration profiles annealed for 30 minutes at 900°C with P_{AsH_3}=0.1, 1 and 10 Torr.

In order to elucidate whether Cr vaporizes from the surface, the substrates were annealed for 15 minutes in pure N_2 ambient in a graphite atomizer. After removal of the substrate, adsorbants on the wall of the atomizer were analysed with the flameless atomic absorption spectrometer. In Figure 4, Cr absorbance is plotted against annealing temperatures. It is evident that Cr vaporizes from the surface, even at high temperatures, where intense thermal dissociation produces Ga droplets at the surface.

From the results, it can be concluded that Cr vaporization is the origin of the rapid Cr diffusion towards the surface.

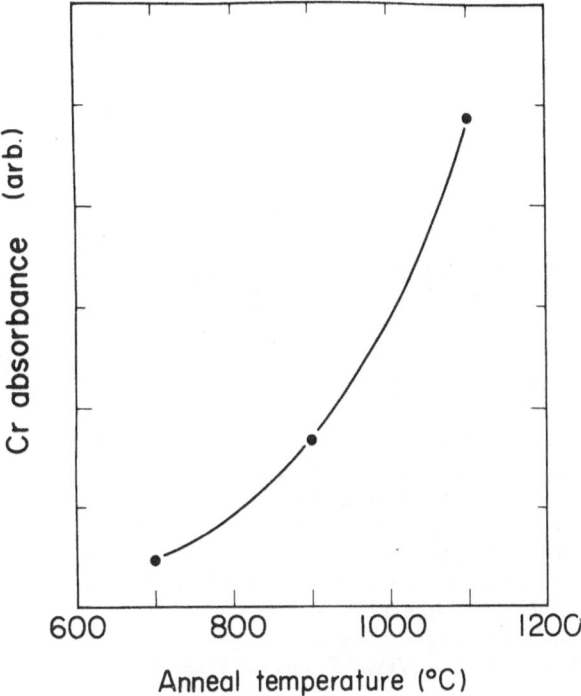

Figure 4 Cr absorbance versus annealing temperature.

Conclusion

Cr-doped semi-insulating GaAs substrates were annealed without encapsulation in pure H_2 or H_2 ambient containing a small amount of AsH_3. Depth profiles of Cr concentration were measured using secondary ion mass spectrometry. After annealing at above 800°C, a region where Cr was depleted was formed underneath the substrate surface. The Cr concentration profile can be explained by the Cr diffusion and vaporization.

Direct evidence for Cr vaporization during annealing was obtained from a flameless atomic absorption spectrometric analysis of adsorbants deposited in the graphite atomizer.

It is concluded that, as a result of Cr vaporization from the surface, Cr diffuses rapidly towards the surface. It is also pointed out that the thermal dissociation of the surface plays an important role in the Cr diffusion. AsH_3 gas addition into the annealing ambient is found to be quite effective in reducing the thermal dissociation and thereby in reducing the enhanced diffusion.

From X-ray microprobe analysis it is found that Cr concentrations in Ga droplets are higher than those in bulk crystals.

ACKNOWLEDGEMENTS

The authors are indebted to Dr Y. Oana for SIMS measurements. They wish to thank Dr M. Ohtomo and Messrs S. Okano and K. Kamei for helpful discussions.

References

1. Barella, J. (1975). *Proc. 5th Biennal Cornell Electrical Engineering Conf.*, p. 135. Ithaca, NY; Cornell UP
2. Crossley, I., Goodridge, I.H., Cardwell, M.J. and Butlin, R.S. (1977). *Inst. Phys. Conf. Ser.*, **33b**, 11
3. Huber, A.M., Morillot, G., Linh, N.T., Favennec, P.N., Deveaud, B. and Toulouse B. (1979). *Appl. Phys. Lett.*, **34**, 858
4. Favennec, P.N. and L'Haridon, H. (1979). *Appl. Phys. Lett.*, **35**, 699
5. Kasahara, J., Arai, M. and Watanabe, N. (1979). *J. Electrochem. Soc.*, **126**, 1977
6. Immorlica, A.A. and Eisen, F.H. (1976). *Appl. Phys. Lett.*, **29**, 94
7. Mattes, B.L., Houng, Y.M. and Pearson, G.L. (1975). *J. Vac. Sci. Technol.*, **12**, 869
8. Cronin, G.R. and Haisty, R.W. (1964). *J. Electrochem. Soc.*, **111**, 874
9. Smith, F.M. and Miller, R.C. (1956). *Phys. Rev.*, **104**, 1242
10. Sato, Y. (1973). *Japan. J. Appl. Phys.*, **12**, 242
11. Kasahara, J. and Watanabe, N. (1980). *Jap. J. Appl. Phys.*, **19**, L151
12. Tuck, B., Adegboyega, G.A., Jay, P.R. and Cardwell, M.J. (1979). *Inst. Phys. Conf. Ser.*, **45**, 114

A MODEL FOR THERMAL CONVERSION OF
SEMI-INSULATING GaAs

JIRO KASAHARA and NAOZO WATANABE
Sony Corporation Research Center,
174 Fujitsuka-cho, Hodogaya-ku, Yokohama, 240 Japan

Abstract

A model for thermal conversion of Cr-doped semi-insulating GaAs is proposed. Residual donors (N_d-N_a) and out-diffusion of originally compensating Cr acceptors are mainly responsible for thermal conversion. The model explained well the experimental carrier concentration profiles in the layers that were thermally converted to n-type under a wide range of annealing conditions, with modification by thermally induced acceptors and donors.

The specification of Cr-doped semi-insulating GaAs for fabricating a MESFET by direct ion implantation was determined, based on the model, from the standpoint of thermal conversion and doping efficiency. The specification is (residual N_d-N_a)/(initial Cr concentration) $\leqslant 0.25$, and (initial Cr concentration) $\leqslant 5\times10^{16}$ cm^{-3} .

1 Introduction

Recent progress in GaAs integrated circuits (GaAs ICs) is remarkable. To fabricate planar ICs, an important technique is a selective implantation directly into semi-insulating (SI) substrates as demonstrated by Eden, Welch and Zucca [1]. Thermal conversion of the SI substrate is fatal for GaAs planar technology, generating an anomalous carrier profile [2].

In our capless annealing experiment, we experienced only n-type conversion. From the investigation of thermal conversion mechanisms, we have found the important role of Cr out-diffusion for the conversion, as suggested by Sato [3] and Favennec and L'Haridon [4]. Considering the out-diffusion of Cr on which we have reported elsewhere [5], we have arrived at a model for thermal conversion by careful observation of carrier concentration profiles induced by annealing. On this model, we give the specification for the substrate for direct ion implantation to fabricate the GaAs planar ICs.

2 Cr out-diffusion

Out-diffusion of Cr from SI GaAs has been observed by SIMS [5]. Depletion of Cr in the vicinity of the surface of the substrate was observed after annealing above 800°C. We did not observe any pile up of Cr near the surface in our capless annealing experiment with AsH₃ [5]. The in-depth profiles of Cr depended on annealing temperature, annealing time, and partial pressure of arsenic (P_{AsH_3}) in the annealing ambient. Arsenic pressure, whose desirable effect on thermal conversion was discussed in a previous paper [2], was also effective in suppressing Cr out-diffusion. Experimentally obtained Cr profiles were well approximated by the following equation (1) derived from a diffusion equation [7]:

$$N(x,t)/N_{Cr} = erf(x/2(Dt)^{1/2}) + [erfc(x/2(Dt)^{1/2} + \alpha(t/D)^{1/2})]$$
$$\times exp(\alpha x/D + \alpha^2 t/D) \tag{1}$$

where N_{Cr} is the initial concentration of Cr (cm⁻³), and α is the interfacial conductance of Cr (cm s⁻¹). The boundary condition is determined by α which depends on P_{AsH_3} and temperature. Through fitting the diffusion constant was determined to be $D = 6.3 \times 10^5$ exp (-3.4 eV kT^{-1}) cm² s⁻¹; and αs at 850°C to be $\geqslant 1 \times 10^{-3}$, 7×10^{-8}, 6×10^{-8} and 4×10^{-8} cm s⁻¹ for $P_{AsH_3} = 0$, 0.15, 3.04 and 7.60 Torr, respectively. For a detailed discussion on Cr out-diffusion see [5].

3   The aspects of thermal conversion

The variation of carrier concentration profiles in the thermally converted layers under widely changed conditions was observed by C-V measurements using an Hg-Schottky barrier contact. We used, in this experiment, (100) oriented Cr-doped GaAs substrates which showed severe thermal conversion for easier investigation. Wafers were annealed by the capless method [6, 8] in H₂ or H₂+AsH₃ flow at temperatures from 800 to 900°C for 15—60 min, the partial pressure of arsenic ranging from 0 to 7.60 Torr.

An n-type conversion layer was observed in samples annealed at temperatures above 830°C for 15 min under $P_{AsH_3} = 0$ Torr. The thermally converted layer extended deeper into the substrate on increasing the annealing temperature from 830 to 900°C, as shown by the dashed lines in Figure 1. The dashed lines in Figure 2 show the annealing time dependence of the profile after annealing at 850°C under $P_{AsH_3} = 0$ and 3.04 Torr, respectively. Under higher P_{AsH_3}, the thermal conversion was less severe and resulted in a shallower conductive layer due to the suppressed Cr out-diffusion [5].

The carrier concentration near the surface was reduced in the samples which showed more severe thermal conversion, implying the existence of acceptors near the surface. Photoluminescence measurements suggested a relationship between the acceptors and arsenic vacancies (V_{As}). Two emission peaks at ~1.41 and ~1.36 eV were observed in addition to the band edge emission. The emission peak at ~1.41 eV, which was associated with V_{As} [8], disappeared from the spectrum after etching off the surface for a few thousand ångströms. The layer thickness where the emission at ~1.41 eV was observed depended on P_{AsH_3}. We may relate the existence of the acceptors to V_{As} near the surface of the annealed sample.

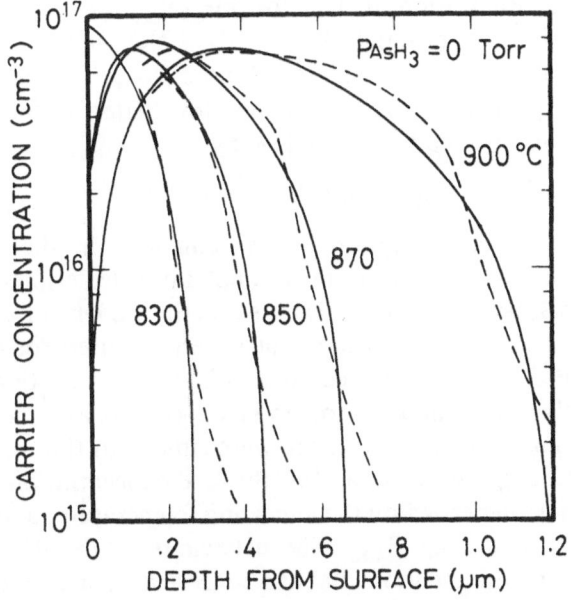

Figure 1 Dependence of carrier concentration profiles on the annealing temperature from 830 to 900°C for 15 min under $P_{AsH_3}=0$ Torr. Solid lines and dashed lines show the calculated profiles with equation (1) and the experimental ones, respectively.

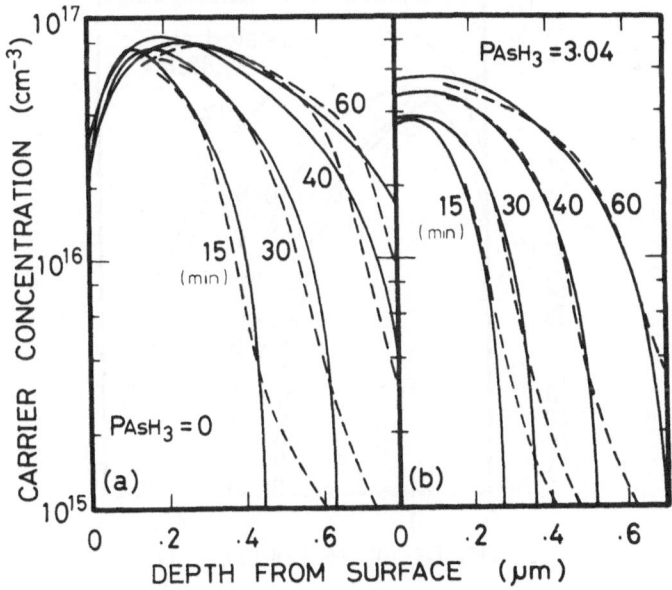

Figure 2 Annealing time dependence of the profiles after annealing at 850°C under $P_{AsH_3}=(a)$ 0 Torr and (b) 3.04 Torr. Solid lines = calculated; dashed lines = experimental.

4   A model for thermal conversion

From the facts described in the previous sections, we arrived at a model for thermal conversion in which we took into account the following parameters: (*a*) initial Cr concentration in the substrate (N_{Cr}); (*b*) out-diffusion of Cr; (*c*) concentration of residual donors (N_d—N_a); (*d*) thermally induced acceptors (N_{TA}); (*e*) thermally induced donors (N_{TD}). The thermally induced donors must be included in the model for better fittings. The carrier concentration at each depth was obtained by

$$n = \{(N_d - N_a) + N_{TD}\} - (N_{DA} + N_{TA}), \qquad (2)$$

where N_{DA} is the concentration of Cr remaining in the substrate after annealing. Figure 4 shows one of the examples of calculation. Each solid line in Figures 1 to 3, being a profile calculated on the model, corresponds to each dashed line. The diffusion constants of Cr and αs used for the calculation were determined from SIMS analysis as discussed in Section 2. Diffusion constants of thermally induced donors and acceptors and their temperature dependences obtained through fitting were similar to those of V_{Ga} and V_{As} obtained by Chiang and Pearson [9]. Surface concentration of the donors and the acceptors depended on the annealing temperature, and the acceptors decreased with increasing P_{AsH_3}. The behaviour of the thermally induced donors and acceptors mentioned above suggested their possible correlation with V_{Ga} and V_{As}, respectively. An important point to notice is that the best

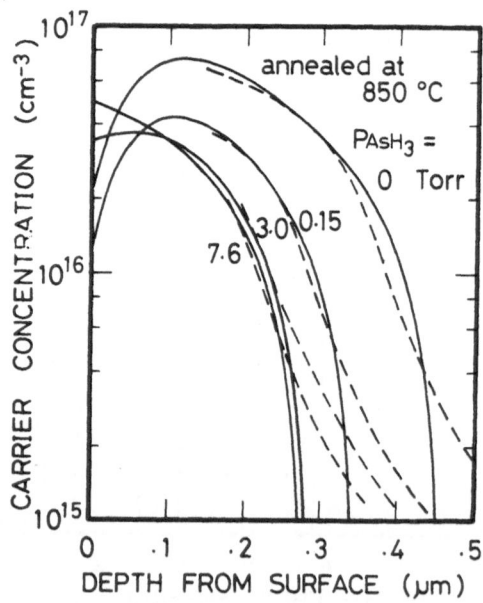

Figure 3   Variation of the profiles in the samples annealed at 850°C for 15 min under $P_{AsH_3} = 0$ to 7.60 Torr. Solid lines = calculated; dashed lines = experimental.

fitting to all the profiles was obtained by only minor changes of the parameters; all the profiles in Figure 2 were calculated with the same set of parameters, changing only the annealing time. We can estimate N_{Cr} and $(N_d - N_a)$ in the substrate by analysing only one carrier concentration profile on the model.

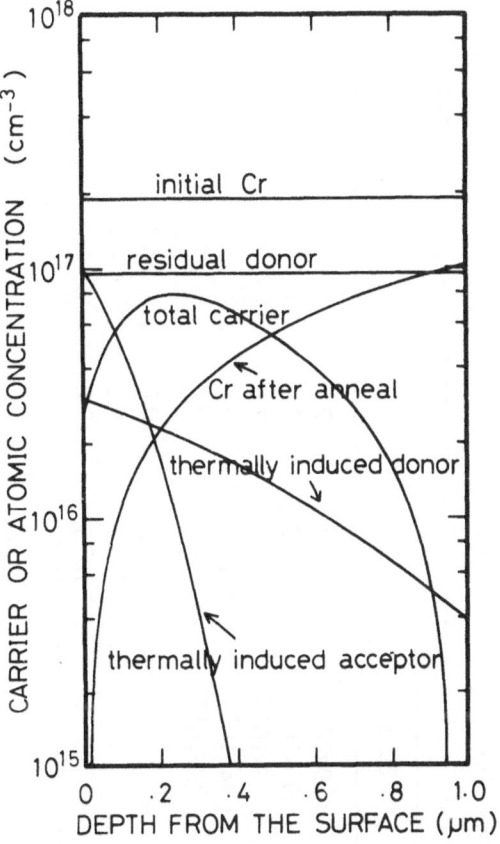

Figure 4 Example of the model calculation.

Thermal conversion of eight ingots from four different suppliers was examined on the model. The ingots of A—E in group I exhibited no thermal conversion after annealing at 850°C for 15 min under $P_{AsH_3} = 0$ Torr but exhibited n-type conductivity after 60 min annealing, as shown in Table 1. The ingots of F—H in group II showed thermal conversion after only 15 min annealing. Analysing the carrier concentration profiles on the model, we determined $(N_d - N_a)$ and N_{Cr} as given in Table 1. The values of N_{Cr} agreed well with those measured by SIMS. In the carrier profiles in the ion implanted layers subsequently annealed at 850°C for 15 min under $P_{AsH_3} = 3.04$ Torr, no anomaly was observed in the wafers from group I, while shoulders appeared in those from group II. To suppress thermal conversion, the specification of $(N_d - N_a)/N_{Cr} \leqslant 0.25$ was required. This specification is not enough, because

Table 1 Summary of the classification of eight ingots

	Ingot	Thermal conversion n_S (cm^{-2})		η (%)	N_d-N_a x10^{16} (cm^{-3})	N_{Cr} x10^{17} (cm^{-3})	$\dfrac{N_d-N_a}{N_{Cr}}$
		15 min	60 min				
I	A	0	8.80x10^{11}	89.1	2.7	1.8	0.15
	B	0	1.00x10^{12}	103.9	0.8	0.5	0.16
	C	0	1.26x10^{12}	66.4	8.0	4.0	0.20
	D	0	1.34x10^{12}	64.6	9.0	4.3	0.21
	E	0	1.46x10^{12}	70.7	9.0	4.0	0.23
II	F	1.00x10^{12}	—	94.3	10.0	3.9	0.26
	G	2.10x10^{12}	—	114.0	9.0	1.9	0.47
	H	3.51x10^{12}	—	167.1	17.0	3.2	0.53

excess N_{Cr} lowers doping efficiency of ion implantation. Considering compensation of implanted atoms by Cr remaining after out-diffusion, we determined the upper limit of N_{Cr}. In a low dose implantation for the MESFET, carrier concentration being $\sim 2 \times 10^{17}$ cm^{-3}, $N_{Cr} \leqslant 5 \times 10^{16}$ cm^{-3} is required to get the doping efficiency above 85%.

5 Conclusion

A model for thermal conversion of Cr-doped semi-insulating GaAs was proposed. Calculated carrier concentration profiles by the model agreed well with those obtained experimentally. The specification of the substrates required for low dose direct ion implantation was determined as $(N_d - N_a)/N_{Cr} \leqslant 0.25$ and $N_{Cr} \leqslant 5 \times 10^{16}$ cm^{-3}. If lower dose implantation is required, N_{Cr} in the substrate must be lowered.

ACKNOWLEDGEMENTS

We would like to acknowledge Dr Morizane for his helpful discussion. We also express our thanks to Mrs Nishiyama and Sakurai for operating ion implantation and Mr Tanigaki for SIMS analysis.

References

1. Eden, R., Welch, B. M. and Zucca, R. (1978). *IEEE:* SC-13, 419
2. Kasahara, J., Arai, M. and Watanabe N. (1979). *J. Appl. Phys.,* 50, 8229
3. Sato, Y. (1973). *Jap. J. Appl. Phys.,* 12, 242
4. Favennec, P.N. and L'Haridon, H. (1979). *Appl. Phys. Lett.,* 35, 699
5. Kasahara, J. and Watanabe, N. (1980). *Jap. J. Appl. Phys.,* 19, L151
6. Kasahara, J., Arai, M. and Watanabe, N. (1979). *J. Appl. Phys.,* 50, 541
7. Carslaw, H. S. and Jaeger, J.C. (1959). *Conduction of Heat in Solids,* 2nd edn, pp. 70—73. Oxford; Clarendon Press
8. Kasahara, J., Arai, M. and Watanabe, N. (1979). *J. Electrochem. Soc.,* 126, 1997
9. Chiang, S.Y. and Pearson, G.L. (1975). *J. Appl. Phys.,* 46, 2986

HYDROGEN ION BOMBARDMENT OF GaAs FOR DEVICE ISOLATION

I.J. SAUNDERS and K. STEEPLES*
Department of Physics, University of Lancaster,
Lancaster LA1 4YB, UK

Abstract

Multiple energy bombardments of n^+ GaAs with the three hydrogen isotopes have shown that deuterons offer great advantages for device isolation. Carrier removal is accomplished at a deuteron dose only 5% of that needed with protons, significantly reducing processing time. Higher doses may be needed for thermal stability, but in all cases deuteron bombardment has been found to be more stable for a given dose. A model for the carrier removal is described.

Introduction

The electrical and optical isolation [1, 2] of devices in GaAs and related compound semiconductors can be satisfactorily performed by proton bombardment. In some cases (e.g. IMPATT diodes and lasers) this has improved the efficiency and reliability of the devices, compared with those isolated by mesa etching, sawing or other techniques. The carrier removal mechanism, however, is not established, and the annealing behaviour shows that defect complexes of several types must be involved. We have studied the effects of the three hydrogen isotopes for such bombardments, namely protons, deuterons and tritons of mass 1, 2 and 3 a.u., respectively.

Except for a few cases discussed later, all the samples of GaAs were from the same region of an n^+ (2×10^{18} cm^{-3}) ingot doped with Si, Bridgman grown by Monsanto Chemical Co. Wafers 0.3 mm thick were sawn at 3° from <100> and chemically polished to a depth of about 50 μm. Deuteron and proton bombardments were performed using multiple energies in 0.1 MeV steps, decreasing from 1.0 MeV to 0.1 MeV. The dose at each energy was the same, and a beam current of 0.1 μA cm^{-2} was used to avoid annealing effects during implantation. Beam energies from 0.7 to 1.3 MeV, increasing in 0.1 MeV steps, were used for the two triton implantations. In most cases suitably energetic diatomic ions (e.g. H_2^+) were employed for convenience and, on impact, these dissociated to give the required particles.

The range of the ions in GaAs was measured by observing the damage

*Now with Texas Instruments, Houston, Texas, USA.

bands produced at each energy. These were revealed by A—B etching of cleaved edges. For ions incident at 1 MeV the proton, deuteron and triton ranges were roughly 12, 10 and 5 μm, respectively, and are discussed elsewhere [3]. The resulting high resistivity layers were about 12, 10 or 7 μm thick but, as low energy tritons could not be used, the top 3 μm of the triton-treated wafers was removed prior to electrical studies. Two terminal, through layer, resistivity measurements were made using ohmic contacts and a Keithley 602 electrometer. The contacts could not be sintered without interfering with annealing studies and, after trials, evaporated indium was found to give the best results, in some cases with a few seconds at 170°C to improve linearity. The bombarded surface had an array of 0.5 mm diameter contacts (produced by in-contact masking), with the other face covered by the base contact. In most cases Ni capping of the contacts was used to improve resistance to probe scratching. Conduction at the various contacts on a given sample and for similar samples compared to within ±10% for high doses, with more scatter for incomplete carrier removal. All samples displayed similar changes for dose variation and annealing.

Results

In Figure 1 the dependence of as-implanted resistivity on ion dose is shown for the three isotopes. The effect of protons is in agreement with previous work on bombardment of n^+ GaAs, from which the high dose behaviour (shown as a broken line) may be deduced. Results for tritons show the anticipated small increase in effectiveness due to the greater defect production of the heavier ion, but deuteron bombardment causes a most surprising carrier removal at unexpectedly low doses. The high resistivity threshold at 10^{13} cm^{-2}/energy corresponds to a removal rate of about 20 electrons per deuteron, which is what would be expected for low doses (see later discussion), but not for those when full compensation is approached. The fall in resistivity above 10^{14} deuterons cm^{-2}/energy is due to the onset of a hopping conduction regime.

Samples were subjected to isochronal annealing in flowing nitrogen without encapsulation. After 30 minutes at each temperature, increasing in 50°C steps from 150°C, the resistivity (and other properties not discussed here) was measured. The three lowest curves in Figure 2 show the resistivity variations for two deuteron implants and one of protons, all at the same dose. The reproducibility and greater thermal stability of the deuteron treatment may be noted. To achieve similar effectiveness at 600°C a proton dose three times greater was required, while a deuteron dose of one quarter gave superior results to the first-mentioned proton treatment (see the upper two curves in Figure 2).

In Figure 3 the full range of deuteron annealing behaviour is illustrated. For doses below about 4×10^{13} cm^{-2}/energy the as-implanted high resistivity

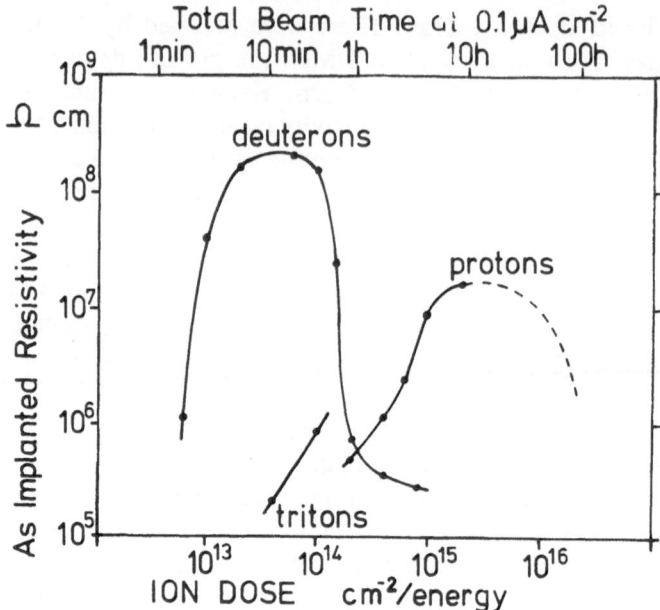

Figure 1   Variation of as-implanted resistivity with dose for the three hydrogen isotopes implanted into n⁺ GaAs.

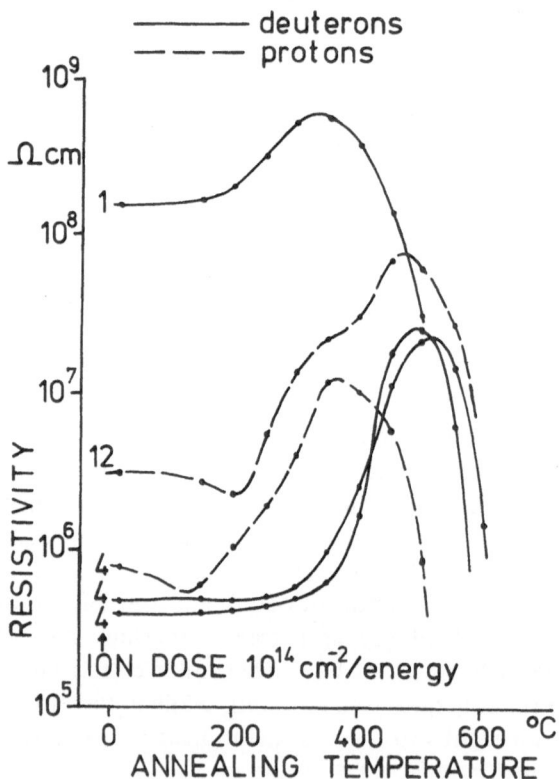

Figure 2   Isochronal annealing for 30 minutes at 50°C steps in flowing nitrogen. Resistivity after each anneal versus annealing temperature for deuteron and proton doses, as indicated.

begins to anneal out at about 200°C. A number of higher temperature annealing stages may be observed in samples given higher doses, indicating a complex defect interaction system.

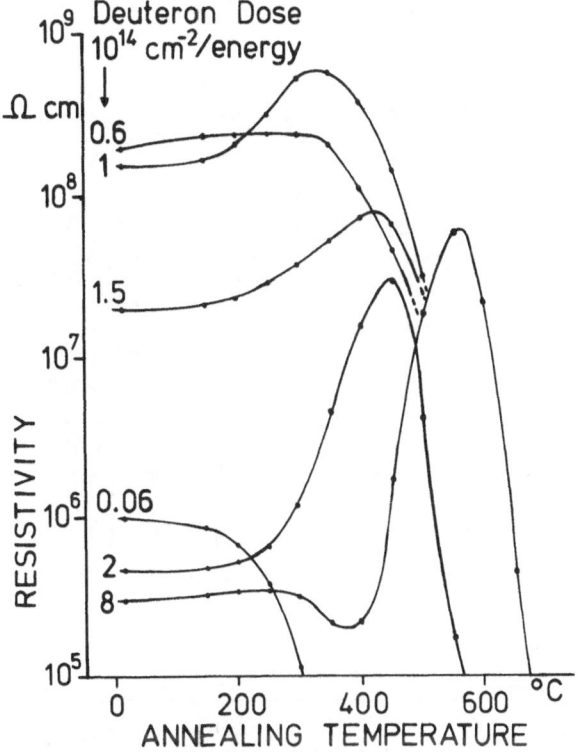

Figure 3   The isochronal annealing behaviour of deuteron bombarded material.

Annealing of samples treated with 2×10^{14} deuterons cm^{-2}/energy was also performed isothermally at 300, 325, 350 and 375°C, with resistivity measurements at intervals of 2 or 5 minutes, depending on the rate of change. Approximately linear relations on an Arrhenius plot (activation energies 0.93 and 0.98 eV) were found for the two resistivity maxima observed, and extrapolation indicates 10^4 hours at 130°C to reach the second of these. However, little reliance can be placed on this extrapolation as the isothermal results predict disappearance of the high resistivity after 30 minutes at 400°C, while the cumulative isochronal treatment produced *maximum* resistivity after this, and after 30 minutes at 550°C there was still 10^5 ohm-cm. It must be concluded that beneficial effects on carrier removal stability are caused by heat treatment of high dose deuteron bombarded samples at 200 to 250°C, and that accelerated ageing studies of devices may give very misleading results.

Figure 4 shows the results of bombardment with 10^{14} deuterons cm^{-2}/energy of material containing various different donors. The effects for S, Sn and Ge are similar to those for Si, but for Se and Te doping the resistivity is sharply reduced by annealing at only 400°C, perhaps indicating that these donors are less suitable for devices isolated by deuteron bombardment.

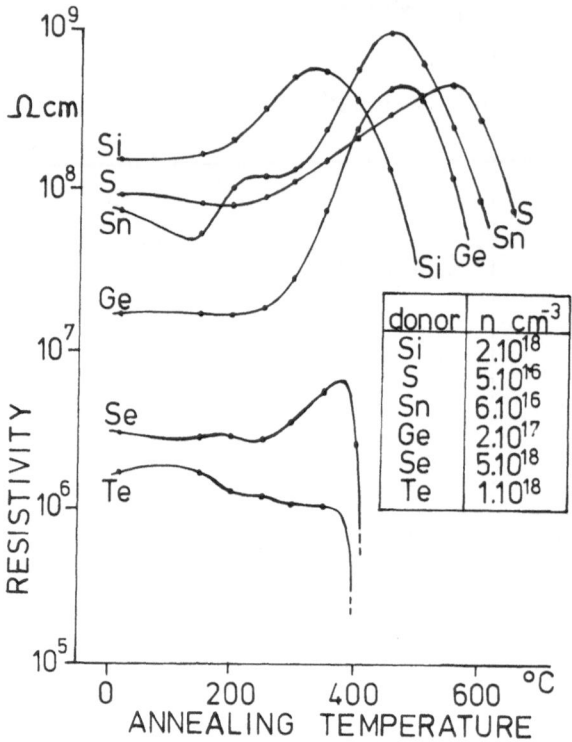

Figure 4 The isochronal annealing of variously doped samples, all implanted with 10^{14} deuterons cm^{-2}/energy.

A range of further studies was performed to find the origin of these effects. Here these can only be summarized. Local mode studies of proton and deuteron implanted VPE layers on semi-insulating substrates [4] showed that the hydrogen isotopes are incorporated into GaAs in the same way, probably on an interstitial site bound to Ga neighbours. The phonon energies of 0.233, 0.168 and 0.138 eV are in the correct ratios (to within 5%) for the masses of protons, deuterons and tritons, respectively. Precision lattice parameter measurements did not reveal anything unusual about the deuteron damage, the lattice dilation being 5, 65 and 330 ppm for 10^{13} deuterons, 10^{14} deuterons and 2×10^{15} protons cm^{-2}/energy, respectively. Nuclear reaction analysis demonstrated the out-diffusion and surface loss of deuterium during annealing at 450°C and over, indicating this as the reason for the final loss of resistivity, although perhaps only part of a complex annealing behaviour.

Discussion

From Figure 3 we see that the high resistivity at low doses is due to an electron trap removed by annealing at 200°C. Such a trap has been observed for various types of radiation treatment of GaAs and is known as E3, being 0.31 eV below the conduction band and very greatly affected by the recombination enhancement of defect reactions [5]. It has been tentatively identified as the gallium vacancy. One can speculate that diffusion of these E3 point defects during low temperature annealing produces some stable complexes with the interstitial deuterium, so that the carrier removal is maintained beyond 300°C if sufficient dose has been given to produce a deuterium concentration similar to the original electron density. Deuterium itself becomes mobile at about 400°C and the resistivity may then fall as deuterium escapes from the GaAs. At the highest doses and annealing temperatures, complexes involving donors form, as observed in luminescence following annealing of radiation damage for donors including Si [5]. However, here the defect mobility near 200°C is also significant, as is shown by the difference in results from isothermal annealing. Protons, being more mobile than the heavier deuterons, permit annealing effects at lower temperatures.

The outstanding problem in interpretation is the high density of point defect electron traps for relatively intense deuteron bombardment. Carrier removal of about 10 per proton at low doses [6] reduces to below 1 per proton as full compensation is approached. If one assumes the electron traps to be of E3 type in all cases before annealing, the yield reduction can be explained as another recombination enhanced effect for these centres. Bombardment creates about 10^4 electron-hole pairs per ion, and for sufficiently low free electron density direct recombination will be reduced, so that it becomes probable that free holes will be captured by the electron traps, releasing over 1 eV to promote diffusion of the defect. The lattice parameter measurements show that simple annihilation does not occur. This suggests that the reduced carrier removal rate results from the formation of clusters, stacking faults, etc.

At doses well below compensation levels, a carrier removal rate roughly proportional to the ion mass may be expected. For example, oxygen has been found to be around ten times more effective than protons [6]. Thus, one estimates a rate of about 20 per deuteron (or 30 per triton) at low doses. The continuation of this rate to high doses of deuterons requires that the defect reactions are suppressed by a modest concentration of deuterium, but not of the other isotopes. As the electronic nature of the hydrogen centres is the same, it appears that the phonon local modes must be involved. During bombardment, the crystal is not at thermal equilibrium and the hydrogen centres can absorb large energies from the shock waves in the lattice as the thermal energy is dissipated following an ion impact. They will reradiate this at multiples of the single phonon energy, namely 0.233, 0.168 and 0.138 eV (at 77 K) for proton, deuteron and triton centres, respectively. Phonon frequency and energy are very high for these lightest of all atoms, and electron-phonon coupling will be strong. The critical energies for the E3 centre are the 0.31 eV needed to release the electron and the residual activation energy for diffusion after recombination enhancement, about 0.35 eV. For protons and tritons the only suitable phonon multiples are over 0.4 eV, and thus the presence of these two isotopes should mainly promote the diffusion and interaction of the defects. However, the two quantum phonon for the deuterium centre is about 0.33 eV, below that needed to promote diffusion but sufficient temporarily to restore to the conduction band electrons trapped at adjacent E3 centres, preventing hole capture and reducing the rate of recombination enhancement of the defect diffusion. Thus a much greater concentration of electron traps may be produced by the deuteron bombardment, since with other ions once compensation is approached the defects become mobile during bombardment and condense into extended forms with less electron absorbing effect.

An alternative model, involving the hydrogen centres as recombination sites with the critical feature being the residual energy after multiphonon emission, has been proposed by one of the authors elsewhere [7] and will not be discussed here.

Conclusions

Although the model of the carrier removal system remains uncertain, it has been clearly established that deuteron bombardment is superior to the established proton treatment for device isolation in GaAs. Implantation time can be reduced and/or thermal stability can be improved. At the reduced doses possible in some cases, lattice damage is considerably reduced, which may be significant for device lifetimes. Lateral spreading of the heavier deuterons under shadow masks should be less than for protons, although this is yet to be confirmed.

ACKNOWLEDGEMENTS

The authors are grateful to AERE, Harwell, and the SRC for supporting this work; to the AERE staff for implantation work, lattice parameter and nuclear reaction measurements, and for much advice and encouragement; to RSRE Baldock for GaAs samples and other assistance; to the Schuster Laboratory, University of Manchester for triton implantations; to STL Ltd for growing epitaxial GaAs for local mode work; and to Professor R.C. Newman, University of Reading, for performing these studies and communicating his results.

References

1. Murphy, R.A., Bozler, C.O., Donnelly, J.P., Laton, R.W., Lincoln, G.A., Sudbury, R.W., Lindley, W.T., Lowe, L.F. and Deane, M.L, (1977). *Inst. Phys. Conf. Ser., 33b*, 210
2. Henshall, G.D., Thompson, G.H.B., Whiteaway, J.E.A., Selway, P.R. and Broomfield, M. (1979). *IEE Solid-State and Electron Devices, 3*, 1
3. Steeples, K. and Saunders, I.J. (1980). Submitted to *Radiation Effects*
4. Newman, R.C. (1979). Private communication
5. Dean, P.J. and Choyke, W.J. (1977). *Adv. Phys., 26*, 1. (This review provides a bibliography for the discussion above)
6. Dietrich, H.B. (1976). *Report of NRL Progress, 1*, 3
7. Steeples, K., Dearnaley, G. and Stoneham, A.M. (1980). *Appl. Phys. Lett.,* in press

INFLUENCE OF CHROMIUM REDISTRIBUTION ON THE ELECTRICAL PROPERTIES OF Se AND Zn IMPLANTED Cr-DOPED SUBSTRATES

P.N. FAVENNEC and H. L'HARIDON
Centre National d'Etudes des Télécommunications, BP No. 40,
22031 Lannion Cedex, France

Abstract

On implanting impurities to give shallow donor or acceptor states in Cr-doped substrates the apparent electrical activity following annealing may be above, below or close to 100%. By supposing that chromium and some other metallic impurity, which acts as an acceptor, redistribute during annealing, the electrical measurements on the implanted SI GaAs substrates can be understood.

It is necessary to protect GaAs during high-temperature annealing to avoid exo-diffusion of arsenic and gallium. A Si_3N_4 cap in the form of a film is usually used for this purpose. Despite this, Cr-doped SI substrates may become n- or p-type or remain semi-insulating near the surface. For fixed experimental conditions the modification of electrical properties is reproducible. It is for this reason that following the annealing of Zn and Se implanted substrates electrical activities of greater than, less than or roughly equal to 100% are obtained.

In Table 1 we show some results obtained for different semi-insulating ingots. The samples were annealed at 900°C for 20 minutes with a Si_3N_4 encapsulation. For each ingot, we have measured the surface carrier concentration after the anneal alone and then after selenium implantations followed by the same anneal. The substrates labelled E, H and J remained stable during the anneal, and therefore their electrical activity after Se implantation was close to 100%. For the substrates labelled A, B and D, a p-type layer appeared near the surface and the apparent electrical activity of Se implantation was lower than 100%. For the substrates labelled C and I, an n-type layer appeared and the apparent electrical activity of the Se implantation was higher than 100%. The same results are obtained for Zn implantations [3]. In all cases, the carrier concentration measured after implantation is found to be the algebraic sum of the implanted carriers and the carriers generated after the anneal alone. The distribution of these carriers is specific to each ingot in given conditions.

Table 1

Ingot	Anneal alone			Implantation + anneal		
	Surface conduction	Surface carrier concentration $(\times 10^{12}$ cm$^{-2})$	Implantation conditions (energy in keV, dose in 10^{12} cm^{-2}, T in °C)		Surface carrier concentration $(\times 10^{12}$ cm$^{-2})$	Apparent electrical activity (%)
E	SI	Unmeasurable	400,	2.5,	200	2.7	108
E	SI	Unmeasurable	400,	2.5,	500	2.45	98
E	SI	Unmeasurable	800,	2.5,	400	2.8	93
H	SI	Unmeasurable	400,	2.5,	400	2.5	100
H	SI	Unmeasurable	400,	6,	400	6.4	106
J	SI	Unmeasurable	400,	2.5,	400	2.7	108
A	p-type	4	400,	6,	400	2.2	30
B	p-type	3	400,	6,	400	3	50
D	p-type	0.5	400,	1.5,	400	1	66
C	n-type	4	400,	6,	400	10	165
I	n-type	8.5	400,	2.5,	400	11	440
I	n-type	8.5	400,	0.25,	400	8.7	3500

A four-level model, including a deep acceptor (Cr), a deep donor, a shallow donor (Si) and a shallow acceptor, could explain rather well the SI behaviour of Cr-doped GaAs [4, 5]. Using this model, we have tried to understand the appearance of conducting layers after annealing. Previously, it has been demonstrated by SIMS analysis that a redistribution of chromium occurs during the heat treatment [6, 7] whereas the silicon concentration remains constant within the depth. By taking into account this Cr depletion, the occurrence of an n-type layer can be understood (we calculate a resulting electron carrier concentration which agrees quite well with the experimental electron density profile [3]), but the occurrence of a p-type layer cannot be explained. To account for the p-type layer, we have to assume that, together with the chromium diffusion, the concentration of a shallow acceptor increases near the surface and exceeds the donor concentration. As yet, we have no clear experimental evidence for the nature of this acceptor, but some experiments indicate that copper could be responsible for p-type conversion.

We have therefore been led to conclude that, at least for doses lower than 2×10^{13} ions cm^{-2}, the carrier profile of implanted and annealed Cr-doped GaAs is the sum of two carrier distributions. The first is induced by the implantation with total electrical activity after 900°C anneal. The second is induced by the redistribution of some impurities, chromium and other acceptor impurities, during the anneal. Metals act generally as acceptors in GaAs. Work is now in progress to shed further light on the role of these metallic impurities in semi-insulating GaAs.

References

1. Eisen, F.H. (1978). *Proc. IBMM Conf.* (Budapest), p. 147
2. Favennec, P.N., Henry, L. and l'Haridon, H. (1978). *Solid St. Electron.*, **21**, 705
3. Favennec, P.N. and l'Haridon, H. (1979). *Appl. Phys. Lett.*, **35**, 699
4. Linquist, P.F. (1977). *J. Appl. Phys.*, **48**, 1262
5. Zucca, R. (1977). *J. Appl. Phys.*, **48**, 1987
6. Huber, A.M., Morillot, G., Linh, N.T., Favennec, P.N., Deveaud, B. and Toulouse, B. (1979). *Appl. Phys. Lett.*, **34**, 858
7. Evans, C.A. Jr, Deline, V.R., Sigmon, T.W. and Lindow, A. (1979). *Appl. Phys. Lett.*, **35**, 291

INFLUENCE OF ANNEALING ON THE ELECTRICAL PROPERTIES OF SEMI-INSULATING GaAs

A. MIRCEA-ROUSSEL, G. JACOB AND J.P. HALLAIS
Laboratoires d'Electronique et de Physique Appliquée,
3, avenue Descartes, 94450 Limeil-Brévannes, France

Abstract

The thermal stability of semi-insulating GaAs crystals is analysed as a function of the heat treatment conditions: the influence of temperature, carrier gas and encapsulation is studied. The annealed wafers are characterized by electrical and photoluminescence measurements. Capless annealed crystals show a p-type conversion above a critical temperature which strongly depends on the carrier gas but only slightly on the as-grown crystal. This conversion is attributed to manganese. Si_3N_4 coated wafers, heat-treated in standard conditions for ion implantation post-annealing, do not show manganese and their electrical behaviour involves the balance of both shallow and deep levels present in the as-grown crystal.

Introduction

Whichever method (i.e. epitaxy or ion implantation) is chosen to obtain a GaAs FET structure, it implies a heat treatment of the GaAs semi-insulating substrate. The performance of the FET thus obtained depends highly on the influence of heat treatment on the electrical properties of the SI substrate.

Various heat treatments have been performed on wafers cut in the (001) plane from Bridgman or LEC grown ingots.

Electrical assessment is achieved by Hall effect on step by step etched samples at temperatures ranging from 77 to 600 K. A rough estimate of the conductivity is also deduced from the spectral photocurrent of the GaAs electrolyte barrier [1]. Front surface photoluminescence spectra are recorded at 4 K.

Capless annealing

Heat treatments were performed on polished wafers, from which at least 20 μm have been chemically etched using sulphuric acid/hydrogen peroxide.

Table 1 summarizes the results obtained for various gas atmospheres and temperatures for 1 hour heat treatment on a chromium-doped SI substrate.

Table 1   **The degree of conversion is expressed by the photocurrent I_{ph} (arbitrary units): for a p-type converted layer, $I_{ph} = 1000$ a.u. corresponds to 10^{16} holes cm^{-3} measured by the Hall effect. The front face (FF) is exposed to the gas flow whereas the rear face (RF) remains on the silica holder.**

Annealing under hydrogen

T (°C)	700		750		800		850	
Face	FF	RF	FF	RF	FF	RF	FF	RF
Type	p/n	p/n	p	p/n	p	p	p	p
I_{ph}	2	2	1000	2	1000	1000	2500	2500 gallium droplets

Annealing under argon

T (°C)	750		800		870		925	
Face	FF	RF	FF	RF	FF	RF	FF	RF
Type	n	n	n	n	n	p	p	p
I_{ph}	2	2	2	2	2	1000	100	2500

Annealing under arsine/hydrogen

T (°C)	750		800		875	
Face	FF	RF	FF	RF	FF	RF
Type	n	n	n	n/p	p	n/p
I_{ph}	10	5	2	2	1000	5

For hydrogen carrier gas the critical temperature, which may change slightly from one sample to another, is about 750°C. Hall data give evidence of a p-type layer, 0.5 μm thick, with a hole concentration of 10^{16} cm^{-3} and an activation energy for the acceptor involved of about 100 meV [2]. In the temperature range 700—730°C, n-type conversion is sometimes observed, Hall measurements show a conducting top layer, about 0.3 μm thick, with donor concentration in the 10^{15} cm^{-3} range and an activation energy of 350—450 meV. This centre could be related to EL6[3].

For argon carrier gas, identical phenomena are observed but the critical temperature is higher.

Heat treatments performed under arsine-hydrogen atmosphere, using a partial pressure of 10^{-3} of arsine, lead to a noticeable decrease of conversion with an upward shift of the critical temperature by 100°C.

Photoluminescence spectra at 4 K for all p-type converted samples are shown in Figure 1. They exhibit supplementary lines as compared to the spectrum of as-grown material: a no-phonon line at 1.409 eV and its four phonon replica. Manganese diffused or implanted samples [4] obtained and measured in our laboratory, definitely show the same spectrum. Chemical analysis by SIMS [5, 6] confirms the presence of manganese in the converted layers. In fact, for some samples the characteristic lines of manganese are observed, although the surface remains semi-insulating: these ingots are highly chromium- and donor-doped [7]. These results rule out other hypotheses which involve a shallow acceptor-vacancy complex centre for the 1.409 eV line. It is worth noting that no PL line due to copper is detected.

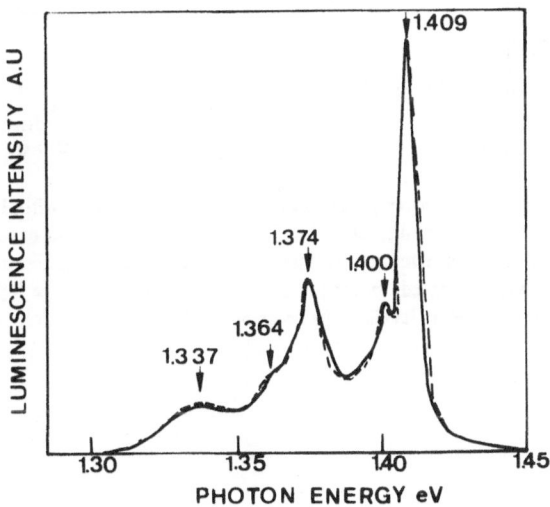

Figure 1 Characteristic 1.4 eV luminescence band at 4 K of (*a*) GaAs after Mn diffusion in vacuum at 750°C (dashed line); and (*b*) GaAs heat-treated in hydrogen flow at 750°C (solid line). Note the identity of the spectra. The control sample heated in vacuum without Mn did not luminesce in this range.

Encapsulated annealings

Samples were cleaned as before and then coated with CVD Si_3N_4. Heat treatments were performed at 875°C for 15 min under argon atmosphere. In these conditions, various situations are found: conversion towards p- or n-type, or stability. In any case, no additional peak occurs in photoluminescence. The observed phenomena can readily be explained on one example using the Shockley diagram shown in Figure 2. The semi-insulating properties of the as-grown ingot are fixed by the respective concentrations of chromium, shallow donors and deep donor (EL2) [8]. Annealing induces changes in concentration of chromium [9] and EL2 [10] (dashed line in Figure 2), which leads to n-type conversion in the example of Figure 2. By analogy, using similar considerations, ingots free of chromium would convert to p-type.

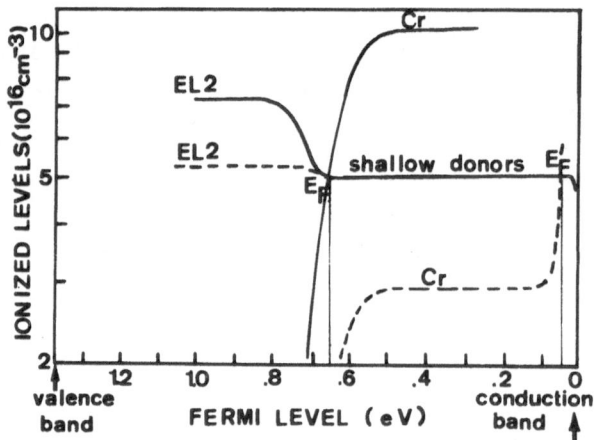

Figure 2   Shockley diagram corresponding to semi-insulating GaAs with 10^{17} cm^{-3} Cr, 5×10^{16} cm^{-3} shallow donors and 2×10^{16} cm^{-3} EL2 (solid line). After encapsulated annealing, shallow donor concentration remains the same while the concentrations of chromium and EL2 are decreased, leading to an n-type converted surface layer (dashed line). E_F and E_F' are the Fermi levels in the as-grown and annealed wafers, respectively.

Conclusions

From a practical point of view, this work confirms the role of manganese in the p-type thermal conversion of capless GaAs. It also shows that the presence of AsH_3 drastically reduces this phenomenon. For encapsulated annealing, the resulting electrical properties apparently do not involve any new impurity and depend only on the as-grown material.

From a more fundamental point of view, it would be interesting to investigate the origin of manganese, its surface dependent mechanism of occurrence and the role played by vacancies in its in- or exo-diffusion.

References

1. Hurtes, Ch., Hollan, L. and Boulou, M. (1979). *Inst. Phys. Conf. Ser.*, **45**, 342
2. Hallais, J., Mircea-Roussel, A., Farges, J.P. and Poiblaud, G. (1977). *Inst. Phys. Conf. Ser.*, **33b**, 220
3. Martin, G.M., Mitonneau, A. and Mircea, A. (1977). *Electron. Lett.*, **13**, 191
4. We acknowledge J. Chaumont for implanting the samples
5. Nordquist, P.E.R. and Klein, P.B. (1979). *Electronic Materials Conference,* July. Boulder, Colorado
6. Clegg, B. Private communication
7. Swiggard, E.M., Lee, S.H. and von Batchelder, F.W. (1979). *Inst. Phys. Conf. Ser.*, **45**, 125
8. Martin, G.M., Farges, J.P., Jacob, G., Hallais, J.P. and Poiblaud, G. (1980). To be published in *J. Appl. Phys.*
9. Huber, A.M., Morillot, G., Linh, N.T., Favennec, P.N., Deveaud, B. and Toulouse, B. (1979). *Appl. Phys. Lett.*, **34**, 858
10. Mitonneau, A. Private communication

REDISTRIBUTION OF S AND Cr IN THERMALLY ANNEALED, SULPHUR IMPLANTED, SEMI-INSULATING GaAs

C.A. EVANS, Jr, C.G. HOPKINS, J.C. NORBERG, V.R. DELINE,
R.J. BLATTNER, R.G. WILSON*, D.M. JAMBA* and Y.S. PARK**
Charles Evans and Associates, 1670 S. Amphlett Blvd, Suite 120,
San Mateo, CA 94402, USA

Abstract

Secondary Ion Mass Spectrometry (SIMS) has been used to study the thermal redistribution of S and Cr in sulphur implanted, semi-insulating GaAs using an SiO_2 encapsulant. Sulphur was observed to diffuse rapidly only at low concentrations while chromium was found to redistribute into regions of implantation damage and lattice stress. The details of Cr redistribution are dependent on the sulphur implantation fluence and energy as well as on the specific annealing conditions.

Considerable attention is being given to the use of direct ion implantation of semi-insulating GaAs substrates for the fabrication of field effect transistors and integrated circuits. Several groups [1—7] have shown that conventional as well as capless annealing techniques [8] can induce Cr depletion in the near surface region. In this depleted region the Cr concentration may drop below the residual donor level (unintentionally incorporated in the material during growth), allowing them to contribute to the electrical profile of the implant.

Using secondary ion mass spectrometry (SIMS), we have determined the atomic redistribution of sulphur and chromium in sulphur implanted semi-insulating GaAs as a function of implantation fluence, energy and annealing temperature, using an SiO_2 encapsulant. Sulphur in-depth distributions have been established using Cs^+ ion bombardment and negative ion spectroscopy while the Cr distributions have been established using O_2^+ ion bombardment and positive ion spectroscopy [9, 10].

Figure 1 shows the as-implanted sulphur profile for a 300 keV, 4×10^{15} cm^{-2} ^{32}S implant into semi-insulating GaAs as well as the ^{32}S profile and ^{52}Cr profile for the annealed material. Although specific to these implantation and annealing conditions, this figure demonstrates the basic results of the study. In order to highlight the sulphur diffusion behaviour, we have subtracted the background which is due to the residual S concentration (1×10^{16} cm^{-3}) in the bulk GaAs and is not due to an instrumental

* Hughes Research Laboratories, Malibu Canyon Road, Malibu, CA 90265, USA
** Air Force Avionics Laboratory, Wright-Patterson AirForce Base, Ohio 45433, USA.

background. (That S is a significant residual donor in all types of GaAs will be discussed in another study [11].)

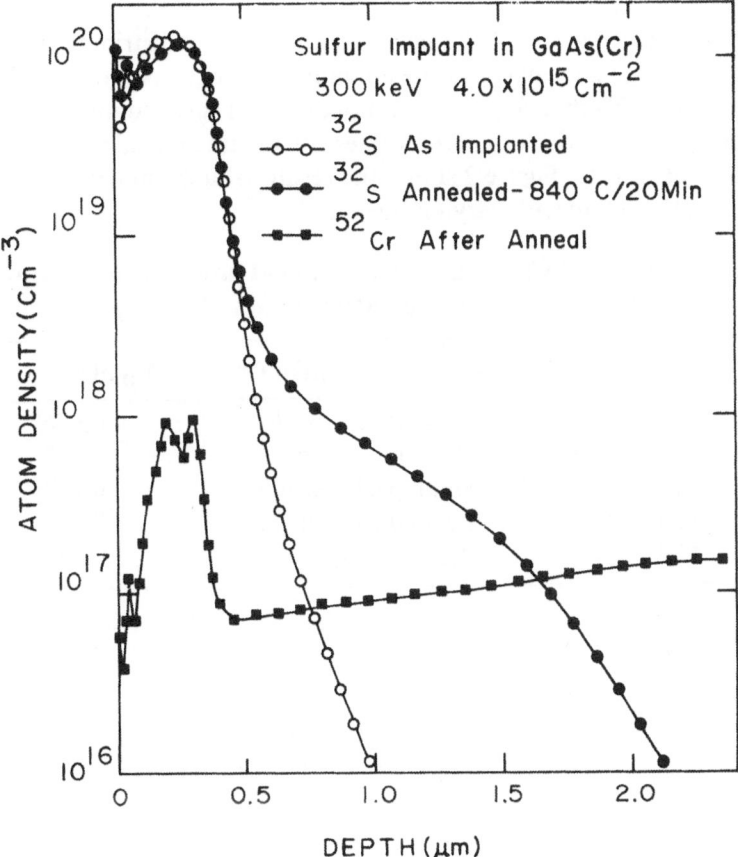

Figure 1 Thermal redistribution of S and Cr in semi-insulating GaAs after a 300 keV ³²S implantation to 4×10¹⁵ cm⁻².

At concentrations above 5×10^{18} cm⁻³, S has not significantly redistributed while diffusion is seen below this concentration. For ³²S implantation we routinely find S diffusion at concentrations below a threshold in the mid 10^{18} cm⁻³ regime.

After thermal processing the Cr exhibits several distinguishable behaviour patterns. First, the Cr depletes in the outer 2 μm of the GaAs and piles up in the encapsulant-GaAs interface, although not for this particular fluence and energy. Moreover, the Cr shows two distinct peaks in and around the projected range of the S implant. As in previous studies [2], we propose that the Cr is being gettered into regions of lattice damage. We see a peak at approximately 0.88 of the ³²S projected range which corresponds to the maxium in the damage profile associated with the implantation. In addition, we see another Cr peak localized at some distance greater than the projected range of the S which we have previously identified in GaAs [2] and in Si [12, 13] as being characteristic of a region of residual implantation damage after annealing. Throughout this study we have seen localization of the Cr in these three different regions.

In order further to detail the phenomenon of Cr redistribution, we have varied the fluence and energy of the implant and the temperature of the post-implant anneal. Table 1 shows the nature of the Cr distribution as a function of ^{32}S energy along with tabulated values for the theoretical projected range of the S implant [14]. Figure 2 shows the results of a 20 min anneal at 840°C for a 300 keV implant with varying fluences.

Table 1 **Nature of the Cr profile as a function of implantation energy (4×10^{13} cm^{-2} with a 840°C anneal for 20 min with an SiO$_2$ cap)**

^{32}S energy (keV)	Nature/location of Cr	R_p of ^{32}S (μm)
75	No apparent peak	0.056
150	No apparent peak	0.113
300	Small peak/$0.2\,\mu$m	0.227
450	Peak/$0.375\,\mu$m	0.345
600	Peak/$0.425\,\mu$m	0.457

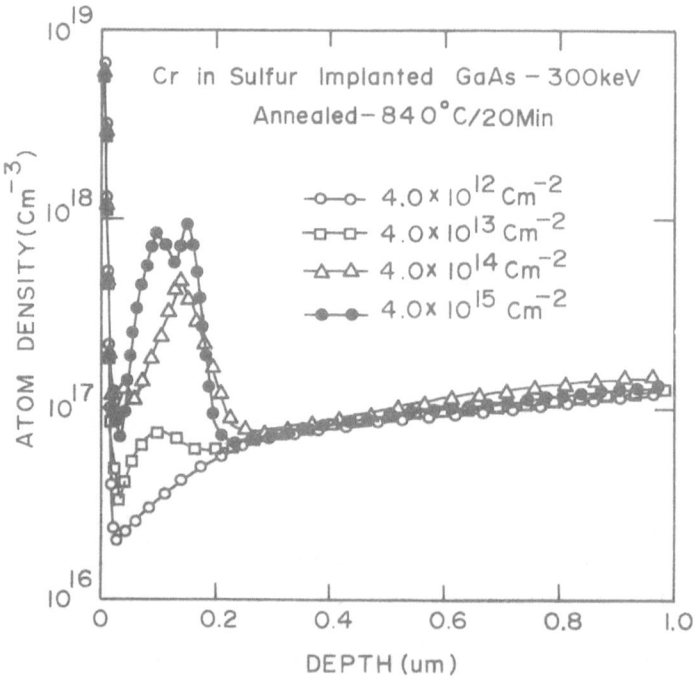

Figure 2   Thermal redistribution of Cr as a function of ^{32}S implantation fluence.

Isochronal annealing for 20 min after a 4×10^{13} cm^{-2} implant at 300 keV was studied in the temperature range of 650—850°C. The mechanism can be visualized as the Cr diffusing from the bulk and gettering into whatever damage may be present in and around R_p. As the temperature is increased, there is an increasing surface concentration and a decreasing amount of Cr near R_p as the damage is removed by thermal processing. Moreover, as the temperature is increased, the interfacial stress is increased, thereby providing more defects in the crystalline lattice which getter the Cr at the surface. Table 2 shows these results in terms of increase in Cr surface concentration as a function of annealing temperature. There may be a systematic error of up to a factor of 2 in these concentrations due to the vagaries in the secondary ionization process in the near surface region. However, the magnitude and trend of the data still support the above conclusions.

Table 2 **Chromium surface concentration as a function of a 20 min anneal with an SiO$_2$ cap of a 300 keV, 4×10^{13} cm^{-2} ^{32}S implant**

Annealing temperature (°C)	Surface concentration (at. cm^{-3})
650	1×10^{17}
700	6×10^{17}
750	2×10^{18}
800	3×10^{18}
850	5×10^{18}

References

1. Huber, A.M., Morillot, G., Linh, N.T., Favennec, P.N., Deveaud, B. and Toulouse, B. (1979). *Appl. Pys. Lett.*, **34**, 859
2. Evans, C.A. Jr, Deline, V.R., Sigmon, T.W. and Lidow, A. (1979). *Appl. Phys. Lett.*, **35**, 291
3. Wilson, R.J., Vasudev, P.K., Jamba, D.M., Evans, C.A. Jr and Deline, V.R. (1980). *Appl. Phys. Lett.*, in press
4. Magee, T.J., Peng, J., Hong, J.D., Deline, V.R. and Evans, C.A. Jr (1979). *Appl. Phys. Lett.*, **35**, 615
5. Asbeck, P.M., Tandon, J., Welch, B.M., Evans, C.A. Jr and Deline, V.R. (1980). *Trans. IEEE Electr. Dev. Lett.*, in press
6. Favennec, P.N. and Haridon, H.L. (1979). *Appl. Phys. Lett.*, **35**, 699
7. Asbeck, P.M., Welch, B.M. and Evans, C.A. Jr. *Electron Density Profiles in Se-Implanted Cr-doped GaAs*, in preparation
8. Anderson, C.L., Vaidyanathan, K.V., Dunlap, H.L. and Kamath, G.S. (1980). *J. Electrochem. Soc.*, in press
9. Storms, H.A., Stein, J.D. and Brown, K.F. (1977). *Anal. Chem.*, **40**, 1399
10. Williams, P., Lewis, R.K., Evans, C.A. Jr and Hanley, P.R. (1977). *Anal. Chem.*, **49**, 1399
11. Evans, C.A. Jr, Norberg, J.C., Hopkins, C.G., Deline, V.R. and Blattner, R.J. *The Identification of Sulfur as a Significant Residual Donor in GaAs*, in preparation
12. Tsai, M.Y., Streetman, B.H., Williams, P. and Evans, C.A. Jr (1978). *Appl. Phys. Lett.*, **32**, 144
13. Tsai, M.Y., Day, D.S., Streetman, B.G., Williams, P. and Evans, C.A. Jr (1979). *J. Appl. Phys.*, **50**, 188
14. Brice, D.K. (1971). Spatial Distributions of Ions Incident on a Solid Target as a Function of Instantaneous Energy. Sandia Labs Report SC-RR-71-0599 Albuquerque, NM

Section 3:
Investigations of
semi-insulating materials

PHOTOCONDUCTIVITY, PHOTOLUMINESCENCE AND ZEEMAN SPECTROSCOPY OF Cr IN GaAs, GaP AND InP

L. EAVES, T. ENGLERT**, T. INSTONE*, C. UIHLEIN**,
P.J. WILLIAMS and H.C. WRIGHT*
Department of Physics, University of Nottingham, University Park,
Nottingham NG7 2RD, UK

Abstract

We report high resolution Zeeman spectroscopy up to 10 T of Cr-associated photoluminescence in GaAs and GaP including the anisotropy of the splitting with B swept in the $(0\bar{1}1)$ plane. The results are interpreted in terms of recombination at an 'isoelectronic molecular' complex involving Cr and a near-neighbour donor. In GaAs the complex has <111> axial symmetry. The anisotropy for GaP is markedly different and corresponds to a defect or complex with principal axes along the <100> directions. The photoconductivity spectra arising from Cr in GaAs, InP and GaP have also been studied. Attention is drawn to the close correspondence between the energies of the zero phonon photoluminescence lines and the thresholds of the Cr-associated photoconductivity peaks in all three systems. The photoconductivity spectra are compared with the results of a new theory for optical cross-sections.

1 Introduction

The precise origins of Cr-associated photoluminescence (PL), photoconductivity (PC) and EPR in III—V compounds are the subject of much controversy despite intensive study. This paper describes PL Zeeman spectra and PC spectra of Cr in the three main III—V compounds with a view to identifying (*a*) the origins of these spectral features, and (*b*) the common features and systematic changes in the spectra as a function of host material and chemical potential. The paper is arranged as follows. Section 2 describes the Zeeman spectra of the well-known sharp PL structure in GaAs and GaP. The results support White's model [1] that the lines do not arise from a Cr^{2+} (^5E—$^5T_2$) transition, but from recombination at a near-neighbour chromium-donor complex. Section 3 presents the PC spectra in GaAs, InP and GaP and compares them with a new theory for optical cross-sections. This section also compares the PC spectra of n-, p- and SI GaAs:Cr. Section 4 summarizes our work.

* Plessey Research (Caswell) Ltd, Towcester, UK.
**Max-Planck Institut Für Festkörperforschung, Grenoble, France.

2   Zeeman anisotropy of PL

The much-discussed, sharp 0.84 eV PL structure in GaAs:Cr was originally thought to arise from an internal ^5E to $^5T_2$ transition of Cr^{2+} ($3d^4$). More recently, the line has been attributed [1] to recombination at a complex involving Cr substituted on a Ga site paired with a near-neighbour donor D (model A of White [1]). This explains the inconsistency between the splitting of the energy levels in the ground state as observed by PL/absorption [2] and EPR [3]. According to White's model, the initial (upper) state involves an electron bound in an almost hydrogen-like orbit to the positively charged state of the complex. This state is denoted by $[Cr^{3+}(3d^3)\text{—}D^+]e$. The final (lower) state $[Cr^{2+}(3d^4)\text{—}D^+]$ has zero overall charge and is an 'isoelectronic molecular' complex similar to Zn—O in GaP [4].

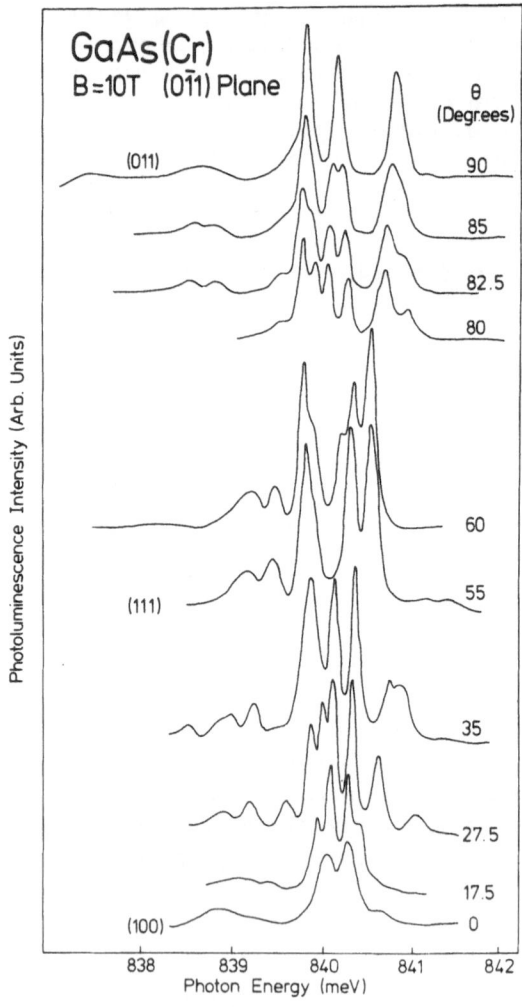

Figure 1   The Zeeman spectra of the 0.84 eV PL zero phonon structure of GaAs:Cr at 2 K with B=10 T rotated in the (0$\bar{1}$1) plane.

The PL data were obtained using the 10 T split-coil superconducting magnet at MPI, Grenoble. Figure 1 shows the Zeeman splitting of the zero-phonon line at 10 T for varying angles of B in the (0$\bar{1}$1) plane. The anisotropy of the main components are plotted in Figure 2. The anisotropy is very similar to that previously observed [4, 5] for GaP:Zn—O and GaP:Li—Li—O and is consistent with recombination at a <111> axial complex.

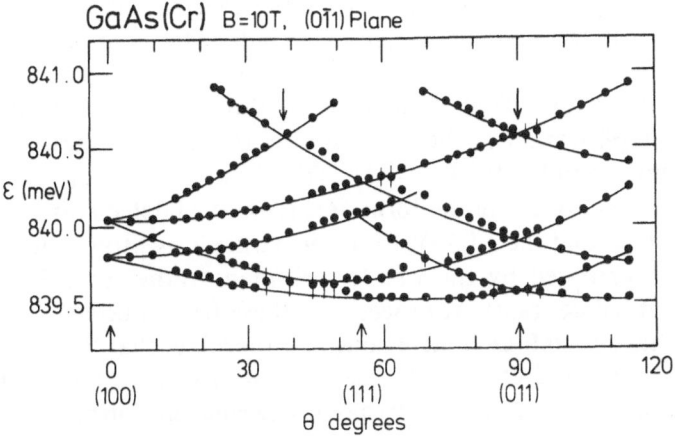

Figure 2 The Zeeman anisotropy at 2 K of the 0.84 eV PL zero phonon structure of GaAs:Cr with B=10 T rotated in the (0$\bar{1}$1) plane.The anisotropy corresponds to a <111> Cr—D axial complex.

For B||<100>, all four <111> axes are equivalent and at high fields the spectrum is dominated by a simple doublet. As B is rotated away from <100> in a (0$\bar{1}$1) plane all but two of the <111> axes become magnetically non-equivalent giving rise to the high multiplicity of lines. An important feature of Figure 2 is the occurrence at the same energy of the crossings, shown by arrows, for B||<211> and B||<011>. In these configurations B is perpendicular in turn to some of the four <111> axes (two for <011> and one for <211>) so that the degeneracy of the states is only partly lifted by the field. As B is rotated away from the perpendicular direction it removes this remaining degeneracy and gives rise to the observed crossings. An effective Hamiltonian, $\mathcal{H}$, of the following form describes the $(Cr^{3+}—D^+)e$ excited state proposed by White:

$$\mathcal{H} = \mu_B J\hat{g}_1 B + \mu_B \sigma\hat{g}_2 B + \tfrac{1}{6}a(J_x^4 + J_y^4 + J_z^4) + D[J_z^2 - \tfrac{1}{3}J(J+1)]$$
$$+ \alpha\sigma_z J_z + \beta(\sigma_+ J_- + \sigma_- J_+)$$

where the terms refer respectively to the Zeeman energies of the Cr-ion and orbiting electron, the cubic and axial fields and the exchange interaction. At present it is not possible to identify the donor involved in the complex, or indeed to rule out another type of axial defect. However, we have found that semi-insulating material thought to contain O as well as Cr gives rise to the strongest 0.84 eV PL. A possible candidate for the axial centre may therefore be $Cr_{Ga}-O_{As}$.

GaP:Cr shows a PL spectrum very similar to that in GaAs. The zero phonon line at 1.03 eV with phonon sidebands to lower energies was first observed by Dean [6] and was subsequently identified with Cr by Kaufmann and Koschel [7]. Just as in GaAs:Cr, the zero phonon line reveals several components under high resolution. At 2 K for B=0, the fine structure consists of a system of three roughly equally spaced central components with two weaker, but similar, components on either side. The Zeeman anisotropy of the principal lines with B=10 T rotated in the (0$\bar{1}$1) plane is shown in Figure 3. The form of the anisotropy is significantly different from GaAs:Cr. There is a great simplification of the spectrum for B∥<111> whereas in GaAs:Cr the spectrum is simplest for B∥<100>. This shows that in GaP, the Cr giving rise to the PL is present in the form of <100> axial defects. It is interesting that Dean [6] found the 1.03 eV PL to be particularly prominent in vapour-grown GaP prepared by the wet H_2 technique rather than by the more conventional halide transport process. As these freely nucleated crystals are known to have a high O content, one can speculate that the complex may again involve Cr and O. In this case, Cr_{GA} with O on a <100> interstitial would have the required symmetry to fit the Zeeman anisotropy.

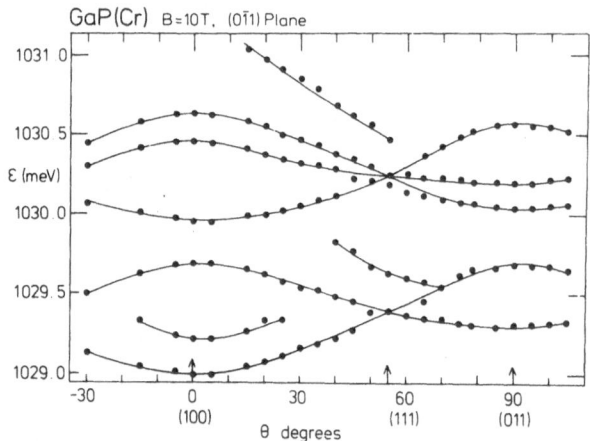

Figure 3   The Zeeman anisotropy at 2 K of the 1.03 eV PL zero phonon structure of GaP:Cr with B=10 T rotated in the (0$\bar{1}$1) plane.

3   Extrinsic photoconductivity

(a) GaAs:Cr

The existence of Cr^{3+} $(3d^3)$ and Cr^{2+} $(3d^4)$ in GaAs is firmly established by EPR [3]. Recently, the existence of Cr^{1+} $(3d^5)$ has been questioned and that of

Cr^{4+} ($3d^2$) proposed instead [8, 9]. The relative concentrations of the various charge states depend on the position of the chemical potential. We have investigated both the d.c. and the 'chopped' PC of a series of n-, p- and SI GaAs. The data, obtained at 2 K using a grating monochromator and lock-in techniques, are strongly dependent on sample doping, as shown in Figure 4. SI material containing approximately equal amounts of Cr and Si (see caption for Figure 4) shows a threshold at 0.8 eV followed by a rather sharp, strong peak at 0.88 eV which has been reported previously (e.g. in [10]) and is generally attributed to

$$Cr^{2+} (3d^4) + h\upsilon (\gtrsim 0.8\,eV) \rightarrow Cr^{3+} (3d^3) + e^-(CB) \tag{1}$$

with the initial Cr^{2+} state 0.8 eV below the Γ-point conduction band edge. In the p-type material this peak is absent as the Cr is in a more positive charge state (presumably Cr^{3+} or Cr^{4+}). In n-type material process (1) occurs either because Cr^{2+} has been produced by photoexcitation of Cr^{1+} [11] or else because Cr^{2+} is the stable charge state in n-GaAs:Cr.

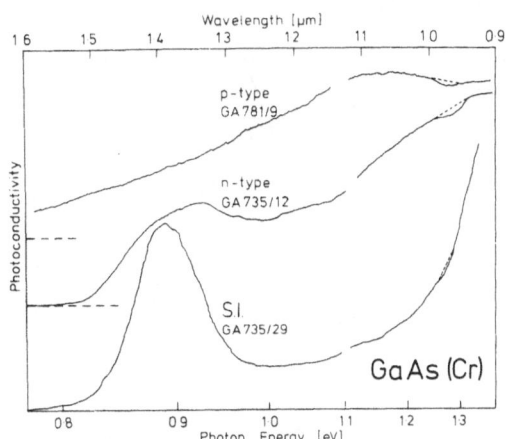

Figure 4   The 'chopped' PC of differently doped samples of GaAs:Cr at 2 K.
Ga 735/29 is SI with $[Cr] = 2 \times 10^{17}$ cm^{-3}, $[Si] = 2 \times 10^{17}$ cm^{-3}.
Ga 735/12 is n-type with n(300 K)$= 4 \times 10^{12}$ cm^{-3},
$[Cr] = 0.5 \times 10^{17}$ cm^{-3}, $[Si] = 1.5 \times 10^{17}$ cm^{-3}.
Ga 781/9 is p-type, Cr- and Zn-doped, with p(300 K)$= 10^{14}$ cm^{-3}.

(*b*)   GaP:Cr

The Cr-associated PC of semi-insulating GaP ($\rho > 10^9$ Ω-cm at 290 K) is shown in Figure 5 over a wide range of temperatures. At low temperatures ($\ll 100$ K) there is a sharp threshold at 1.2 eV; at higher temperatures, a shoulder appears at slightly lower energy with a threshold at around 1.0 eV. This behaviour suggests that there is a ground state (*a*) in the middle of the gap, 1.2 eV below the X-point conduction band minima with an excited state (*b*) about 0.2 eV below the band edge. Thus, at low temperatures the photoconductivity threshold corresponds to a transition from the ground state to the conduction band edge. At higher temperatures a photothermal ionization process sets in whereby an intra-impurity transition (*a* to *b*) followed by thermal excitation into the conduction band gives rise to the lower energy threshold. Gloriozova and Kolesnik [12], using photocapacitance techniques, also conclude that such a two-stage process occurs although they place the excited state within only 30 meV of the conduction band edge.

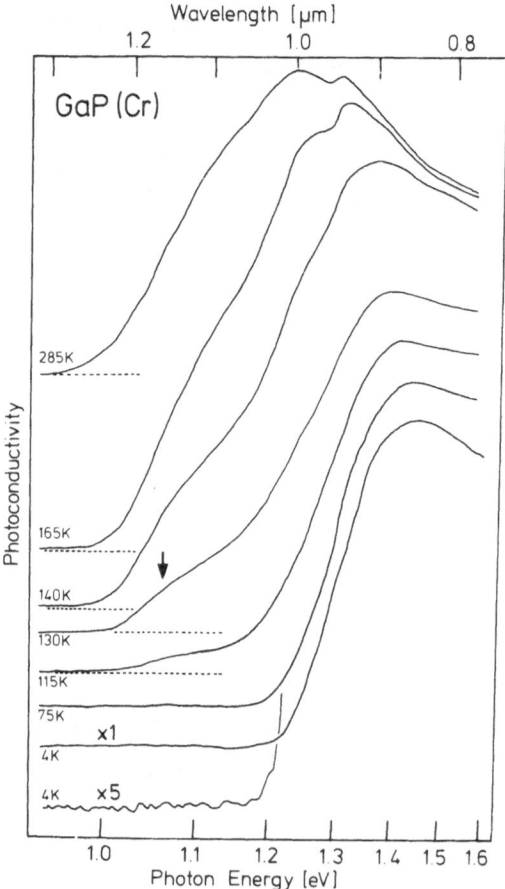

Figure 5 The chopped PC of high resistance, ρ (300 K) $\sim 10^9$ Ω -cm, GaP:Cr as a function of temperature.

(c) InP:Cr

The amount of PC data on this system is rather limited. A typical spectrum for a high resistance sample, shown in Figure 6, is dominated by a strong peak at 1.05 eV. It is rather similar in form to that found in SI GaAs:Cr. This suggests that it arises from the type of process given by equation (1). The weaker threshold at 0.77 eV is not present in all the Cr-doped samples we have investigated and may not necessarily be due to Cr. The InP:Cr spectrum is discussed with that of the other two materials in the next section.

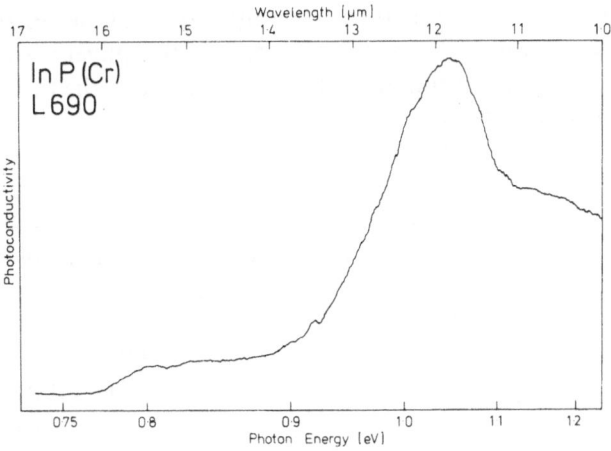

Figure 6   The chopped PC of high resistance, ρ (300 K) $\sim 10^8 \Omega$-cm, InP:Cr at 2 K.

(d) PHOTOCONDUCTIVITY — COMPARISON WITH THEORY

The forms of the PC spectra of InP:Cr and GaAs:Cr shown in the above sections are in good agreement with a new method for calculating deep trap optical cross-sections using evanescent waves associated with the Γ-point and indirect extrema to describe the deep levels [13, 14]. In contrast to the Lucovsky theory [15] this method, which takes full account of band non-parabolicity, predicts that the optical cross-section for the transitions to the Γ-point minimum peaks at only about 0.1 ev above threshold [13]. This is in contrast to the slow rise from threshold predicted by simple Lucovsky theory [15]. Thus the two sharp peaks at 0.88 and 1.05 eV, which are observed in GaAs and InP, can be explained in terms of a single transition from a bound state to the Γ-point conduction band edge. The further rise in the PC at 1.1 eV in GaAs:Cr is about 0.3 eV above the lower edge threshold and may correspond to transitions to the higher conduction band minima. In GaP:Cr , we observe that the photoconductivity peak is much broader than in the direct gap materials. This agrees with Burt's calculated cross-section for transitions from a deep level to the indirect conduction band minima [13].

4 Conclusion

The Zeeman spectroscopy and anisotropy confirm White's model of recombination at an axial complex or defect to explain the strong Cr-associated PL in III—V materials. The previous concensus of opinion was that the GaAs:Cr PL arose from a ^5E—$^5T_2$ transition of Cr^{2+}. This view, combined with energy level diagrams for Cr^{2+} obtained from extrinsic PC data, led to the suggestion that the upper ^5E state was near, or possibly in resonance with, the conduction band edge (see [16] for a discussion of this). It now seems that the dominant PL and PC peaks in III—V materials arise from quite different processes, the former involving complexes and the latter simple centres. However, the results presented in this paper show that for all three materials, the zero phonon PL line is very close to the threshold in PC. This is illustrated in Table 1.

Table 1

Material	Zero phonon PL (eV)	PC threshold (eV)
GaAs	0.84	0.8
InP	0.84 [17]	0.9
GaP	1.03	1.2 (T < 100 K)
		1.0 (T > 100 K)

We have also found a similar correspondence for Fe in InP. An explanation of the coincidence in energy awaits a quantitative theory on the energy levels of transition metal ions and their complexes in III—V materials.

ACKNOWLEDGEMENTS

We acknowledge P.J. Dean, R.G. Humphreys, M.S. Skolnick and A.M. White (RSRE, Malvern), M.J. Cardwell, P.R. Jay and D.J. Stirland (Plessey, Caswell), and G. Landwehr (Grenoble) for advice. This work was supported in part by the SRC.

References

1. White, A.M. (1979). *Solid St. Commun.*, **32**, 205
2. Lightowlers, E.C., Henry, M.O. and Penchina, C.M. (1978). *Proc. Int. Conf. on Physics of Semiconductors*, Edinburgh, p. 307
3. Krebs, J.J. and Stauss, G.H. (1977). *Phys. Rev. B*, **16**, 971
4. Henry, C.H., Dean, P.J. and Cuthbert, J.D. (1968). *Phys. Rev.*, **166**, 754
5. Dean, P.J. (1971). *Phys. Rev. B*, **4**, 2596
6. Dean, P.J. (1973). *J. Luminescence*, **7**, 51

7. Kaufmann, U. and Koschel, W.H. (1978). *Phys. Rev. B,* **17,** 2081
8. Kaufmann, U. and Schneider, J. To be published
9. Stauss, G.H., Krebs, J.J., Lee, S.H. and Swiggard, E.M. To be published
10. Lin, A.L. and Bube, R.H. (1976). *J. Appl. Phys.,* **47,** 1859
11. Eaves, L. and Williams, P.J. (1979). *J. Phys. C: Solid St. Phys.,* **12,** L725
12. Gloriozova, R.I. and Kolesnik, L.I. (1978). *Sov. Phys. Semicond.,* **12,** 66
13. Burt, M.G. (1980). *J. Phys. C: Solid St. Phys.,* **13,** in press
14. Inkson, J.C. (1980). *J. Phys. C: Solid St. Phys.,* **13,** 369
15. Lucovsky, G. (1965). *Solid St. Commun.,* **3,** 299
16. Stocker, H.J. and Schmidt, M. (1976). *Proc. Int. Conf. on Physics of Semiconductors,* Rome, p. 611
17. Koschel, W.H., Bishop, S.G. and McCombe, B.D. (1976). *Proc. Int. Conf. on the Physics of Semiconductors,* Rome, p. 1065

PHOTOCONDUCTIVITY STUDIES OF
Cr DEEP ACCEPTORS IN InP

S. FUNG and R.J. NICHOLAS
Clarendon Laboratory, Parks Road, Oxford OX1 3PU, UK

Abstract

Photoconductivity has been observed from Cr-doped InP at temperatures between 6 and 300 K. Photoconductive onsets have been observed at 0.47 eV at 6 K (0.40 eV at 300 K), and at 0.93 eV at both 6 and 300 K. The structure above 0.93 eV is found to be suppressed by compensation of the Cr with shallow Sn donors introduced by nuclear transmutation doping. The two onsets are attributed to photoexcitation into and out of the dominant Cr trap located 0.40 eV below the conduction band at 300 K. An additional broad structure is observed at 0.8 eV which is thought to be due to an internal transition of the Cr ion.

Introduction

It has been known for some time that InP doped with a sufficiently high Cr concentration can become semi-insulating, although for a given doping level InP:Cr has a lower resistivity than InP:Fe. Very little experimental work has been done to investigate the deep Cr acceptor levels in this material. Voltage-current measurements on Cr-doped material have indicated energy levels at 0.2 and 0.54 eV [1], and it was suggested that the 0.54 eV level was due to Cr, while low frequency current oscillation measurements [2] have indicated the presence of a level at 0.62 eV. A broad photoluminescence band at 0.8 eV has also been reported [3] and was attributed to an internal transition between the crystal field split levels of Cr^{2+} ions.

This paper reports photoconductivity measurements on Cr-doped InP which show that the dominant level occurs at 0.40 eV (0.47 eV) below the conduction band at 300 K (6 K). The Cr ions are assumed to be introduced substitutionally at the In site in an analogous manner to GaAs:Cr [3]. Thus, the Cr^{3+} ($3d^3$) is the neutral state and the Cr^{2+} ($3d^4$) and Cr^{1+} ($3d^5$) are the one and two electron traps, respectively. Photoconductive signals are found due to both electron and hole excitation, and a photoresponse edge at 0.74 eV and a peak at 0.81 eV are probably associated with an intracentre transition of either Cr^{2+} or Cr^{1+} ions.

Experimental

Both a.c. and d.c. photoconductivity methods are used, as described in [5]. Spectra of a range of samples are taken between 6 and 300 K. In the d.c. experiment each spectrum is scanned over a period of 60 minutes. The signal response time is found to be rather short, approximately 300 ms, as opposed to the long response time observed in some samples of InP:Fe [5] and GaAs:Cr [6]. The d.c. photoconductivity is up to two orders of magnitude greater than the dark value at temperatures below 77 K.

Two samples have been irradiated by thermal neutrons in order to introduce 2×10^{16} cm^{-3} Sn donors. The In undergoes a nuclear transmutation to Sn which acts as a shallow donor on the In site. This method has been proved very successful in controlling Sn donors in InSb [7, 8]. The room temperature resistivity of these samples is reduced by up to two orders of magnitude indicating a considerable compensation of the deep Cr acceptors. The experiments were repeated after adjustment of the compensation in this way. The resistivity of the samples used, both before and after irradiation, is shown in Table 1.

Table 1 **Sample characteristics**

Sample number	Resistivity (Ω-cm) (300 K)	Resistivity after irradiation (Ω-cm) (300 K)
L630	5×10^3	100
L690	10^6	10^5
L587	10^5	

Results

A.C. PHOTOCONDUCTIVITY

The a.c. photoresponses of three bulk InP:Cr samples have been taken. Typical recordings of samples L630, L690 and L587 at 6 K are shown in Figure 1. The samples L690 and L587 are more semi-insulating than L630. All the three recordings exhibit an onset of photoconductivity at 0.47 eV. Similar spectra taken at 300 K indicate an onset at 0.40 eV. This shift with temperature is almost equal to that of the fundamental gap. The photoresponse increases steadily with photon energy and subsequently levels off at approximately 0.70 eV. There appears to be another onset at 0.74 eV and the signal reaches a maximum at about 0.81 eV. The samples L690 and L587 show another edge at 0.93 eV and a maximum at 1.01 eV before decreasing to a low value below the fundamental gap. It is clear that the intensity in the spectrum for the more strongly compensated L630 sample is smaller at 1.01 eV.

Figure 2 shows spectra of the sample L630 taken at different temperatures. These indicate that there is a considerable response at 1 eV at 300 K and this decreases with temperature so that the response at 0.80 eV appears as a peak.

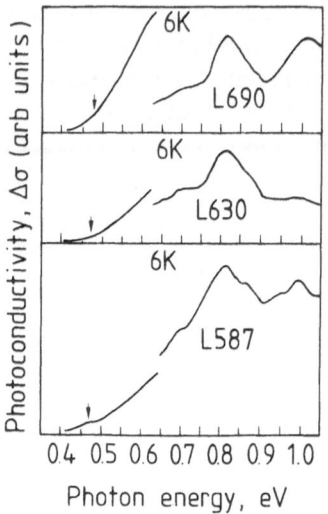

Figure 1 Experimental recordings of the a.c. photoresponse of three samples of InP:Cr as a function of photon energy. The spectra are taken at 6 K and the sample numbers are shown. The observed photoconductive signals have been normalized to the incident light intensity. The spectra are taken in two sections using different infra-red filters in order to eliminate any spurious second order effects. The arrows show the position of the low energy photoconductive onset as determined by a fit to the photoionization cross-section.

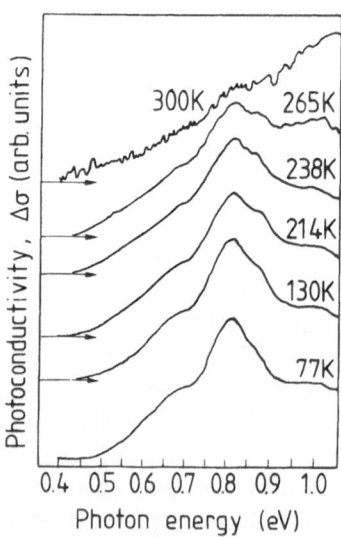

Figure 2 Experimental recordings of the a.c. photoresponse of sample L630 taken at lattice temperatures between 77 and 300 K. The spectra are normalized to the incident light intensity.

D.C. PHOTOCONDUCTIVITY

Both samples L690 and L630 have very high resistivity when they are cooled to below 200 K. At 6 and 77 K the resistivity exceeds 10^{12} ohm-cm, and it is possible to take d.c. photoresponse spectra with unchopped light. Spectra taken this way are very similar to those of the a.c. results.

RESULTS FROM IRRADIATED SAMPLES

Irradiated samples L690 and L630 (IRL690 and IRL630) have been investigated by the a.c. method. There is no detectable change in the spectrum of IRL630 while IRL690 exhibits features similar to those of the more compensated L630 before irradiation. The response peak at 1.01 eV has now disappeared in the IRL690 spectrum whereas other features remain unaltered.

The resistivity of IRL630 at low temperatures is strongly reduced by the irradiation indicating that almost all of the deep Cr traps have been filled by the shallow donor compensation, while that of IRL690 remains very high showing that not all traps have been filled.

Interpretation

It is to be expected that the Cr centres in InP will behave in a similar way to GaAs:Cr, where the Cr^{3+} is the neutral charge state and both one (Cr^{2+}) and two (Cr^{1+}) electron trap states have been observed [4]. A Cr^{2+} EPR signal has been observed in InP:Cr [9]; however, no signal was detected from either the Cr^{3+} or the Cr^{1+} states. The Fermi level in semi-insulating Cr-doped material is typically in the region of 0.4—0.5 eV from the conduction band edge, as has been found for the samples used in the present investigation, so that it is to be expected that a Cr acceptor energy is in this region. The photoconductive onset at 0.47 eV (6 K) is thought to result from excitation from a Cr acceptor to the conduction band in one of the following two processes:

$$Cr^{2+} \rightarrow Cr^{3+} + e^- \text{ (conduction band)} \tag{1}$$

or

$$Cr^{1+} \rightarrow Cr^{2+} + e^- \text{ (conduction band)}. \tag{2}$$

It is not possible to determine unambiguously from the present measurements which process is occurring, although it is thought to be more likely that only the one electron trap (equation 1) is involved. The trap energy (0.47 eV at 6 K) was determined by fitting the observed photoresponse using the method described by Fung *et al.* [5].

The complementary processes of either:

$$Cr^{3+} + e^- \text{ (valence band)} \rightarrow Cr^{2+} \qquad (3)$$

or

$$Cr^{2+} + e^- \text{ (valence band)} \rightarrow Cr^{1+} \qquad (4)$$

are thought to result in the photoresponse observed in the region above 0.9 eV. The strong reduction in the relative intensity of this region of the spectrum produced by increased compensation with shallow donors shows that the initial state for this transition must be a Cr trap in a lower state of occupation than for the rest of the spectrum. The transition energy for process (3)/(4) is estimated to be 0.93 eV ($\pm$ 0.02 eV), with some uncertainty due to the contribution to the signal from other features in the spectrum. The sum of the two transition energies for processes (1)/(2) and (3)/(4) are thus 1.33 eV (300 K) and 1.40 eV (6 K), in good agreement with the band gap energies of 1.35 eV and 1.42 eV, as would be expected for an impurity which is coupled only weakly to the lattice.

The strong temperature dependence of the signal from transition (3)/(4) is due to the fall in the Cr^{2+}/Cr^{3+} population with falling temperature. Thermal excitation of either electrons into the conduction band or holes into the valence band no longer becomes sufficient to provide a significant population in the high charge state which is the initial state for the transitions (3)/(4).

The origin of the structure in the region 0.7—0.9 eV is not completely clear. The data taken with different samples with different levels of compensation indicate that this structure is associated with Cr in the same initial state as gives rise to the 0.47 eV edge. The origin of this structure may thus be:

1. Due to an internal transition between the crystal field levels ($^5T_2 \rightarrow ^5E$) of Cr^{2+}. This would conform to the assignment of the photoluminescence band which gives a peak at 0.8 eV [3], which is interpreted as the level separation. With the Cr^{2+} ground state 0.47 eV below the band edge, this would put the upper level in the conduction band and hence would explain the strong photoresponse and broad luminescence.
2. A similar picture to (1) with the ground state as a Cr^{1+} level which is crystal field-split.
3. A transition from the Cr^{1+} or Cr^{2+} level to a higher conduction band minimum. This would require a satellite valley located ~ 0.3 eV above the Γ-point, which is rather less than that estimated by Pitt [10].

It is thus concluded that no evidence has yet been found for more than one trap state from InP:Cr, and the level scheme and optical transitions observed may be represented by Figure 3.

The Cr^{2+} acceptor ground state lies 0.47 eV (0.40 eV) below the conduction band edge at 6 K (300 K) with the upper level 0.34 eV above the band edge. It is possible that the transitions and levels in Figure 3 could result from Cr^{2+} and Cr^{1+} states, with the Cr^{2+} one electron trap not making any contribution to the photoconductive signal. If this were the case, however, it would be expected that the one electron trap should lie significantly below the double trap and hence enable Cr-doped material to be made with a significantly lower Fermi energy and hence more strongly semi-insulating.

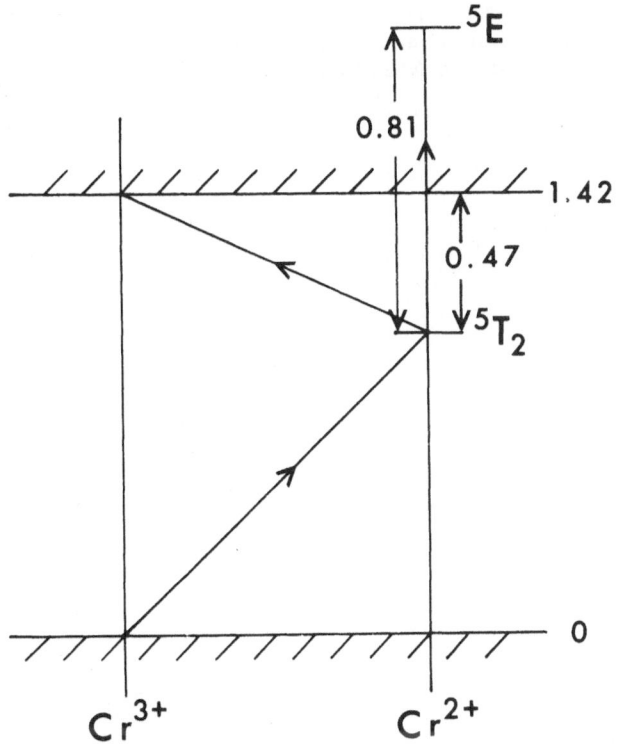

Figure 3 A schematic view of the possible energy levels and separations for Cr in InP at 6 K. The arrows show the electron and hole excitation processes (1) and (3) as well as the internal Cr^{2+} transition. The energy levels and separations are shown in eV.

ACKNOWLEDGEMENTS

We would like to thank Dr B. Cockayne of the RSRE, Malvern, for provision of the InP:Cr samples, and Mr Izatt and the staff of the Scottish Universities Research and Reactor Centre, East Kilbride, for performing the neutron irradiation.

References

1. Pande, K.P. and Roberts, G.G. (1976). *J. Phys. C: Solid St. Phys.*, **9**, 2899
2. Kovalevskya, G.G., Alynshina, V.I. and Schlobodchikov, S.V. (1975). *Sov. Phys. Semicond.*, **9**, 1385

3. Koschel, W.H., Bishop, S.G. and McCombe, B.D. (1976). *Proc. 13th Int. Conf. on Physics of Semiconductors,* p. 1065. Rome; Tipografia Marves

4. Lin, A.L. and Bube, R.H. (1976). *J. Appl. Phys.,* **47**, 1859

5. Fung, S., Nicholas, R.J. and Stradling, R.A. (1979). *J. Phys. C: Solid St. Phys.,* **12**, 5145

6. Eaves, L. and Williams, P.J. (1979). *J. Phys. C: Solid St. Phys.,* **12**, L725

7. Kuchar, F., Fantner, E. and Bauer, G. (1974). *Phys. Stat. Solidi (a),* **24**, 513

8. Clark, W.G. and Isaacson, R.A., (1967). *J. Appl. Phys.,* **38**, 2284

9. Stauss, G.H., Krebs, J.J. and Henry, R.L. (1977). *Phys. Rev. B,* **16**, 974

10. Pitt, G.D. (1973). *J. Phys. C: Solid St. Phys.,* **9**, 2899

PHOTOLUMINESCENCE STUDIES OF DEEP IMPURITY STATES IN Fe-DOPED InP

S.G. BISHOP, P.B. KLEIN, R.L. HENRY and B.D. McCOMBE
Naval Research Laboratory, Washington, DC 20375, USA

Abstract

Photoluminescence, photoluminescence excitation, and optical absorption measurements have been carried out on semi-insulating and n-type InP:Fe. Interband excitation of the ~ 0.35 eV intracentre Fe^{2+} transition is interpreted in terms of electron capture by an Fe^{3+}, resulting in an Fe^{2+} centre in the $^5T_2$ excited state, which relaxes radiatively to the ^5E ground state. An extrinsic band in the photoluminescence excitation spectrum of the ~ 0.35 eV luminescence exhibits an onset at $\sim 1.13-1.15$ eV, which is attributed to transitions from the valence band to the $^5T_2$ excited state of Fe^{2+}. This assignment is found to be consistent with the energy level scheme for InP:Fe proposed by Fung, Nicholas and Stradling.

Introduction

In recent years iron-doped InP has been the subject of investigation by several spectroscopic techniques, including optical absorption [1, 2], photoluminescence [2—4], photocapacitance [5], photoconductivity [6—8], and electron spin resonance (ESR) [9]. These studies have attempted to determine the microscopic physical characteristics of the iron impurity centres and to fix the energies of the iron deep acceptor levels within the InP gap. It has been established [1, 2] that iron occupies the cation site in InP and exists in both the isoelectronic (neutral) Fe^{3+} ($3d^5$) state and the one electron trap state Fe^{2+} ($3d^6$). The relative proportion of these two charge states is determined by the degree of compensation of the deep iron acceptor levels by shallow donor impurities.

Recently, Fung et al. [8] proposed an energy level scheme for Fe^{2+} in InP which was based upon their photoconductivity results in conjunction with the results of several other workers [1—3, 5, 6, 9—11]. In this paper we present optical absorption, photoluminescence (PL) and photoluminescence excitation spectra (PLE), which tend to confirm their proposed model and to clarify the detailed interpretation of some of their photoconductivity spectra in terms of the model.

Experimental

Luminescence and optical absorption experiments were carried out on semi-insulating and n-type bulk InP:Fe. The crystals were grown by the liquid encapsulation technique in a pyrolytic boron nitride crucible. Four samples were studied: two semi-insulating (SI) Fe-doped samples (1-91-H and 1-74-H), one n-type Fe-doped sample (1-66-H) and one undoped sample (1-81-H).The characteristics of the samples are summarized in Table 1.

Table 1   **Characteristics of the InP samples that were studied**

Sample	Dopant	Fe concentration (cm^{-3})	Donor concentration (cm^{-3})	Conductivity
1-81-H	Undoped	$<10^{15}$	$\sim 3\times10^{15}$	n
1-91-H	Fe	1×10^{16}	$\sim 3\times10^{15}$	SI
1-74-H	Fe	1×10^{17}	$\sim 3\times10^{15}$	SI
1-66-H	Fe, Sn	1×10^{16}	$\sim 4\times10^{16}$	n

Photoluminescence was excited by the 6471 Å emission of a Kr^+ laser and detected by a PbS cell at 77 K and a phase-sensitive detection system. A Spex 3/4 meter grating monochromator was used both to analyse the PL and to disperse light from a 150 W tungsten-halogen lamp for use as a source for the PLE spectroscopy and the optical absorption measurements. In the latter cases, light exciting the monochromator was focussed directly on to the sample, which was immersed in liquid helium. Through the use of appropriate filters, either the integrated PL signal or the transmitted light was focussed on to the PbS detector by a spherical mirror. By scanning the monochromator wavelength, PLE or absorption spectra were obtained. All spectra were corrected for the wavelength response of the system.

Results and discussion

Figure 1 shows a comparison of the near IR PL spectra obtained in the SI sample 1-91-H, the n-type sample 1-66-H and the undoped sample 1-81-H. All three crystals exhibit a nearly identical 4-line spectrum, which has been attributed to ~ 0.35 eV transition from the $^5T_2$ excited state to the ^5E ground state of the ^5D term of Fe^{2+} $(3d^6)$ ions in the cubic crystal field of the InP lattice. The ~ 14 cm^{-1} separations of the four sharp lines are attributed to spin-spin and second order spin-orbit splitting of the ^5E ground state [2]. As expected, the intensity of the Fe^{2+} PL is much weaker in the undoped 1-81-H

($\ll 10^{15}$ cm^{-3} Fe) than in the iron-doped 1-91-H crystal ($\sim 10^{16}$ cm^{-3} Fe). However, the Fe^{2+} PL intensity is also strongly quenched in the n-type crystal 1-66-H ($\sim 10^{16}$ cm^{-3} Fe). ESR studies [9] of SI crystals of InP:Fe with low donor concentrations have shown that most of the iron is in the Fe^{3+} charge state. It is expected that back doping InP:Fe with shallow donors will convert some of the Fe^{3+} to Fe^{2+}, the one electron trap state, and that n-type crystals would contain mostly Fe^{2+} and Fe^{1+}. In the case of 1-66-H, the iron was confirmed to be predominantly in the Fe^{2+} state by optical absorption measurements at 3.5 μm; the sharp Fe^{2+} absorption was found to be about four times as strong as that in 1-91-H. This is consistent with the relative Fe^{2+} concentrations expected from Table 1.

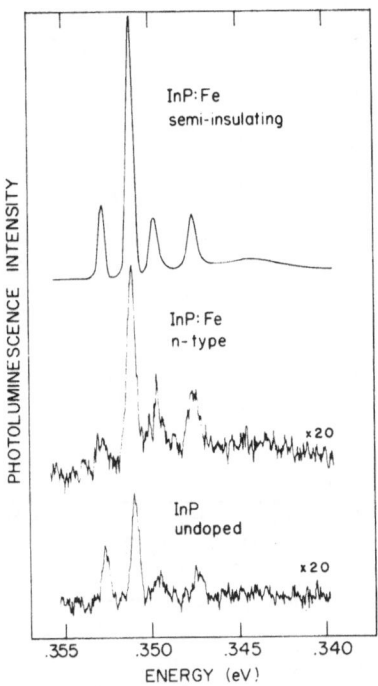

Figure 1 Photoluminescence spectra, at 4.2 K, attributed to Fe^{2+} in semi-insulating and n-type InP:Fe and in undoped InP.

Because of the strong quenching of the 0.35 eV PL in the n-type sample, it would appear that the strength of the Fe^{2+} PL is correlated with the equilibrium concentration of Fe^{3+}. This suggests an excitation mechanism [4] in which interband photoexcitation of free electrons and holes is followed by capture of a conduction band electron by an Fe^{3+} ion and results in the formation of Fe^{2+} in the $^5T_2$ excited state. The centre then relaxes to the ^5E ground state of Fe^{2+} by emission of a ~ 0.35 eV photon. This mechanism is analogous to that suggested [12] for the excitation of Cr^{2+} for PL in GaAs:Cr containing Cr^{3+} in equilibrium.

The PLE spectra for the 0.35 eV PL band in the SI crystals 1-91-H and 1-74-H are shown in Figure 2. The prominent peaks in these spectra just above 1.4 eV represent the onset of efficient interband excitation of PL by the mechanism just described. The steep rise just below 1.4 eV corresponds to the rising band edge absorption, while the fall off in efficiency at high photon energies (and high absorption coefficients) is presumably attributable to surface recombination. Both spectra exhibit a band of extrinsic excitation extending from a low energy onset of ~1.13—1.15 eV up to the band edge absorption. The strength of this excitation band for the Fe^{2+} PL is correlated with the concentration of Fe^{3+}; the intensity of the extrinsic PLE band relative to the band edge excitation peak is stronger in the 1-74-H crystal ($\sim 10^{17}$ cm^{-3} Fe) than in the 1-91-H crystal ($\sim 10^{16}$ cm^{-3} Fe).

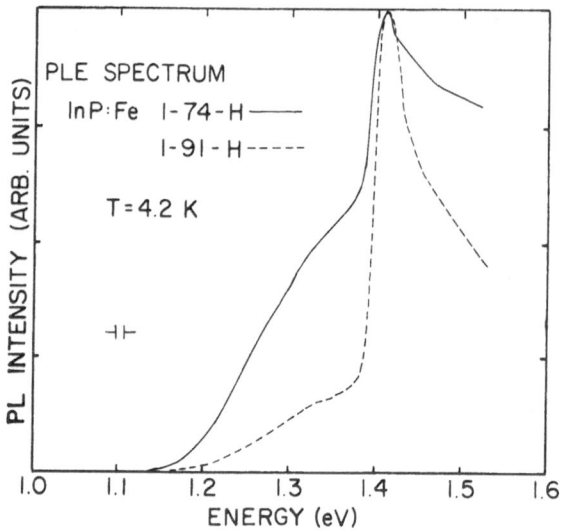

Figure 2   Photoluminescence *excitation* spectra of the ~ 0.35 eV Fe^{2+} luminescence band in semi-insulating InP:Fe.

On the basis of these results, it is suggested that the extrinsic excitation of Fe^{2+} PL may be attributed to a ligand-to-metal charge transfer transition in which a valence electron from an adjacent phosphorus atom is transferred to an Fe^{3+} ion, thereby creating an Fe^{2+} ion in the excited $^5T_2$ state. This excited Fe^{2+} ion then relaxes to the ^5E ground state by the emission of a ~ 0.35 eV photon. This interpretation is quite consistent with the model suggested by Krebs and Stauss [9] to explain their ESR results, which has subsequently been incorporated in the energy level scheme used by Fung *et al.* [8] in the analysis of their photoconductivity spectra. This level scheme, shown in Figure 3, was established on the following basis. Krebs and Stauss [9] observed two onsets in the near IR quenching spectrum of the Fe^{3+} ESR signal, at ~ 0.75 eV and ~ 1.1 eV. Noting that the separation of these onsets is about ~ 0.35 eV, they suggested that the onsets were due to the transfer of electrons from the valence band to Fe^{3+}, which converts the centre to the ^5E ground state of Fe^{2+} in the case of ~ 0.75 eV or to the $^5T_2$ excited state of Fe^{2+} for the ~ 1.1 eV onset. In each case the Fe^{3+} population would be decreased, as observed. This scheme places the ground state of the Fe^{2+} about 0.75 eV above the top of the valence band. Fung *et al.* [8] subsequently suggested that the 0.64—0.66 eV onset, which they observe in photoconductivity spectra for SI InP:Fe, could result from the excitation of electrons from the ground state of the one electron trap Fe^{2+} state to the Γ point conduction band minimum. Consistent with this interpretation, the 0.64 edge was found to be dominant in their more strongly Fe^{2+} sample and is only weakly observed in their predominantly Fe^{3+} sample. In addition, Fung *et al.* [8] observed quenching features in the photoconductivity spectra at 0.78 eV and 1.13 eV which they associate with the valence band to Fe^{3+} electron transfer transitions suggested by Krebs and Stauss [9], which produce the ^5E ground state and $^5T_2$ excited state of Fe^{2+}.

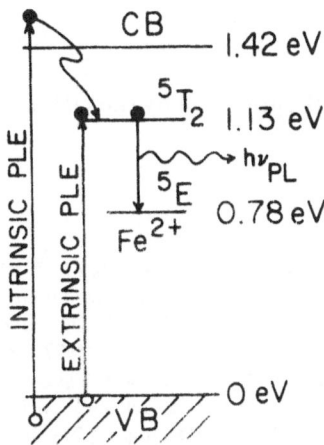

Figure 3   Proposed energy level scheme for Fe in InP [8, 9]. Intrinsic excitation involves capture of a conduction band electron by an Fe^{3+} ion to form the $^5T_2$ excited state of Fe^{2+}.

The PLE spectra of Figure 2 and our interpretation of the extrinsic PLE band are consistent with the energy level scheme in Figure 3. The 1.13—1.15 eV onset of extrinsic excitation of Fe^{2+} PL closely matches the 1.13 eV separation of the $^5T_2$ excited state of Fe^{2+} from the top of the valence band, as shown in the inset in Figure 3. While the features at ~ 1.1 eV in the ESR and photoconductivity quenching spectra prompted the original assignment of this energy level, the PLE spectra provide more direct evidence for this assignment since the PLE spectrum directly associates this threshold with the excitation of the Fe^{2+} PL band.

It should be pointed out that any extrinsic absorption process with threshold at ~ 1.1 eV, which can excite electrons into the conduction band, could give rise to the 0.35 eV PL through the Fe^{3+} capture mechanism that we have proposed. However, such an absorption mechanism would require the presence of a very deep (~ 1.1 eV) neutral donor state. No such deep donors have been reported for semi-insulating InP:Fe and the assignment of the 1.1 eV PLE threshold to a valence band-to-$Fe^{2+}(^5T_2)$ transition seems far more likely.

Summary

A model was proposed for the excitation mechanism of Fe^{2+} PL in InP which involves the capture of an electron by an Fe^{3+} centre, resulting in Fe^{2+} in the $^5T_2$ excited state. The PL transition then proceeds between the $^5T_2$ level and the ^5E ground state. The results of optical absorption and PLE measurements tend to confirm the energy level scheme for Fe^{2+} in InP suggested by Fung *et al.* [8].

ACKNOWLEDGEMENT

This research was partially supported by the Office of Naval Research.

References

1. Ippolitova, G.K., Omel'yanovskii, E.M., Pavlov, N.M., Nashel'skii, A.Ya. and Yakobson, S.V. (1977). *Sov. Phys. Semicond.*, **11**, 773 (1977, Fiz. Teckh. Poluprovodn., **11**, 1315)
2. Koschel, W.H., Kaufmann, U. and Bishop, S.G. (1977). *Solid St. Commun.*, **21**, 1069
3. Koschel, W.H., Bishop, S.G. and McCombe, B.D. (1977). *Proc. 13th Int. Conf. on Physics of Semiconductors* (Ed. F.G. Fumi), p. 1065. Rome; Tipografia Marves
4. Bishop, S.G., Henry, R.L., Klein, P.B. and McCombe, B.D. (1978). *Extended Abstracts of the Electrochem. Soc. Meeting*, Seattle, Washington, 21—26 May, vol. 78—1, 320
5. Grushko, N.S. and Gutkin, A.A. (1975). *Sov. Phys. Semicond.*, **8**, 1179 (1974, Fiz. Teckh. Poluprovodn., **8**, 1816)
6. Kovalevskya, G.G., Alyushina, V.I. and Slobodchikov, S.V. (1975). *Sov. Phys. Semicond.*, **8**, 1217 (1975, Fiz. Teckh. Poluprovodn., **9**, 2125)
7. Look, D.C. (1979). *Phys. Rev. B*, **20**, 4160
8. Fung, S., Nicholas, R.J. and Stradling, R.A. In press
9. Krebs, J.J. and Stauss, G.H. (1977). *Phys. Rev. B*, **16**, 974
10. Mitzuno, O. and Watanabe, H. (1972). *Electron. Lett.*, **11**, 118
11. Pande, K.P. and Roberts, G.G. (1976). *J. Phys. C: Solid St. Phys.*, **9**, 2899
12. Instone, T. and Eaves, L. (1978). *J. Phys. C: Solid St. Phys.*, **11**, L771

PHOTOLUMINESCENCE AND PHOTOCONDUCTIVITY STUDY OF THE 1.10 eV ENERGY LEVEL IN Fe-DOPED InP

PHIL WON YU

Physics Department, University of Dayton, Dayton, Ohio 45469, USA

Abstract

Photoluminescence and photoconductivity measurements in Fe-doped high-resistivity InP have been performed in the temperature range 4—300 K. Particular emphasis was placed on the explanation of the 1.06 eV emission. The emission spectrum has phonon structure at lower temperatures. The observed band shape and peak position are well explained by the configuration co-ordinate model. A nearest neighbour molecular-like centre $(Fe)_{In} — (V)_P$ is responsible for the 1.06 eV emission as well as the ~ 1.10 eV emission present in doped and undoped InP. Both d.c. and a.c. photoconductivity spectra show a strong photoconductivity onset at 1.1—1.2 eV for T=7—300 K. This onset can be ascribed to the defect centre $(Fe)_{In}—(V)_P$ which has a carrier trapping level of 0.24 eV.

Introduction

Iron-doped InP has recently proved very useful as a substrate material for microwave devices due to its high resistivity. The optical and electrical properties of Fe-doped InP have been the subject of many investigations. It has been shown [1, 2] that Fe occupies the cation site and exists in both Fe^{2+} ($3d^6$) and Fe^{3+} ($3d^5$) states. Electrical measurements [1, 3] show the presence of a deep level at 0.64—0.68 eV. Sharp zero phonon lines near 0.35 eV in low-temperature absorption and luminescence spectra have been interpreted [1, 2] as intracentre transitions ($^5E—^5T_2$) of Fe^{2+}. A broad absorption band near 0.44 eV at 300 K has also been observed [1] and has been studied [4] in detail for undoped and Fe-doped samples. Furthermore, a photoconductive band [5] near 0.44 eV at 300 K has been observed and interpreted as resulting from Fe^{2+} intracentre optical transition followed by a thermal excitation to the conduction band.

A broad photoluminescence band at 1.1—1.2 eV has been previously observed [6—11] at 77 K in undoped and doped crystals. The origin of the emission was tentatively ascribed to an association with $(V)_p$ (phosphorus vacancy) [6] or Cu [8]. Demberel et al. [10] explained the 1.12 eV band at 77 K present in Fe-doped crystal as an isoelectronic complex formed with Si and

Fe. In this work, we have made photoluminescence and photoconductivity investigations on Fe-doped high-resistivity substrates ($\rho_{300} \geqslant 10^7$ Ω-cm), with particular emphasis on the energy level near 1.10 eV. We explain the emission band at 1.06 eV in Fe-doped high resistivity substrates by a configuration co-ordinate model. The defect centre responsible for the 1.06 eV emission is shown to be associated with Fe. Also, the strong photoconductive onset at 1.1—1.2 eV in Fe-doped high resistivity substrates is discussed in terms of the presence of the defect centre.

Experimental

The 6471 Å line of a Kr laser with a maximum intensity of 200 mW was used as an excitation source for photoluminescence measurements. The emission spectra were analysed with a 3/4 m monochromator with a 600 grooves per mm grating blazed at 1.24 μm and were detected by a liquid nitrogen cooled PbS cell. Both d.c. and a.c. (14 Hz) PC measurements were made with an apparatus capable of measuring resistances up to 10^{14} Ω. The photoexcitation was provided by a monochromatic light beam with a light intensity of 5×10^{15} photon cm^{-2} s^{-1} in the wavelength regions of 0.7—2.55 μm and 2.9—3.5 μm. The intensity is lower near 2.8 μm due to a strong atmospheric absorption. However, the PC signal was assumed to be linear with the light intensity.

Results and discussion

1.06 eV EMISSION

Figure 1 shows the temperature dependent emission spectra obtained from Fe-doped high resistivity crystals. Three emissions at 1.42, 1.38 and 1.06 eV are seen at 5 K. The 1.42 eV emission is due to exciton complexes. The band near 1.38 eV can be ascribed to a donor-acceptor pair transition, with a typical acceptor being Zn. The relative emission versus excitation intensity for two bands at 1.06 and 1.42 eV follows approximately a linear relationship at 4.2 K over the excitation intensities of ~2—200 mW. The relative emission intensities between the two bands at 1.06 and 1.42 eV, i.e. $I_{1.06}/I_{1.42}$, range from 15 to 0.3 with an excitation intensity of ~200 mW at 4.2 K for the samples obtained from four different sources. However, the peak position and the phonon structure of the 1.06 eV band are the same for all the samples measured. The phonon structure disappears with an increase in temperature, as shown at 46 and 100 K in Figure 1. Figure 2 shows the 4.2 K emission spectra near 1.10 eV obtained from (*a*) n-type conducting bulk material, and (*b*) Fe-doped high resistivity substrate material (a detailed explanation of the 1.10 eV band will be published elsewhere). As shown in Figure 2 we find that the peak energy of the broad band near 1.1 eV depends on the samples, and phonons are clearly involved in the emission band. We also find that the zero phonon energy E_0 is at 1.18 eV and that the coupled phonon energy $\hbar\omega$ is 38 ± 2 meV. This phonon energy is the TO phonon energy at the Γ point [12].

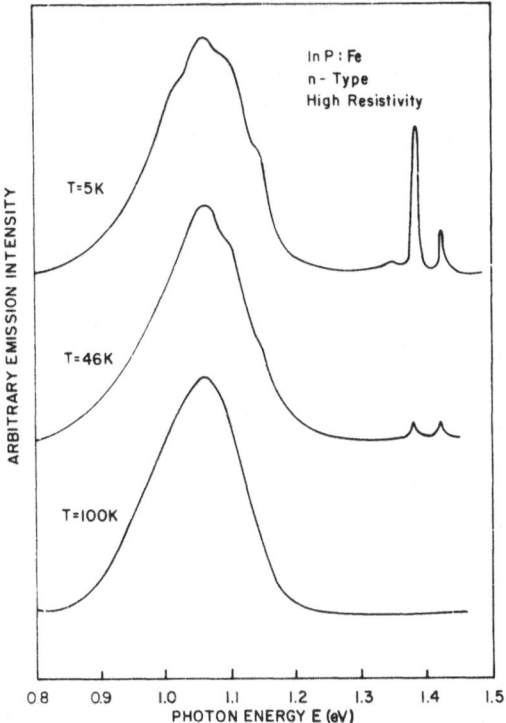

Figure 1   The emission spectra obtained from an Fe-doped high resistivity crystal at T=5, 46 and 100 K. The phonon structure of the 1.06 eV band is seen clearly at T=5 K but disappears with an increase in temperature. The 1.42 eV and 1.38 eV bands also disappear with an increase in temperature.

In the optical transition involving a deep centre the excitation of phonons through the strong lattice localized centre interaction results in phonon side bands. The theoretical spectra of the phonon side bands in such a case were calculated using the configuration co-ordinate model [13]. Let us take the simple case when the non-degenerate electron states at the localized centre are coupled linearly to a single normal co-ordinate of the lattice. At T=0 K, the position of the n-th phonon side band is given by $E_n = E_0 - n\hbar\omega$, in luminescence. The transition probability of the n-th phonon is given by $I_n = \exp(-S)S^n/n!$, where S is a measure of the strength of the lattice-localized centre interaction.

This model is applied to explain the 1.06 eV emission present in Fe-doped high resistivity materials, together with the parameters $E_0 = 1.18$ eV and $\hbar\omega = 38$ meV, which are obtained from the similar emission spectra near 1.10

eV in Figure 2. The observed band shape can be fitted to the theoretical line spectrum with a proper adjustment of the coupling strength, S. The value of S for Fe-doped materials was found to be 3.5. The less distinct phonon structure in Fe-doped samples is due to a large value of S compared to that in the sample of Figure 2a. The difference of the peak energy is due to the difference of S, which depends on the charge density difference between the two electronic states. Also, we find that E_0 is at 1.18 eV, regardless of the samples. Consequently, all the emission bands near 1.10 eV in InP can be attributed to the same radiative centre.

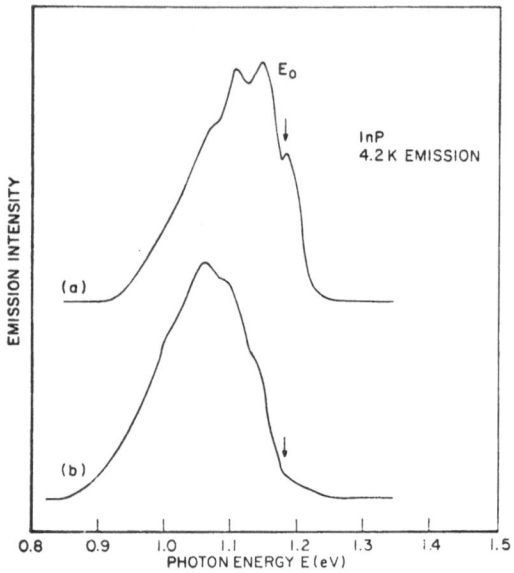

Figure 2   4.2 K emission spectra near 1.10 eV obtained from (*a*) n-type conducting bulk material, and (*b*) Fe-doped high resistivity substrate material. The arrow indicates the zero phonon line energy E_0 (=1.18 eV). The zero phonon line in (*b*) is not distinct due to a large value of S. However, the presence of a peak is clear by comparison with other spectra with S≼3. One illustration is made in (*a*).

The ~1.10 eV band is tentatively ascribed to the nearest neighbour centre $(Fe)_{In}^- (V)_P$ for the following reasons:

1. The 0.35 eV emission due to the intracentre transition $(^5E—^5T_2)$ of Fe^{2+} [1, 2] appears in all samples showing the ~1.10 eV band, which indicates that Fe is a major contaminant even in undoped bulk and epitaxial materials.
2. Cr-doped crystals do not show the ~1.10 eV bands.
3. The 1.06 eV band is dominant in Fe-doped materials together with the 0.35 eV emission.
4. The emission intensity of the ~1.10 eV band is relatively lower in VPE samples compared to that in LPE samples, which is consistent with the photocapacitance measurements by Chiao and Antypas [9].

PHOTOCONDUCTIVITY SPECTRA

Figures 3 and 4 show, respectively, d.c. and a.c. photoconductivity spectra obtained from an Fe-doped high resistivity InP sample at T=7, 90, 223 and 300 K. The a.c. method was necessary to find those signals that are much weaker than the dark conductivity. The a.c. and d.c. photoconductivity spectra at 7 K show an initial onset of the photoresponse at ~0.60 eV, a maximum near 0.78 eV, and a subsequent quenching of the photoresponse at energies higher than 0.78 eV. Another strong photoconductive onset occurs at 1.1—1.2 eV with the continuous increase of photon energy before the band gap region is reached. At T>7 K, the characteristics of the a.c. and d.c. photoconductivity spectra can be described as follows: (*a*) the PC maximum present at 7 K disappears; (*b*) the PC response has an onset near 0.60 eV, increases up to ~0.8 eV, has a weaker onset near 0.8 eV, changes little at 0.8<hν<1.1 eV, and shows a strong onset at 1.1—1.2 eV. Particularly, the a.c. spectra at T=223 and 300 K show a broad band peaking at 0.44 eV in the energy range of 0.38—0.60 eV.

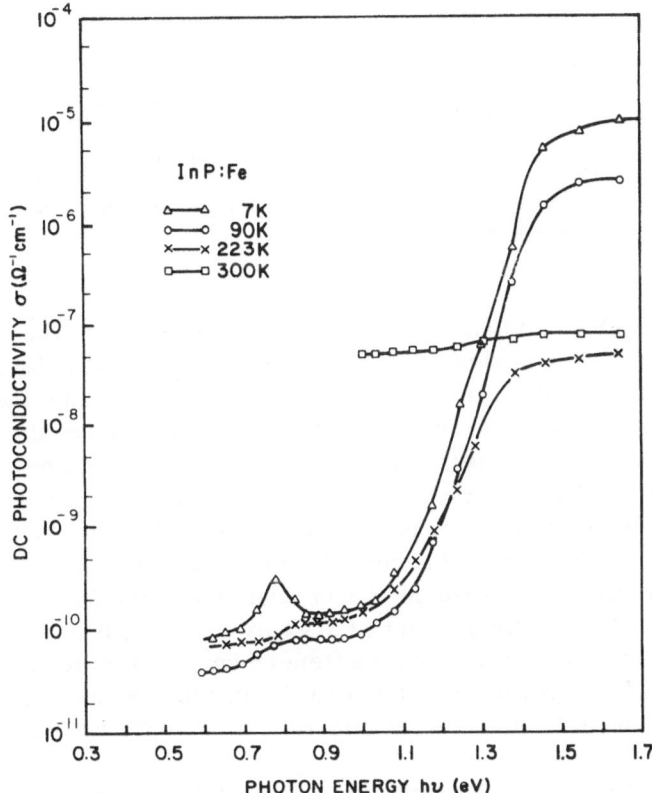

Figure 3 D.c. photoconductivity (σ) spectra at T=7, 90, 223 and 300 K, obtained from an Fe-doped high resistivity sample.

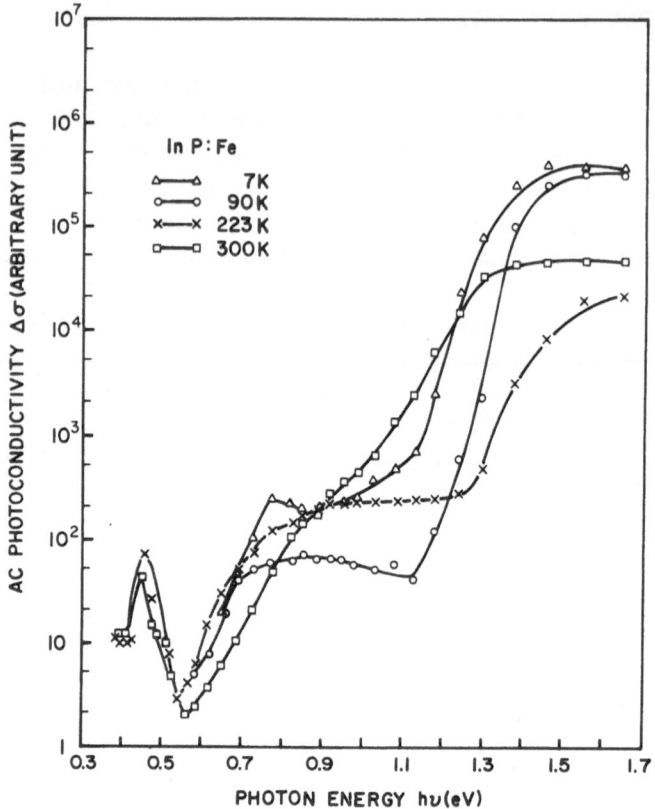

Figure 4   A.c. photoconductivity ($\Delta\sigma$) spectra at T=7, 90, 223 and 300 K, obtained from the same sample as in Figure 3.

The 0.60 eV onset can be taken to result from the transition of $Fe^{2+}(^5E)\rightarrow$the conduction band, as discussed by Fung, Nicholas and Stradling [14]. They ascribed the PC maximum at 0.78 eV and the subsequent quenching effect to the strong reduction in the lifetime of the photoexcited carriers when the transition valence band$\rightarrow Fe^{2+}(^5E)$ becomes possible with $h\nu\gtrsim0.78$ eV. The competition between VB$\rightarrow Fe^{2+}(^5E)$ and $Fe^{2+}(^5E)\rightarrow$CB is likely to be responsible for the PC maximum at 7 K and the slowly varying features at $0.8<h\nu<1.1$ eV in our experiment. However, this PC response will be influenced by the Fermi level which determines the fractional occupation of $Fe^{2+}(^5E)$. The broad band at 0.44 eV was ascribed to the resonance photoconductivity [5] as resulting from an Fe^{2+} intracentre optical excitation followed by a thermal excitation to the conduction band. This PC band corresponds to the 0.35 eV emission due to Fe^{2+} intracentre transition in the PC.

From both photoluminescence and photoconductivity experiments we know that the zero phonon line of the 1.06 eV band is at 1.18 eV and the strong PC onset occurs at 1.1—1.2 eV. Furthermore, the plot of the relative emission intensity of the 1.10 eV band versus $10^3 \, T^{-1}$ yields an activation energy of 0.24 ± 0.02 eV, which corresponds to the thermal release of a carrier bound to the defect centre $(Fe)_{in}^- \, (V)_P$. The zero phonon energy therefore involves a transition to the defect centre relative to the band edge.

We can consider that the PC onset is initiated from the zero phonon energy of the 1.06 eV band. The peak energy of the 1.06 eV band remains almost at the same energy for T=7—100 K as shown in Figure 1, and shifts to a lower energy with the increase in temperature. However, we note that $Fe^{2+}(^5T_2)$ is at $E_v+1.13$ eV, near T=7 K, when $Fe^{2+}(^5E)$ is located at $E_v+0.78$ eV. The transition $VB \rightarrow Fe^{2+} \, (^5T_2)$ will enhance hole excitation, which also gives a PC onset at 1.13 eV. In this case the large difference of the photoionization cross-section between the transitions $VB \rightarrow Fe^{2+}(^5E)$ and $VB \rightarrow Fe^{2+}(^5T_2)$ remains to be answered.

ACKNOWLEDGEMENTS

We would like to thank D.C. Reynolds and D.C. Look for many useful discussions. This work was performed at the Avionics Laboratory, Wright Patterson AFB, under contract No. F33615—76—C—1207.

References

1. Ippolitova, G.K., Omel'yanovskii, E.M., Pavlov, N.M., Nashel'skii, A.Ya. and Yakobson, S.V. (1977). *Sov. Phys. Semicond.,* **11**, 773
2. Koschel, W.H., Kaufmann, H. and Bishop, S.G. (1977). *Solid St. Commun.,* **21**, 1069
3. Mizuno, O. and Watanabe, H. (1975). *Elctron. Lett.,* **11**, 118
4. Iseler, G.W. (1979). *Inst. Phys. Conf. Ser.,* **45**, 144
5. Look, D.C. (1979). *Phys. Rev. B,* **20**, 4160
6. Mullin, J.B., Royle, A., Straughan, B.W., Tufton, P.J. and Williams, E.W. (1972). *J. Crystal Growth,* **13/14**, 640
7. Williams, E.W., Elder, W., Astles, M.G., Webb, M., Mullin, J.B., Straughan, B.W. and Tufton, P.J. (1973). *J. Electrochem. Soc.,* **12**, 1741
8. Roder, O., Heim, U. and Pilkuhn, M.H. (1970). *J. Phys. Chem. Solids,* **31**, 2625
9. Chiao, S.H. and Antypas, G.A. (1978). *J. Appl. Phys.,* **49**, 466
10. Demberel, L.A., Papov, A.S., Kushev, D.B. and Zheleva, N.N. (1979). *Phys. Stat. Sol. (a),* **52**, 341
11. Kawamura, Y., Ikeda, M., Asahi, H. and Okamoto, H. (1979). *Appl. Phys. Lett.,* **35**, 481
12. Borcherds, P.H., Alfrey, G.H., Saunderson, D.H. and Woods, A.D.B. (1975). *J. Phys. C: Solid St. Phys.,* **8**, 2022
13. Keil, T.H. (1965). *Phys. Rev. A,* **140**, 601
14. Fung, S., Nicholas, R.J. and Stradling, R.A. (1979). *J. Phys. C.:Solid St. Phys.,* **12**, 5145

THE CATHODOLUMINESCENCE OF Cr-DOPED GaAs PREPARED BY MOCVD EPITAXY

D.R. WIGHT, I.D. BLENKINSOP and S.J. BASS
RSRE, London Road, Baldock, Hertfordshire, UK

Abstract

The cathodoluminescence of p-type, n-type and SI epitaxial layers of Cr-doped GaAs has been studied in the temperature range 10—300 K. Four luminescence bands have been detected which are associated with the presence of chromium in the crystals. Correlation with deep level assessments on similar MOCVD material undertaken by other workers in collaboration with this research, has indicated that two of the bands are due to radiative hole capture at defined deep level states. Band A with threshold near 1.2 eV (10 K) is associated with a state which may be the Cr substitutional acceptor occupied by two electrons (Cr^{1+}). The band with threshold near 0.8 eV (10 K) is associated with recombination at the better established, single electron occupied (Cr^{2+}) substitutional acceptor which is considered to be a dominant Cr species, controlling the electrical properties of the crystals. The latter assignment is successfully tested by comparing the temperature dependence of the luminescence and photocapacitance thresholds. A third band, band D, with threshold also near 1.2 eV is associated with the formation of the neutral (Cr^{3+}) state of the same centre by electron capture from the conduction band. The occurrence of two of the luminescence bands is in substantial agreement with results obtained by other workers on similarly counter-doped, ingot materials.

Introduction

Cr-doped GaAs is an important part of GaAs technology because it is universally used to obtain the highly insulating material, which is one of the unique advantages of GaAs over Si technology. Considerable efforts have been made to generate a firm scientific basis for this semi-insulating (SI) behaviour but no fully satisfactory understanding has been achieved to date. ESR studies show that the chromium may exist in four charge states [1] and that the third charge state (Cr^{2+}) is consistent with that predicted for chromium substituting for gallium [2]. The electronic properties of these deep level states are not well established and different authors present various schemes to explain the observable experimental properties [3—5]. This has, in part, been due to the fact that the ingot materials commonly available are not well characterized, and contain many unwanted impurities or defects which can complicate the issue. Also, there has been a tendency to attempt to interpret results obtained using only one or perhaps two experimental techniques on such materials, so rather incomplete assessments have been made. This paper relates some of the results of an attempt to improve this situation by using higher quality epitaxial material whose active chromium concentration has been deduced from electrical compensation measurements, and whose deep level properties have been appraised using photocapacitance [6, 7], DLTS [7], photoconductivity [8] and luminescence techniques as a co-operative exercise with other workers. The material production, and study of electrical and luminescence properties have been undertaken at RSRE and are the subject of this paper.

Experimental

Two sets of epitaxial materials were grown using the MOCVD technique employing hexacarbonyl chromium as the doping gas [9]. The first set was grown on to n-type and semi-insulating GaAs substrates and the second set was grown on to GaAlAs alloy layers, previously grown to enable the

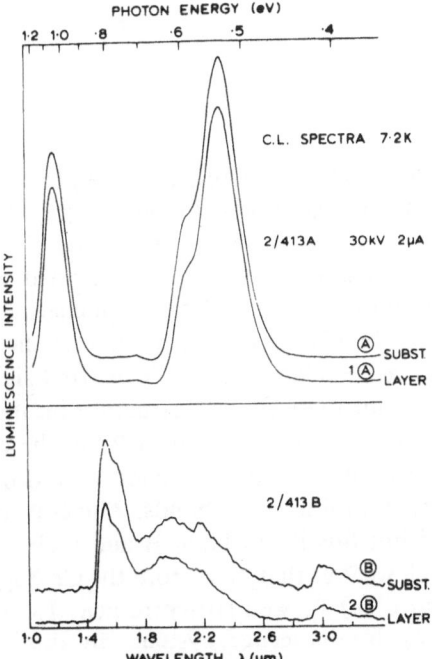

Figure 1   CL spectra [2, 3] taken from two samples from an epitaxial growth on two different substrates (A) n-type and (B) SI. Also shown are spectra obtained after the layers had been removed by etching, confirming the dominance of the substrates in the spectra. An Si filter was employed but no further order sorting filters were used so second order structure is seen.

epitaxial GaAs:Cr layer to be removed by etching techniques [10]. This was important because of the substrate interference in luminescence and photoconductivity experiments. An example of this is shown in Figure 1, where the luminescence from epitaxial layers grown concurrently on to two substrates is shown to be dominated by substrate luminescence. The exciting electron beam did not penetrate through the epitaxial layer, and so it appears that the edge luminescence generated in the epitaxial material was generating deep level photoluminescence in the substrate. Both sets of epitaxial materials studied contained Cr-doped layers and non-Cr-doped layers as controls. Counter-doping with Zn and S was used to obtain p-type and n-type Cr-doped layers, SI material being obtained when Cr doping alone was employed. The carrier and chromium concentrations were estimated from previous calibrations [9] and the samples removed from substrates were selected for cathodoluminescence (CL) studies in the temperature range 10—300 K. A chopped $4\mu A$ beam of 60 kV electrons was focussed on to the sample, which was mounted in indium on a continuous flow cryostat. A.c. detection techniques and a grating spectrometer with order sorting filters were used to obtain the luminescence spectra. The detector was a liquid nitrogen cooled PbS element sensitive to wavelengths (λ) out to ~3.5 m.

Results

The growth and electrical data from the samples studied are shown in Table 1, and the 10 K spectra obtained from n-type, SI and p-type GaAs:Cr material are shown in Figure 2. Spectra from the control samples are also included in Figure 2. Where the traces are extended to λ values less than 1.0 μm (Si filter) the emission spectra have been checked using a GaAs filter. It can be seen that four bands can be associated with the presence of Cr in the samples and these have been labelled A—D. At first sight the bands labelled A and D look similar but there is weak structure on the high energy side of band D which is absent in band A, and at 77 K the difference in the line shapes and peak positions are quite pronounced (Figure 3), so they are thought to be different luminescence bands. The control sample (Figure 2e) has one broad band but this is not band B and it should be noted that the intensity is very much weaker than that from the Cr-doped sample. Also, the control sample (Figure 2a) showed structure near 1.0 μm; this is the tail of strong higher energy luminescence systems in this sample, which have apparently been quenched by the addition of chromium (Figure 2b). Band C is the well-known Cr sharp line emission [11] which is seen here at low resolution, so the lines are not resolved; it is noticed that this emission is absent in both p- and n-type Cr-doped material. The presence of band B in n-type material and band C in SI material has been previously reported by Instone and Eaves [12] in their studies of counter-doped (Si) ingot material of known electrical properties. However, the persistent band B is not reported by these authors, presumably because they did not study the spectra to longer wavelengths. We have studied the spectra from conducting n-type GaAs:Cr in the temperature range 10—300 K in order to seek correlations with the collaborative deep level studies and these results are shown in Figure 4. Both bands persist and broaden with the threshold and peak wavelengths of band B moving to higher photon energy.

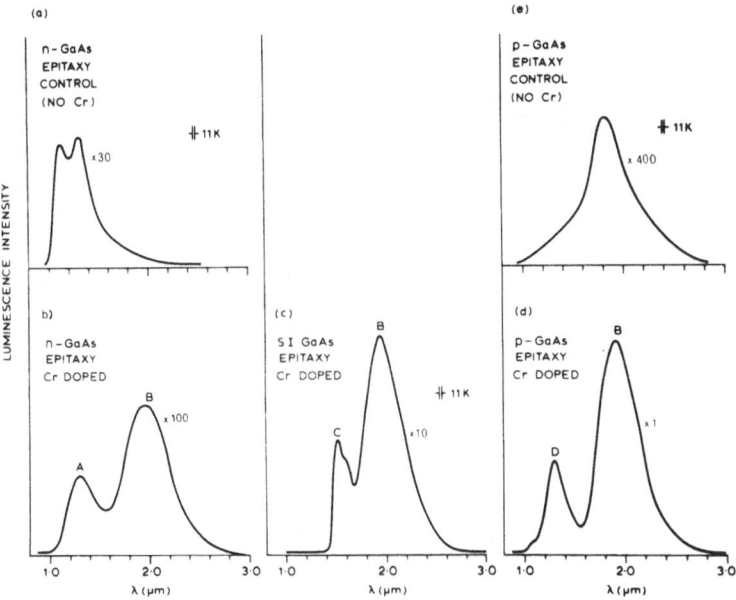

Figure 2   CL spectra obtained from a range of epitaxial Cr-doped layers removed from their substrates. Spectra from non-Cr-doped control samples are also included. Sample details are given in Table 1. The relative recording gain is shown for each spectrum.

Table 1 **The estimated active chromium concentrations and electrical properties of the MOCVD samples used for CL studies (Figure 2)**

Sample:	655A	633A	628	651A	653A
Dopants:	S	S, Cr	S, Cr	Zn, Cr	Zn
$N_D - N_A$ ($\times 10^{16}$ cm^{-3})†	+20	+20	+1.0	−100	−100
[Cr]* ($\times 10^{16}$ cm^{-3})	−	1.9	1.9	1.9	−
Type	n	n	SI	p	p
Layer thickness (μm)	17.7	15.0	15.0	18.5	19.5

* Estimated assuming that each chromium atom removes two electrons in n-type material and that the Cr-doping efficiency in n-type SI and p-type material is constant.
† Estimated from doping data and ignoring the chromium content.

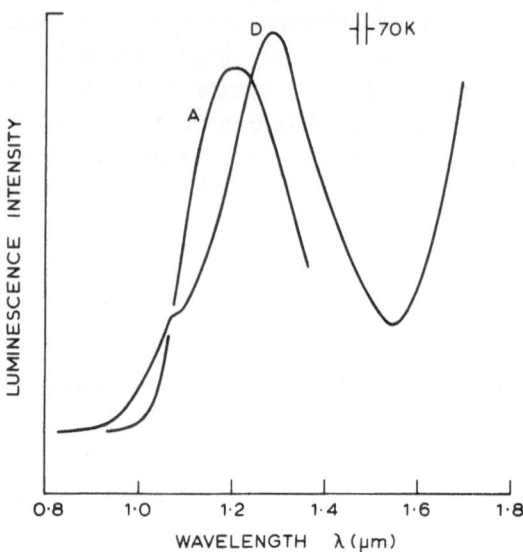

Figure 3 CL spectra taken at ~70 K showing the difference between bands A and D (see Figure 2).

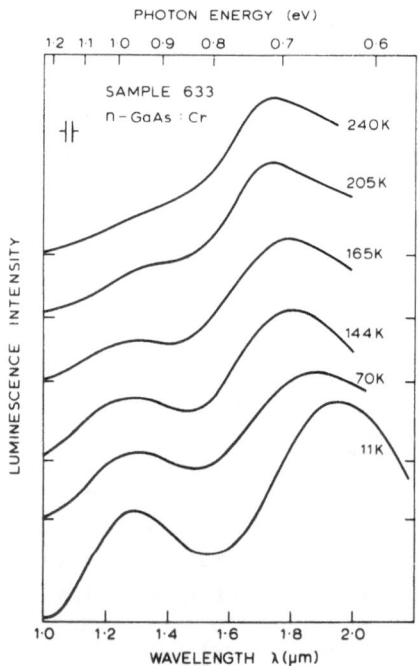

Figure 4 CL spectra from an n-type GaAs:Cr, S epitaxial layer at sample temperatures in the range 10—240 K.

Discussion

We turn immediately to the collaborative deep level studies where Szawelska and Allen [6], Chandler *et al.* [7] and Amato *et al.* [8] have studied similar samples. They all confirm a dominant deep level state of concentration close to that of the chromium concentration with an energy position ∼0.75 eV from the conduction band, detecting the state by optical and thermal excitation to the conduction band in SI and n-type layers and by optical and thermal excitation from the valence band in p-type material. This state is thought to be the Cr^{2+} substitutional Cr acceptor. Szawelska and Allen [6] report a second state ∼0.4 eV from the conduction band and suggest that this is a second, Cr^{1+}, charge state of the chromium substitutional acceptor, the presence of which is also indicated by the results of Brozel *et al.* [13] and ESR data. The work of Chandler *et al.* [7] detects a similar state but their initial analysis produces some doubts about this interpretation.

In p-type GaAs:Cr material, Chandler *et al.* [7] report a state ∼0.4 eV from the valence band detected using DLTS techniques, which may be the Cr^{3+} state. The deep level transitions concerned are depicted in Figure 5, using the notation described by Szawelska and Allen [6].

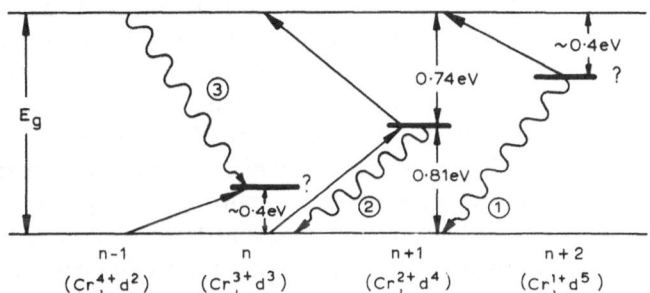

Figure 5   The (80 K) electronic transitions (→) discussed in the text and the luminescence transitions (curly arrows) [2—4] assigned to bands A, B and D, respectively.

In comparing these deep level results with our luminescence data, there is an obvious association of band A with the luminescence emitted when the shallower (Cr^{1+}) electron state recombines with a free hole ((1) in Figure 5), because the threshold energy is about E_g —0.4 eV, and it is only observed in n-type material. This assignment deserves further, detailed testing but since there is some doubt as to the nature of this state, no further discussion will be indulged in here. There appears to be a strong concensus, however, about the Cr^{2+} state and that it arises from the dominant chromium species controlling the electrical properties of these and probably other GaAs:Cr materials. The threshold position of band B associates it with the luminescence emitted when the Cr^{2+} electron recombines with a free hole ((2) in Figure 5) but its presence in n-, p-type and SI material deserves discussion. In SI material the Cr^{2+} state can be considered largely occupied in the unexcited crystal, so hole recombination can be expected. In n-type material the presence of strong internal edge luminescence ($\lambda \sim 0.82$ μm) was confirmed by measurements with the PbS detector and an S1 photomultiplier, and it is expected that such optical pumping will generate a depopulation of the Cr^{1+} state by the process depicted in Figure 5, thus creating Cr^{2+} states to enable transition (2) to be detected. It should be noted that band B was very weak in the n-type sample suggesting a small Cr^{2+} concentration, consistent with the above idea. In excited p-type material, high concentrations of the Cr^{3+} species are expected because of the positive charge on the Cr^{4+} centre. One only needs a fast electron capture process at this (neutral) Cr^{3+} centre to explain the observation of an intense luminescence signal in band B.

If we associate band D with radiative electron capture at the Cr^{4+} state recently detected in ESR studies on unexcited p-type GaAs:Cr [1], the threshold energy is again in rough agreement with DLTS data ($E_g - 0.4$ eV). It is also apparent that band B should not be detected in p-type samples unless band D is also present, as is observed in the spectra (Figure 2d).

In order to test the association of band B with the $Cr^{2+} \rightarrow Cr^{3+}$ deep level transition we will consider the temperature dependence of the threshold energies. Taking the photocapacitance data of Chandler et al. [7], the threshold for the transition to the conduction band is seen to move to longer wavelengths as the temperature is increased from 80 K, such that the energy separation from the valence band increases only slightly. In view of the facts that the no-phonon features cannot be located and that phonon broadening further confuses the issue, no rigorous comparison of transition energies is feasible without further analysis for which no accurate parameters are available. We therefore have chosen simply to extrapolate the lines of maximum slope to the background intensity level to obtain luminescence threshold energies. For the photocapacitance threshold energies we have the 80 K values for the conduction and valence band transitions obtained by Szawelska and Allen [6] ($E_{TC} = 0.74 \pm 0.01$ eV, and $E_{TV} = 0.81 \pm 0.01$ eV, respectively). By finding the 80 K value of the photocapacitance cross-section at 0.74 eV in the work of Chandler et al. [7] and locating the photon energies at which this value of cross-section occurs at different temperatures, the temperature dependence of E_{TC} was deduced. These values were then subtracted from the band gap energy to obtain the separation from the valence band ($E_g - E_{TC}$). Since the local lattice relaxation energy, which occurs when the centre becomes occupied, is small [6] and unlikely to distort the temperature dependence, we have also normalized the ($E_g - E_{TC}$) values to agree with the 80 K E_{TV} value of 0.81 eV, to predict the temperature dependence of E_{TV} itself. The luminescence thresholds and the values for ($E_g - E_{TC}$) and E_{TV} are compared in Figure 6 where a surprisingly good agreement is seen which, despite the assumptions made, adds weight to the proposed interpretation.

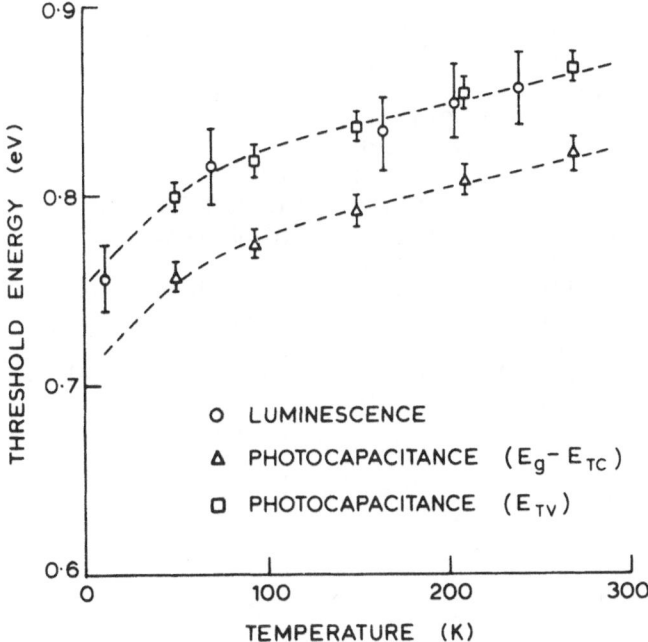

Figure 6 The temperature variation of the luminescence (valence band) and photocapacitance (conduction band) transition thresholds for the Cr^{2+} state in n-type GaAs:Cr MOCVD material. The photocapacitance data have been converted to show the energy separation from the valence band for comparison with the luminescence. The luminescence thresholds were obtained by extrapolating the lines of maximum slope to the half intensity position near the threshold turning point in the spectra.

Conclusions

The luminescence behaviour of Cr acceptors in MOCVD GaAs is in substantial agreement with that observed in carefully controlled ingot materials. Evidence has been presented which associates a dominant luminescence band with hole recombination at the Cr^{2+} charge state of the major, substitutional chromium species. Two other chromium-associated bands have been tentatively associated with deep level states, which may be (a) the neutral state (Cr^{3+}), and (b) the doubly charged state (Cr^{1+}) of the same centre.

ACKNOWLEDGEMENT

We are indebted to P.J. Dean for his advice on the status of the ESR work.

References

1. Kaufmann, U. and Schneider, J. (1980). *Appl. Phys. Lett.,* in press
2. Stauss, G.H., Krebs, J.J., Lee, S.H. and Swiggard, E.M. (1979). *J. Appl. Phys.,* **50,** 6251
3. Lin, A.L. and Bube, R.E. (1976). *J. Appl. Phys.,* **47,** 1859
4. Look, D.C. (1977). *Solid St. Commun.,* **24,** 825
5. Stocker, J.H. (1977). *J. Appl. Phys.,* **48,** 4583
6. Szawelska, H.R. and Allen, J.W. (1979). *J. Phys. C,* **12,** 3359
7. Chandler, T., Brunwin, R.F. and Hamilton, B. (1980). This volume
8. Amato, M.A., Arikan, M.C. and Ridley, B.K. (1980). This volume
9. Bass, S.J. (1978). *J. Cryst. Growth,* **44,** 29
10. Griffiths, R.G., Blenkinsop, I.D. and Wight, D.R. (1979). *Elect. Lett.,* **15,** 629
11. Lightowlers, E.C. and Penchina, C.M. (1978). *J. Phys. C,* **11,** L405
12. Instone, T. and Eaves, L. (1978). *J. Phys. C,* **11,** L771
13. Brozel, M.R., Butler, J., Newman, R.C., Ritson, A., Stirland, D.J. and Whitehead, C. (1978). *J. Phys. C,* **11,** 1857

TRUE MOBILITIES IN SEMI-INSULATING
O- AND Cr-DOPED GaAs

DAVID LOOK
Physics Department, University of Dayton,
Dayton, Ohio 45469, USA

Abstract

It is well known that an analysis of the magnetic field dependences of the resistivity ρ and Hall coefficient R, in nearly intrinsic material, will uniquely yield μ_n, μ_p, n, and p as long as *single*-carrier magnetic field dependences are not important. Unfortunately, recent data indicate that such effects are often important in semi-insulating GaAs and thus a new approach is necessary. We suggest an approach that requires only the usual, low field Hall effect and resistivity parameters, R_0 and ρ_0, but that also assumes knowledge of the intrinsic carrier concentration n_i, and the hole mobility μ_p as a function of the electron mobility, i.e. $\mu_p = f(\mu_n)$. We describe analyses which yield $n_i \simeq (2.6 \pm 0.5) \times 10^6$ cm^{-3}, and a *tentative* relationship for μ_p namely, $\mu_p^{-1} \simeq 9 \times 10^{-4} + 13 \mu_n^{-1}$. Curves of μ_n versus R_0/ρ_0, with ρ_0 as a parameter, are calculated by using these relationships. For $\rho_0 \leqslant 4 \times 10^8$ Ω-cm, the only solutions are $\mu_n \approx R_0/\rho_0$ which means that simple Hall measurements will suffice for nearly all O-doped and undoped semi-insulating GaAs. For higher ρ_0, the solutions are double-valued, and means of selecting the proper solution are discussed. A simple compensation-distribution model is presented which yields the positions of the 'O' and Cr energy levels.

Introduction

The importance of semi-insulating GaAs as a substrate material for GaAs FETs and logic circuits is well established. Thus, much time and effort have been spent in the development of effective characterization techniques. Two of the most relevant properties to be measured are the resistivity and mobility, the latter giving an indication of the crystal quality and impurity content. In this paper we discuss the problems associated with mobility measurements in semi-insulating GaAs, and suggest an approach that will overcome many of them.

Magnetic field dependences of ρ and R

In the presence of mixed conduction the Hall coefficient and conductivity may be written [1] as

$$R = \frac{R_n\sigma_n^2 + R_p\sigma_p^2 + R_nR_p\sigma_n^2\sigma_p^2(R_n+R_p)B^2}{(\sigma_n+\sigma_p)^2 + \sigma_n^2\sigma_p^2(R_n+R_p)^2B^2} \tag{1}$$

$$\sigma = \frac{(\sigma_n+\sigma_p)^2 + \sigma_n^2\sigma_p^2(R_n+R_p)^2B^2}{\sigma_n(1+R_p^2\sigma_p^2B^2) + \sigma_p(1+R_n^2\sigma_n^2B^2)} \tag{2}$$

where $R_n = r_n/en$, $R_p = r_p/ep$, $\sigma_n = e\mu_n n$, $\sigma_p = e\mu_p p$ and B is the magnetic field strength. These equations yield the relationship

$$\frac{1}{B^2} + \mu_n^2 Y = S_{\rho,m}\frac{\rho_0}{\Delta\rho} = S_{R,m}\frac{R_0}{R_0-R} \tag{3}$$

where

$$S_{\rho,m} = \frac{c(1+b)^2}{b(1+bc)^2}\mu_n^2 \tag{4}$$

and

$$S_{R,m} = \frac{c(1+b)^2(1-c)}{(b^2c-1)(1+bc)^2}\mu_n^2 \tag{5}$$

Here $c \equiv n/p$ and $b \equiv \mu_n/\mu_p$, and it is clear from equation (3) that $S_{\rho,m}$ and $S_{R,m}$ are the slopes of $1/B^2$ versus $\rho_0/\Delta\rho$ and $1/B^2$ versus $R_0/(R_0-R)$ plots, respectively. The subscript '0' denotes a measurement at zero magnetic field, and we have assumed unity Hall factors, although they can be carried along explicitly if desired [2]. As shown elsewhere, it is possible uniquely to determine μ_n, μ_p, n and p from measurements of R_0, ρ_0, $S_{\rho,m}$, and $S_{R,m}$, as long as *all* of the magnetic field dependence in R and σ is due to mixed-carrier effects, i.e. if R_n, R_p, σ_n and σ_p are themselves independent of magnetic field [2, 3].

In recent months it has become clear to us and others [4] that single-carrier magnetoresistive effects cannot be ignored, especially for O-doped and undoped semi-insulating material, which usually has higher conductivity than Cr-doped material. In particular, we have measured S_ρ, and S_R in several conductive samples ($c \doteq 0$, or $c \doteq 1$), which should have negligible mixed-carrier contributions, according to equations (4) and (5). It may be noted that our measured values, designated $S_{\rho,s}$ and $S_{R,s}$, are typically much larger than those determined by Willardson and Duga [5]. (The work by these authors has been used by us, in the past, as a basis for ignoring single-carrier effects.)

To add single-carrier effects to equations (1) and (2) we can set

$$\sigma_n = \sigma_{no}(1 - q\sigma_n \mu_n^2 B^2)$$

and

$$R_n = R_{no}(1 - qR_n \mu_n^2 B^2)$$

with similar equations for σ_p and R_p. Such a relationship for σ has good theoretical and experimental justification [1], while the form of R is primarily empirical. The insertion of these equations into equations (1) and (2) greatly complicates the solutions, of course, but in order to preserve the *form* of equation (3), which obtains even for most of our conductive samples, terms of higher order than B^2 must be eliminated. With this restriction we can show that $S_\rho = S_{\rho,m} + S_{\rho,s}$ and $S_R = S_{R,m} + S_{R,s}$, where

$$S_{\rho,s} = \frac{bc}{1+bc} q\sigma_n \mu_n^2 + \frac{1}{1+bc} q\sigma_p \mu_p^2 \tag{6}$$

$$S_{R,s} = \frac{b^2 c}{b^2 c - 1} qR_n \mu_n^2 - \frac{1}{b^2 c - 1} qR_p \mu_p^2 + \frac{2bc(bc+1)}{(b^2 c-1)(bc+1)} (q\sigma_n \mu_n^2 - q\sigma_p \mu_p^2). \tag{7}$$

Equations (6) and (7) were recently derived by Betko and Merinsky [4], although they did not identify the above restriction. They analysed several Cr-doped GaAs samples under the assumption $S_{\rho,s} \hat{=} 4\times 10^4$ cm^4 V^{-2}s^{-2}, a value much lower than what we normally measure (typically, $10^6 - 10^7$ cm^4 V^{-2}s^{-2}) on *conductive* samples. In any case, if the single-carrier effects are comparable to the mixed-carrier effects we can no longer solve uniquely for the electrical parameters, and a different approach is necessary.

An approach to the problem

It would be of great value if the results of a simple (low field) Hall measurement, namely R_0 and ρ_0, would be sufficient to determine μ_n, μ_p, n and p, even in the presence of mixed conduction. Rigorously, this is impossible of course, since we have only two equations (equations (1) and (2), with B=0), and four unknowns. But n and p are related by $np = n_i^2$, and it should also be possible, in principle, to relate μ_p and μ_n since the carrier scattering mechanisms are similar. From scattered data in the literature, and some of our own, all on *conductive* samples, we have deduced the following *tentative* relationship for μ_p and at room temperature: $\mu_p^{-1} = 9\times 10^{-4} + 13\mu_n^{-1}$. (This relationship can hopefully be refined as more data become available.) Thus, with the proper values of n_i and $\mu_p = f(\mu_n)$, equations (1) and (2) (at B=0) can be solved iteratively for μ_n and n, and then $S_{\rho,m}$ and $S_{R,m}$ can be calculated according to equations (4) and (5), respectively. In addition, if the experimental slopes S_ρ and S_R have been determined, then the single-carrier contributions $S_{\rho,s}=S_\rho - S_{\rho,m}$, and $S_{R,s}=S_R - S_{R,m}$ can also be calculated.

Before using the above approach routinely it is necessary to determine a good value of n_i. To do this the solutions of some 55 Cr-doped, and 17 O-doped (or undoped) semi-insulating GaAs crystals were examined, using various forms of $\mu_p = f(\mu_n)$. The Cr-doped samples ranged in resistivity from 6×10^7 to 1.7×10^9 Ω-cm, while the undoped samples ranged from 3×10^7 to 3×10^8 Ω-cm. Of the 55 Cr-doped crystals, 11 were rejected beforehand as having results which varied with different surface preparation techniques, or were clearly bad in some other respect. The other samples were analysed by allowing n_i to vary from $(1.0$ to $15)\times10^6$ cm^{-3}, and the functional form of $\mu_p = f(\mu_n)$ to vary within reasonable limits, with $\mu \leqslant 8000$ cm^2 V^{-1} s^{-1}, the lattice-limited mobility. The results are as follows.

1. The variations of n_i and μ_p had little effect on the solutions of any sample having $\rho_0 \lesssim 4\times10^8$ Ω-cm, which included *all* of the O-doped or undoped crystals.
2. Nine of the Cr-doped samples had *no* solutions for $n_i > 4\times10^6$ cm^{-3}, and two had no solutions for $n_i > 3\times10^6$ cm^{-3}.
3. With the additional (reasonable) constraint, $\mu_n \gtrsim 1000$ cm^2 V^{-1}s^{-1}, seven of the samples had no solutions for $n_i < 1.5\times10^6$ cm^{-3}, and three had no solutions for $n_i < 2.3\times10^6$ cm^{-3}.

This process was refined by using the 'best' empirical relationship, $\mu_p^{-1} \simeq 9\times10^{-4} + 13\ \mu_n^{-1}$. The result was that the least number of 'violations' (namely 1), occurred at $n_i = 2.6\times10^6$ cm^{-3}. Thus, this value, with an error bar of $\pm 0.5\times10^6$ cm^{-3}, is taken to be the best measured value of n_i at this time.

Data analysis and discussion

Curves of μ_n versus R_0/ρ_0 are presented in Figure 1. The values of n_i and $\mu_p = f(\mu_n)$ given above were used in the calculations, along with unity Hall factors. As stated earlier, a good approximation for $\rho_0 \lesssim 4\times10^8$ Ω-cm is $\mu_n \approx R_0/\rho_0$ (and also, $n \approx (eR_0)^{-1}$). Thus, simple Hall measurements should suffice for nearly all O-doped or undoped semi-insulating GaAs crystals. For Cr-doped samples, on the other hand, there will usually be two possible solutions. The lower-valued solution will always have $c \equiv n/p > 1$, and the upper-valued one, $c < 1$, so that a thermoprobe measurement should be able to distinguish between them, at least if the inequality is large enough. (Usually, thermoprobe measurements in Cr-doped GaAs [6] have indicated p-type material, i.e. $c < 1$.) Another help is a measurement of S_ρ, and then a calculation of $S_{\rho,s}=S_\rho - S_{\rho,m}$, which should be positive, according to equation (6) (since $q\sigma_n$ and $q\sigma_p$ are positive). A strong negative value of $q\sigma_n$, say $q\sigma_n < -0.1$, indicates a bad or very questionable solution. A third possible aid in distinguishing between solutions is an independent measurement to indicate crystal quality and purity, e.g. trace element analysis. This type of aid is helpful when the two solutions are quite different, i.e. one very high and the other very low.

Of the 44 Cr-doped samples examined, 14 admitted of only one solution, and 30 of two solutions. Of the latter 30, 24 clearly had one bad solution, as indicated by a negative $S_{\rho,s}$, or a value of $c \equiv n/p \gg 1$, which is inconsistent with thermoprobe measurements, when $\rho_0 \gtrsim 5 \times 10^8$ Ω-cm. Thus, only 6 of the 44 samples had ambiguous solutions. (Typically, such samples had $R_0/\rho_0 \simeq 2 \times 10^3$ cm^2 V^{-1}s^{-1}, with possible solutions $\mu_{n1} \simeq 2 \times 10^3$, and $\mu_{n2} \simeq 4 \times 10^3$ cm^2 V^{-1}s^{-1}.) The O-doped (or undoped) semi-insulating samples all had only one solution, as discussed earlier.

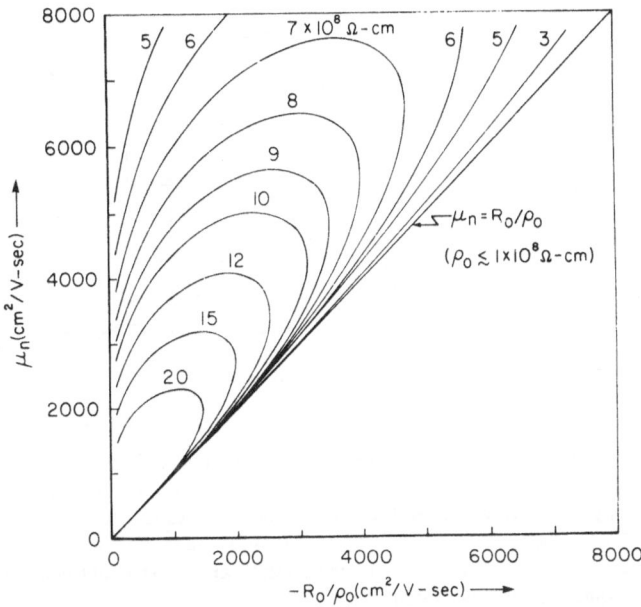

Figure 1 Plots of μ_n versus R_0/ρ_0 for various values of ρ_0 (units of 10^8 Ω-cm). The assumptions are $n_i = 2.6 \times 10^6$ cm^{-3}, $\mu_\rho^{-1} = 9 \times 10^{-4} + 13\mu_n^{-1}$, and $r_n = r_p = 1$. These curves hold only for samples with negative (n-type) Hall coefficients, the usual case.

The room-temperature Fermi levels for the 38 Cr-doped and 17 O-doped samples were calculated from $\epsilon_C - \epsilon_F \simeq 0.0255$ $\ln (4.37 \times 10^{17}/n)$ eV, and the distributions, $N_O(\epsilon_F)$ and $N_{Cr}(\epsilon_F)$, are plotted as histograms in Figure 2. It is clearly evident that the Fermi levels for O-doped samples lie above those for Cr-doped samples. (There will naturally be some overlap in the data because many of the samples will contain both Cr and O.) The theoretical curves in Figure 2 result from a simple analysis much like that described in [3]. The energy levels which fit the data are $\epsilon_C - \epsilon_0 \simeq 0.59 \pm 0.03$ eV, and

$\epsilon_C - \epsilon_{Cr} \simeq 0.68 \pm 0.02$ eV. It is interesting that the N_{Cr} (ϵ_F) curve begins to increase again above $\epsilon_C - \epsilon_F \simeq 0.68$ eV, suggesting a deeper Cr level, perhaps at about 0.75 eV. This *deeper* Cr level does not appear in the analysis if the lowest valued solution of μ_n is *always* chosen, but this assumption is not supported by thermopower measurements, as discussed above.

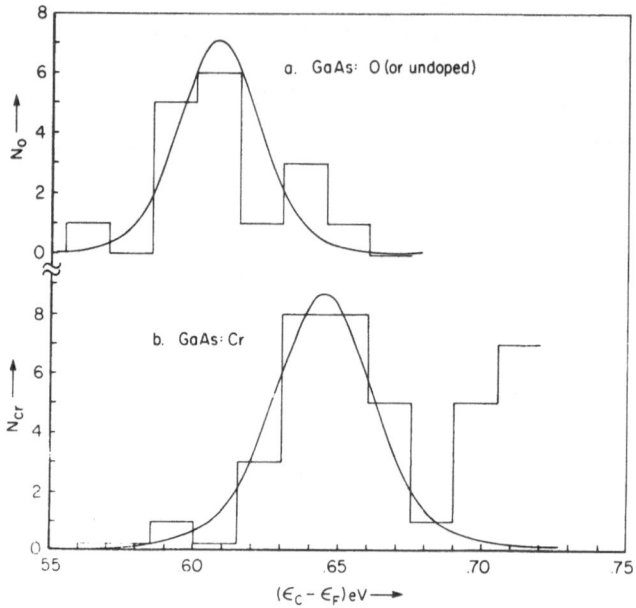

Figure 2 Histograms of number of samples (in a 0.015 eV energy interval) versus Fermi energy, at room temperature. Curve (*a*) is for 17 O-doped (or undoped) semi-insulating GaAs crystals, curve (*b*) is for 26 Cr-doped crystals. The solid lines are a theoretical fit (see text) with parameters $g_0 = \frac{1}{2}$, $\epsilon_C - \epsilon_O = 0.59$ eV, and $g_{Cr} = 4$, $\epsilon_C - \epsilon_{Cr} = 0.68$ eV. A deeper level, comprising 12 Cr-doped crystals, also appears in the Cr histogram, but a fit to these data was not attempted.

It is difficult to verify the derived energy levels by other means. Room temperature photoconductivity data in semi-insulating GaAs crystals always show thresholds in the 0.5—0.6 eV region, and sometimes there appears to be a shoulder near 0.7 eV. At this high a temperature, however, such data generally do not yield precise numbers. Furthermore, data which result from thermal activation processes in most cases give an energy at T=0, which can be quite different from the room-temperature value. Nevertheless, we feel that our results are basically correct, although subject to refinement as further data become available. Such refinement will probably come either from a more careful determination of $\mu_p = f(\mu_n)$, or from the incorporation of thermopower measurements into the analysis.

References

1. See, e.g., Putley, E.H. (1968). *The Hall Effect and Semiconductor Physics,* ch. 3. New York; Dover
2. Look, D.C. (1975). *J. Phys. Chem. Solids,* **36,** 1311
3. Look, D.C. (1977). *J. Appl. Phys.,* **48,** 5141
4. Betko, J. and Merinsky, K. (1979). *J. Appl. Phys.,* **50,** 4212
5. Willardson, R.K. and Duga, J.J. (1960). *Proc. Phys. Soc. (London),* **75,** 280
6. Cronin, G.R. and Haisty, R.W. (1964). *J. Electrochem. Soc.,* **111,** 874

MAGNETO-OPTICAL MEASUREMENTS ON THE 0.839 eV EMISSION IN GaAs:Cr

N. KILLORAN, B.C. CAVENETT and W.E. HAGSTON
Department of Physics, University of Hull, Hull, UK

Abstract

The origin of the 0.839 eV emission in semi-insulating GaAs:Cr has been investigated by optically detected magnetic resonance (ODMR) and Zeeman spectroscopy on the zero phonon line and satellites. The Zeeman measurements have been carried out for $B \parallel < 111>$, $<110>$ and $<100>$ as well as angular dependence of the splittings in the (110) plane. The results show that the centre has a symmetry axis along the $<111>$ direction, with a small orthorhombic distortion and these conclusions are in agreement with the ODMR signal angular dependence. The experiments show that the emission does not arise from the $<100>$ Jahn-Teller distorted Cr^{2+} centre observed in EPR but is due to a $<111>$ centre such as that suggested by White where an exciton recombines at a $(Cr^{2+} - D^+)$ pair.

The role of chromium in semi-insulating GaAs is far from being well understood, despite much activity to determine the nature of the compensation process. Krebs and Stauss [1] were the first to observe magnetic resonance of isolated chromium ions in the charge states Cr^{1+}, Cr^{2+} and Cr^{3+} corresponding to the configurations $3d^5$, $3d^4$ and $3d^3$, respectively. Excitation of the samples with light changed the relative strengths of the resonances and, in particular, radiation with energy greater than 0.9 eV increased the Cr^{2+} and Cr^{1+} concentration and decreased the Cr^{3+} concentration. More recently, the dynamics of these processes have been investigated by White, Krebs and Stauss [2], who concluded that the transient behaviour governing the conversion between Cr^{2+} and Cr^{1+} is determined by electron tunnelling to a deep hole trap. Far IR magnetic resonance measurements by Wagner and White [3] confirmed the Krebs and Stauss [1] conclusions that chromium is present in the material as isolated substitutional ions. However, not all of the evidence points to such simple defects. For example, infra-red luminescence with a zero phonon line (ZPL) at 0.839 eV was reported by Koschel, Bishop and McCombe [4] and Stocker and Schmidt [5] and, although this has been generally attributed to an internal transition of the substitutional Cr^{2+} ion, high resolution measurements of both the luminescence and the absorption by Lightowlers *et al.* [6, 7] showed the existence of at least 13 lines in the zero phonon region.

These authors suggested an energy level scheme with five levels in the ground state and seven levels in the excited state. Clearly, the dilemma has been to reconcile these emissions with the transitions within the Cr^{2+} level scheme based on the static Jahn-Teller model discussed in [1] and based on the analogous model for Cr^{2+} in the II—VI compounds suggested by Vallin *et al.* [8]. This discrepancy has led White [9] to suggest that the 0.839 eV emission may be the result of exciton recombination at a chromium donor isoelectronic defect in analogy with the well-known exciton recombination observed in GaP at Zn—O and Cd—O donor acceptor pairs [10]. In the case of GaAs, White [9] has suggested that the hole is localized in the 3d shell and that transitions $(Cr^{3+}-D^0)$ to $(Cr^{2+}-D^+)$ take place. The principal axis of symmetry would be <111> for nearest neighbour pairs, and <100> or <110> for silicon on the nearest and next nearest gallium sites, respectively. We have investigated the 0.839 eV emission by optically detected magnetic resonance (ODMR) and Zeeman spectroscopy on the zero phonon line components and indeed show that the centre is axial along <111> with a small orthorhombic distortion, analogous to the nearest neighbour DA pairs observed in ZnS [11, 12].

The high resolution spectroscopy and Zeeman measurements were carried out with the samples and superconducting magnet immersed in liquid helium at 2 K. The sample was excited using the 752 nm or 799 nm lines from a Coherent krypton 3000 K laser and luminescence was detected by a North Coast Ge detector via a Spex 0.75 m monochromator. Care was taken to mount the samples in a strain-free way.

Spectra taken at zero magnetic field comparing the ZPL components of Plessey and Sumitomo material are shown in Figures 1a and 1b, respectively. These spectra agree with the measurements of Lightowlers *et al.* [7] at 6.5 K, although in our measurements fewer lines were observed due to thermalization. However, some of the lines appeared to have more than one component. Figure 2 shows the splittings of the Sumitomo ZPL components in a magnetic field of up to 6.5 T for B∥<111>, k∥B. At the highest resolution only the splitting of the more intense lines (F and G) could be followed with accuracy due to the complexity of line splittings and the reduction in line intensities resulting from thermalization at high magnetic fields. Note that the intensities of the F line Zeeman components increase with respect to the intensities of the G line components as B increases. This effect can be ascribed to magnetic field-induced mixing of the wavefunctions of the states involved in transitions F and G. Figure 3 shows the angular dependence of the Zeeman splittings at 6 T in the (110) plane. For B∥<100> the lines F and G split into two components and as B moves away from <100> each component splits into three. This implies that the defect axis is along <111> since for B∥<100> all centres are at 54° 44′ to <100> resulting in a simple pattern for this direction.

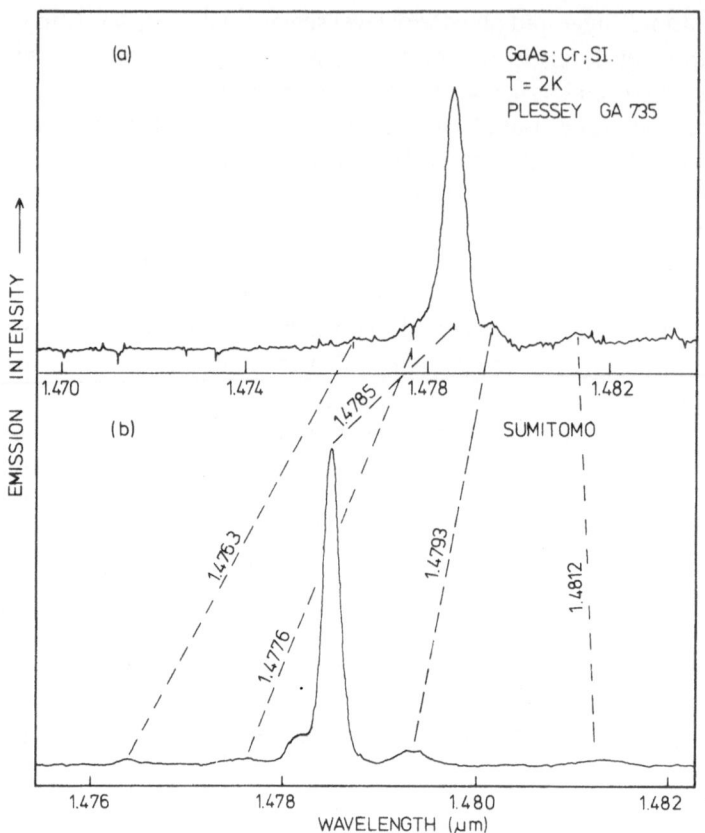

Figure 1 High resolution emission spectra taken at 2 K for (*a*) Plessey GA 735 GaAs:Cr, and (*b*) Sumitomo material. The emission intensity in (*b*) is approximately ten times greater than in (*a*).

The lower energy lines in Figure 3 are weak and difficult to follow. If we are correct in our assignments we must ascribe orthorhombic symmetry to the centre. Then the magnetic data must be described by three g-values, g_1, g_2 and g_3 corresponding to $<111>$, $<\bar{1}10>$ and $<\bar{1}\bar{1}2>$ directions, respectively (for a $<111>$ defect axis), where g_2 and g_3 are similar in magnitude. This model accounts for the turning points in the angular dependence of the Zeeman splittings at $<112>$, $<111>$ and $<110>$. Further, the splittings at $<112>$ and $<110>$ imply that g_2 and g_3 are slightly different in magnitude for both the excited and ground states of the optical transitions. The effective spin Hamiltonian applicable to this model is:

$$\mathcal{H} = \mu_B \mathbf{B} \cdot \mathbf{g} \cdot \mathbf{S} + D[S_z - \tfrac{1}{3}S(S+1)] + E(S_x^2 - S_y^2).$$

A quantitative determination of the parameters in the above Hamiltonian, together with the factors which could account for the asymmetry of the Zeeman splittings about the zero field line positions, is in progress.

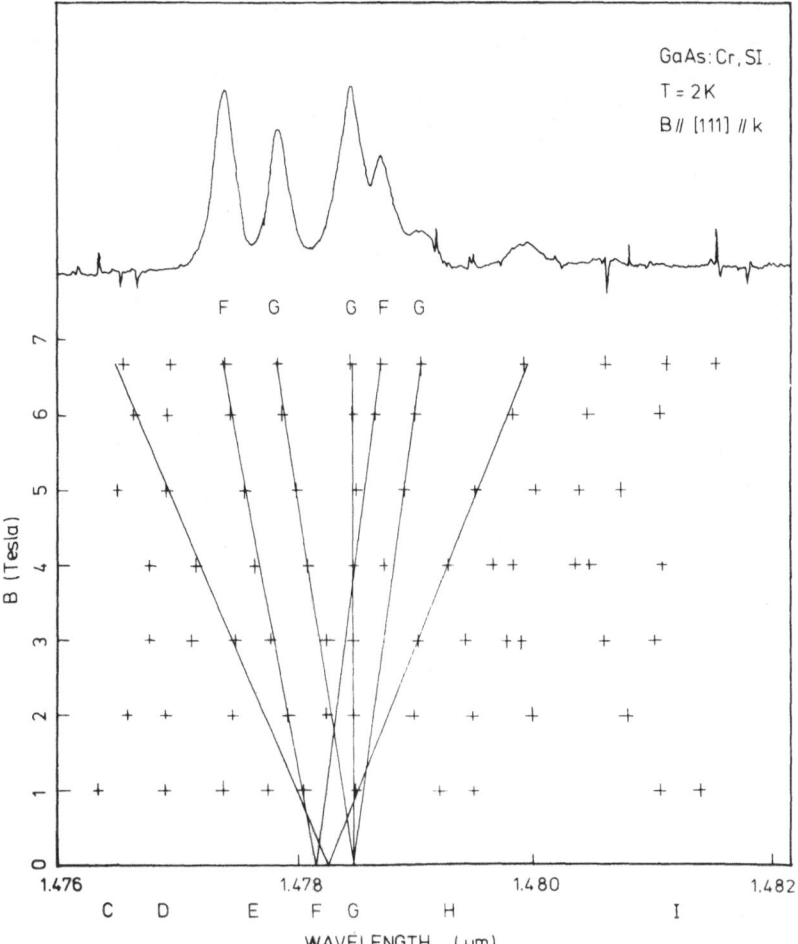

Figure 2   Zeeman splitting for the Sumitomo material for B∥<111>∥k showing the linear splittings with magnetic field. The Zeeman spectrum for B=6.5 T shows the thermalization of the high energy lines corresponding to the excited state g-value being less than the ground state g-value for a doublet to doublet transition.

In order to understand the electronic structure of the excited state we have investigated the emission by the technique of optically detected magnetic resonance, which has been described by us before [13, 14]. The samples were placed either in an X-band (9.4 GHz) or a J-band (16.6 GHz) cavity situated at the centre of a superconducting magnet, all immersed in liquid helium at 2 K. The luminescence excited by the krypton laser was detected using the Ge detector and a Hilger and Watts 0.33 m spectrometer. Both the X-band and the J-band microwave systems incorporated a travelling wave tube amplifier providing up to 2 W of microwave power and a PIN diode switch for chopping the microwaves at 125 Hz. A lock-in detector operating at this frequency was used to detect the microwave-induced emission changes at resonance. Figure 4 shows the ODMR spectra for B‖<100> at X-band with 500 mW of 799 nm σ^+ and σ^- circularly polarized excitation. Clearly, the ODMR signals are very sensitive to spin polarization in excitation but no emission polarization dependence of the ODMR signals has been obtained. Spectral dependence measurements of the two intense resonances showed that all of the 0.839 eV emission contributed to the signal.

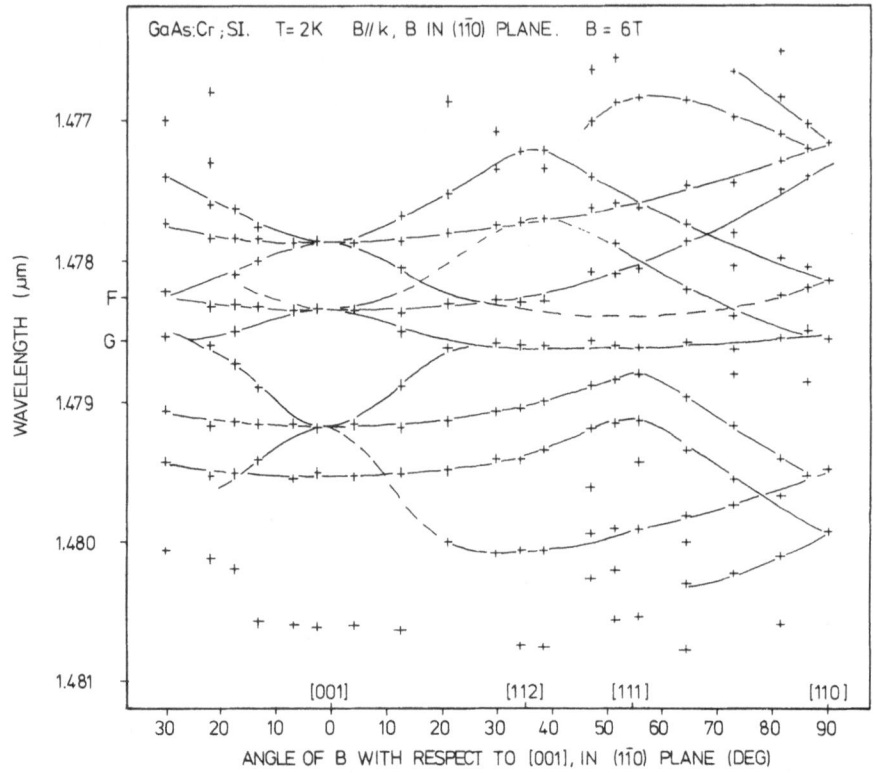

Figure 3   Angular dependence of the Zeeman splittings for rotation of B in the (110) plane for B=6 T.

An angular dependence measurement for B rotated in the (110) plane shows that the resonance signal at 0.20 T in Figure 4 has a dependence consistent with the orthorhombic symmetry (<111> principal symmetry axis) already deduced from the Zeeman data. The signal at 0.33 T is clearly isotropic and may be associated with Cr^{1+} and Cr^{4+}. Its appearance on the 0.84 eV luminescence may be due to a spin dependent non-radiative process coupled to the excited state of the luminescence. The signal at 0.46 T is usually very weak; therefore, any splitting of this signal could not be observed and, in addition, its peak was isotropic. The circular polarization excitation dependence suggests that the partial thermalization in the excited state is enhanced by optical pumping with σ^+ light.

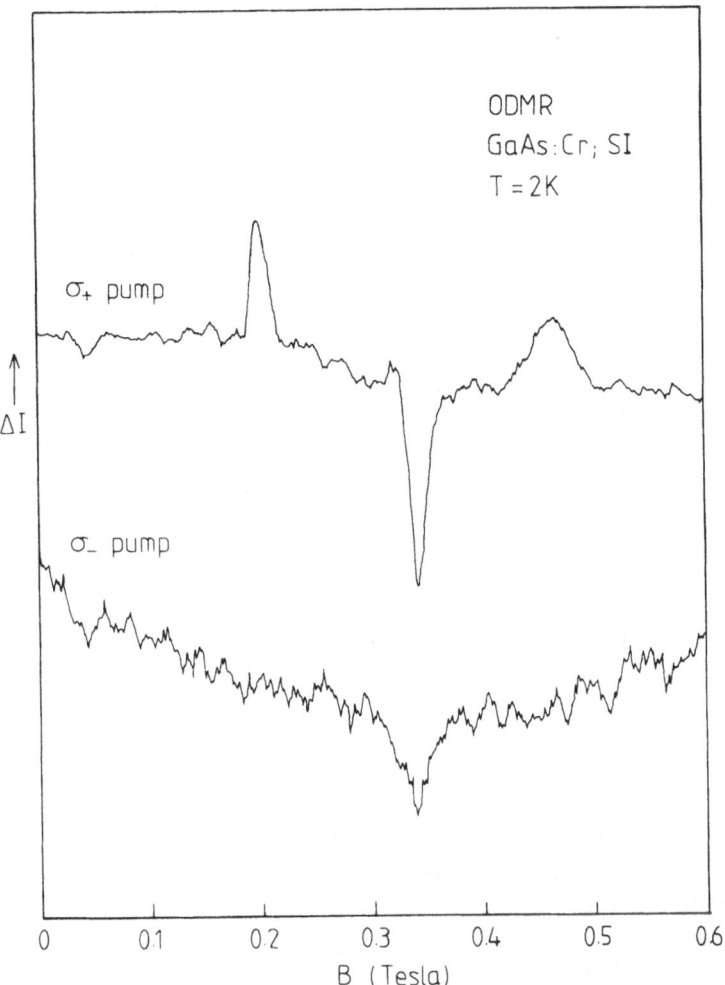

Figure 4   ODMR signals for B‖< 100> with circularly polarized excitation corresponding to an
S=2 thermalized spin system.

We propose that line F is due to a spin doublet to spin doublet transition
and line G is due to a spin triplet to spin triplet transition, as shown
schematically in Figure 5. These optical transitions are consistent with the
model proposed by White [9]. In order to account for the thermalization
observed in Figure 2, where the high energy lines are more intense than the
low energy transitions, we have the g-value of the ground state greater than
the g-value of the excited state. Consequently, the Zeeman splittings observed
are effectively due to the differences between excited and ground state g-
values.

We conclude that the magnetic data for the 0.839 eV emission show that
the principal axis of the centre involved is < 111> and so the luminescence is

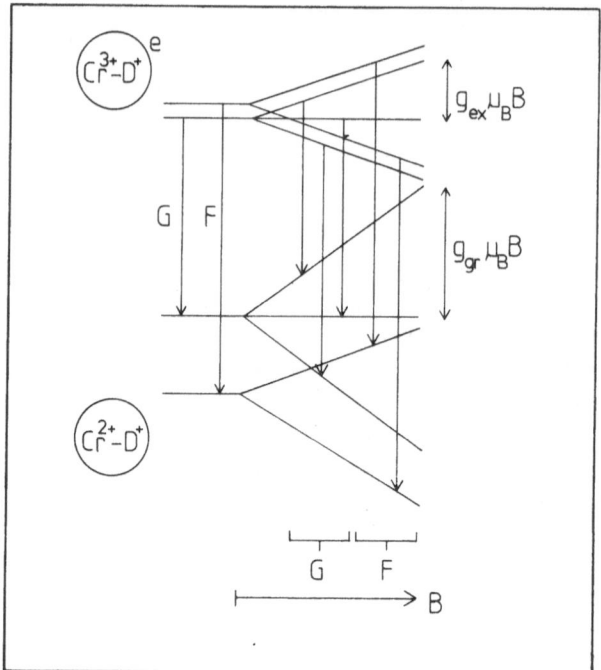

Figure 5   Schematic energy level model for transitions F and G.

not due to an internal transition of the isolated Cr^{2+} ion which has $< 100>$ symmetry. The results are consistent with the exciton model suggested by White where recombination takes place at a $(Cr^{2+}-D^+)$ pair.

ACKNOWLEDGEMENTS

We gratefully acknowledge valuable discussions with Drs A.M. White, E.C. Lightowlers, P.J. Dean, L. Challis and L. Eaves. We thank Dr White for encouraging us to carry out the Zeeman measurements. One of us (N.K.) is grateful to the SRC for a Studentship. We wish to thank The Royal Society, SRC and the Ministry of Defence for generous financial support.

References

1. Krebs, J.J. and Stauss, G.H. (1977). *Phys. Rev. B,* **16**, 971
2. White, A.M., Krebs, J.J. and Stauss, G.H. (1980). *Solid St. Commun.,* in press
3. Wagner, R.J. and White, A.M. (1979). *Solid St. Commun.,* **32**, 399
4. Koschel, W.H., Bishop, S.G. and McCombe, B.D. (1976). *Solid St. Commun.,* **19**, 521
5. Stocker, H.J. and Schmidt, M. (1976). *J. Appl. Phys.,* **47**, 2450
6. Lightowlers, E.C. and Penchina, C.M. (1978). *J. Phys. C,* **11**, L405
7. Lightowlers, E.C., Henry, M.O. and Penchina, C.M. (1978). *Inst. Phys. Conf. Ser.,* **43**, 307

8. Vallin, J.T., Slack, G.A., Roberts, S. and Hughes, A.E. (1970). *Phys. Rev. B,* **2,** 4313
9. White, A.M. (1979). *Solid St. Commun.,* **32,** 205
10. Henry, C.H., Dean, P.J. and Cuthbert, J.D. (1968). *Phys. Rev.,* **166,** 754
11. Title, R.S. (1967). *The Physics and Chemistry of II—VI Compounds* (Eds M. Aven and J.S. Prener), ch. 6. Amsterdam; North Holland
12. Nicholls, J.E., Davies, J.J., Cavenett, B.C. and Dunstan, D.J. (1979). *J. Phys. C,* **12,** 361
13. Cavenett, B.C. (1978). *Luminescence Spectroscopy* (Ed. M.D. Lumb), ch. 5. London; Academic Press
14. Cavenett, B.C. *Advances in Physics,* to be published

CHARGE-CARRIER TRANSPORT IN SEMI-INSULATING InP SURFACE LAYERS

L.G. MEINERS and H.H. WIEDER
Electronic Material Sciences Division, Naval Ocean Systems Center, San Diego, California 92152, USA

Abstract

The surface charge-carrier transport and electronic properties of n-type, Fe-doped, <100>-oriented, Czochralski-grown, semi-insulating InP were investigated by means of field effect controlled galvanomagnetic measurements. It is shown that the surface Fermi level is pinned 0.16 eV below the conduction band, that such surfaces are accumulated, that typical surface electron densities are of the order $n_s \sim 10^{11}$ cm^{-2} and corresponding mobilities are of the order $\mu_s \sim 5 \times 10^2$ cm^2 V^{-1}s^{-1} and that they are functions of the oxide deposition process. Optimum procedures yield $n_s \sim 10^{12}$ cm^{-2} and $\mu_s \sim 9 \times 10^2$ cm^2 V^{-1} s^{-1}.

Semi-insulating (SI) InP is used [1—3] for making enhancement type metal-insulator-semiconductor field effect transistors (MISFETs). Such n-channel devices have a large dynamic range and exhibit [1] microwave power gain. They are intended to become the elementary components of a monolithic microwave integrated circuit technology based on SI InP. The synthesis and electrical properties of Fe-doped bulk crystalline InP have been investigated by Mizuno and Watanabe [4] and by Iseler [5]. The bulk crystals are n-type, the free electron density is a function of the mole fraction of Fe added to the stoichiometric melt and of the residual donor and acceptor concentrations. The activation energy of electrons from the Fe deep acceptor levels to the conduction band is between 0.6 and 0.7 eV, typical resistivities are $\rho \geqslant 10^7$ ohm-cm, electron densities $10^7 < n < 10^9$ cm^{-3} and mobilities $10^2 < \mu < 3 \times 10^3$ cm^2 V^{-1}s^{-1}.

The specific field effect controlled charge-carrier transport mechanism involved in the SI-MISFET has not been identified, as yet, nor have the electrical and galvanomagnetic properties of the SI InP surfaces been determined previously. Of particular interest are the position of the pinned surface Fermi level, E_F^* in comparison with those of undoped and n-doped InP, the surface potential dependence of the surface density, n_s, and the mobility, μ_s, of the surface charge carriers. In order to determine these parameters experimentally we have made field effect controlled conductivity and Hall measurements of <100> oriented Fe-doped bulk crystalline SI InP which had an electron density, $n=2.25\times10^8$ cm^{-3} and mobility $\mu=3040$ cm^2 V^{-1} s^{-1} at ~300 K.

Haeusler and Lippmann [6] have shown that for a symmetrical cross-shaped Hall generator the magnetoresistance of geometrical origin introduces an error of less than 0.3% in the expected linear relationship between the Hall voltage V_h and the transverse magnetic field B in the range between B=0 and B=1 T. De Mey [7, 8] and Versnel [9], respectively, made an extended theoretical analysis of the symmetrical cross by a conformal representation. De Mey found that the error in linearity does not depend on the Hall mobility, the magnetic field, or the extent of the contacts made to the cross arms. Furthermore, the broad contacts allow a higher current to be used than point contacts without introducing thermal and thermoelectric errors in the Hall measurements. The symmetrical cross is, in fact, a variant of the clover-leaf specimen configuration used for van der Pauw type [10] resistivity and Hall measurements. The applicability of van der Pauw's relationships to specimens which are inhomogeneous in their thickness dimensions has been analysed theoretically by Pauwels [11]. To first order, the perturbation introduced by small-scale inhomogeneities can be neglected. An insulated gate controlled symmetrical cross structure, such as shown in Figure 1, was used for the subsequently described electrical and galvanomagnetic measurements. The SI InP specimens were cleaned in a 49% KOH solution heated to just below its boiling point, in order to remove their surface oxides. This was followed by a rinse in 18 MΩ H$_2$O, a light etching in a 2% (by volume) solution of bromine in methanol and sequential rinsing cycles of methanol and H$_2$O. Electrodes, such as shown in Figure 1, were made by the vacuum deposition of a w/o Au and 20 w/o Sn alloy followed by a 30 minute 325°C alloying cycle. The SiO$_2$ dielectric layers were deposited in a plasma-assisted chemical vapour phase reactor utilizing the pyrolysis and reaction of silane and nitrous oxide. HCl was used in the reaction chamber in order to provide predeposition etching of some InP specimens and was used for other specimens during the SiO$_2$ deposition. Smooth, glassy layers were formed reproducibly with bulk resistivities greater than 10^{16} ohm-cm. A gate electrode in the shape of a cross was vacuum-deposited on the oxide, as shown in Figure 1, and for the length-to-width ratio of ~9 used for this geometry, the estimated errors of geometrical origin are less than 0.1%. Van der Pauw type measurements were used to determine the gate voltage dependent resistivity and Hall coefficient of each specimen using the technique and apparatus described by Hemenger [12] for making such measurements on high resistance samples. This scheme employs high input impedance unity gain amplifiers to drive a guard for each of the sample leads. In this manner leakage currents are reduced and the stray capacitance of the leads is eliminated. The Hall parameters of the devices were measured and the calculations performed in the manner specified by ASTM F76—78 to minimize the effect of contact non-linearities. Approximately 5 minutes were required at each value of gate voltage to complete the Hall measurements and record the data. The devices were also measured in the pulsed mode with a 60 Hz curve tracer in order to compare the Hall and field effect mobilities.

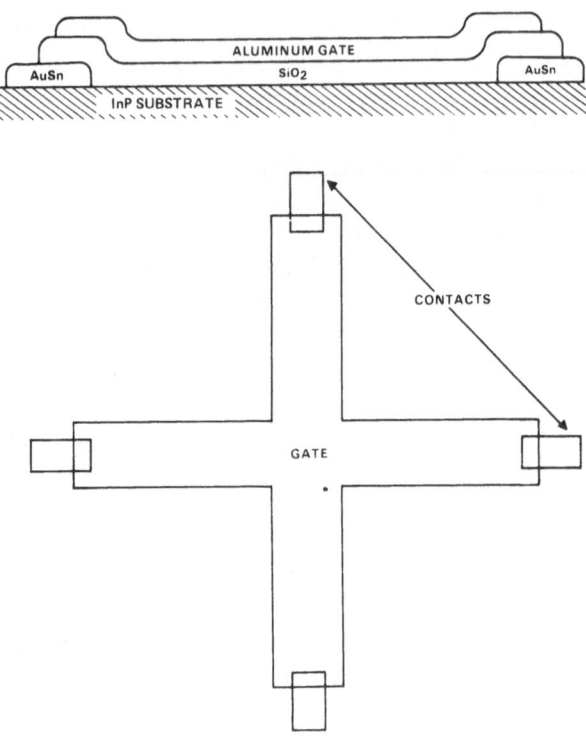

Figure 1   Symmetric cross van der Pauw structure.

Figure 2 shows the gate voltage dependence of the surface electron concentration determined from Hall measurements and Figure 3 shows the corresponding mobility dependence, determined from Hall and resistivity measurements made on the corresponding samples. The measured surface electron densities and mobilities are dependent on the surface preparation and treatment. Those specimens whose surfaces are cleaned only in KOH and bromine methanol prior to the deposition of the SiO_2 layers have much lower peak surface electron densities and mobilities than those which were subjected to gaseous HCl present in the reaction chamber during the deposition of the SiO_2 layers in $\sim 0.03\%$ concentration of the total gas flow. Such devices had surface electron densities $> 2 \times 10^{11}$ cm^{-2} and corresponding mobilities of $\sim 9 \times 10^2$ cm^2 V^{-1} s^{-1}.

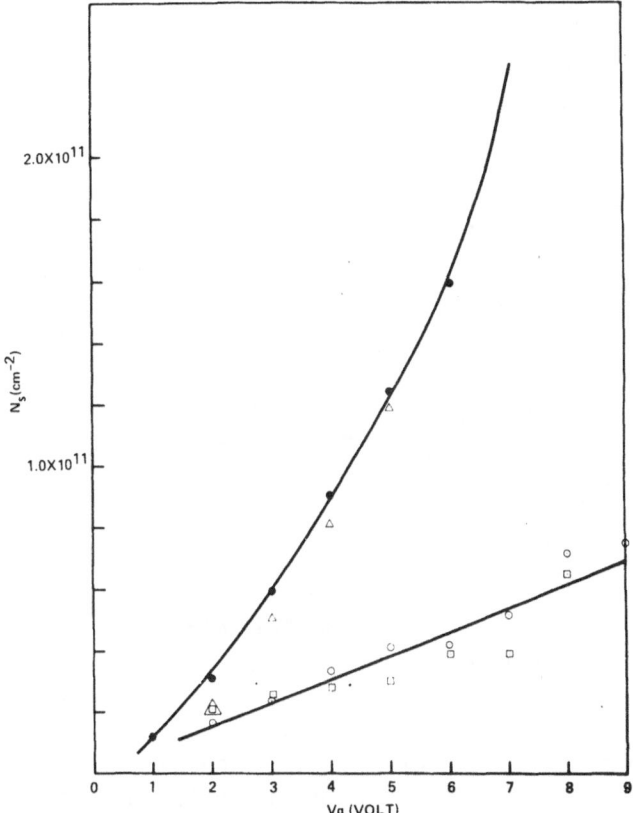

Figure 2   Surface electron density of Fe-InP ($n=2.2\times10^8$ cm^{-3}) as a function of gate voltage. The open circles (sample A) are for a sample etched in bromine methanol; the open squares (sample B) are for a sample exposed to HCl plasma etch and with 0.5% HCl in the gas stream during the SiO$_2$ growth; the open triangle and the solid circles (samples C and D) are for devices etched in bromine methanol and with HCl in the gas stream only during SiO$_2$ growth.

The field effect mobility of these devices was measured by obtaining the common source I-V characteristics of the devices and was calculated from

$$\mu_{FE} = \frac{L}{Z}\left[\frac{g_m}{C_{ox}(V_g - V_{th})}\right] \tag{1}$$

where L/Z is the length-to-width ratio of the channel, g_m is the transconductance, C_{ox} is the oxide capacitance per unit area and V_{th} is the threshold voltage. Mobilities of 98 and 84 were obtained for device A and B, respectively, and mobilities of 416 and 541 cm^2 V^{-1} s^{-1} were obtained for devices C and D. The field effect mobility of each device was roughly 0.5 of its Hall mobility. The results of the electrical and galvanomagnetic measurements suggest that the effect of a positive gate voltage applied to the gate is to drive the surfaces of the InP specimens into accumulation. The

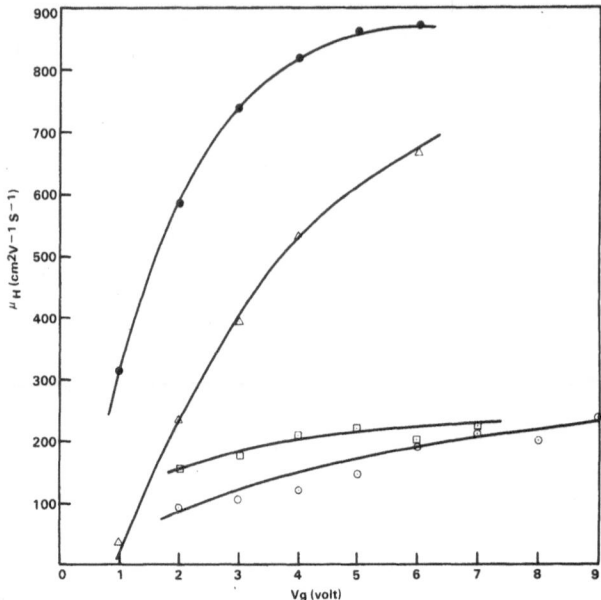

Figure 3   Surface electron mobility as a function of gate voltage. The sample identification is the same as for Figure 2.

surface electron density n_s and surface charge concentration $Q_s = qn_s$ can be related to the surface potential V_s following the formalism of Seiwatz and Green [13]. For an n-type semiconductor in the accumulation regime:

$$Q_s = \left(\frac{2kT}{\epsilon_s}\right)^{1/2}\left[N_c\left\{F_{3/2}\left[q/kT(E_F-E_c+V_s)\right] - F_{3/2}\left[q/kT(E_F-E_c)\right]\right\} + \right.$$

$$\left. + N_D \ell n \left[\frac{2+\exp(q/kT(E_c-E_D-E_F-V_s))}{2+\exp(q/kT(E_c-E_D-E_F))}\right]\right]^{1/2} \quad (2)$$

where ϵ_s is the semiconductor dielectric constant, k is Boltzmann's constant, T is the temperature, N_C is the effective density of states in the conduction band, E_C, E_F and E_D are, respectively, the energies of the conduction band edge, of the Fermi level and of the donor level. $F_{3/2}(x)$ are the Fermi-Dirac integrals as defined and tabulated by Blakemore [14]. The total charge on the InP surface is

$$Q_T = C_{ox}V_{ox} = C_{ox}(V_T-V_s) = Q_s + Q_{ss} \quad (3)$$

where V_{ox} is the potential drop across the SiO_2 capacitance, C_{ox} and V_T is total potential drop across the specimen; Q_T includes the surface charge density Q_s and the charge in the surface states Q_{ss}. The interface state density is

$$N_{ss} = \frac{\partial Q_{ss}}{\partial V_s} = \frac{\partial}{\partial V_s}\left[C_{ox}(V_T-V_s)-Q_s\right]. \quad (4)$$

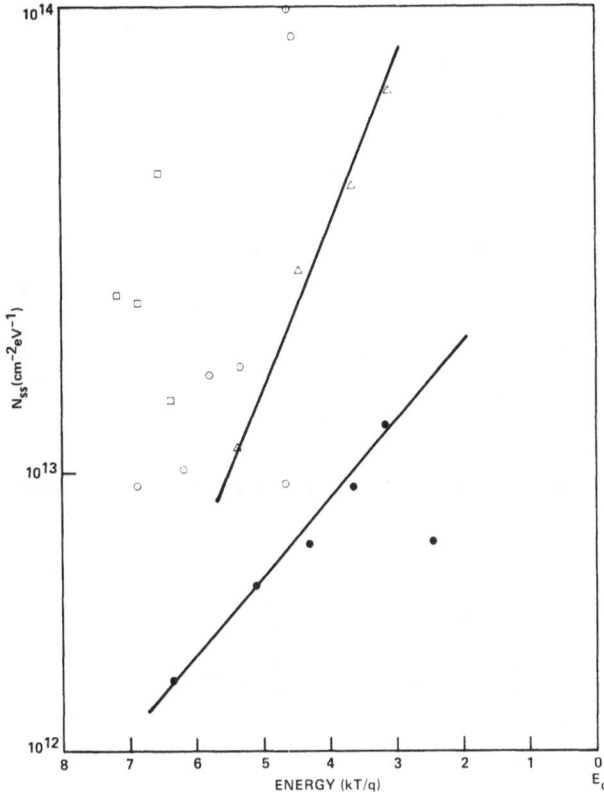

Figure 4 Interface state density as a function of the energy below the conduction band edge. The sample identification is the same as for Figure 2.

Equation (4) was used to calculate the surface potential dependence of N_{ss} (Figure 4), which shows it to be dependent on surface preparation prior and during oxide deposition as well as on V_s. For the optimum surface treatment (using HCl during the deposition of the SiO_2 layer) $N_{ss} \approx 8 \times 10^{12}$ eV^{-1} cm^{-2} within 0.1 eV of E_C; this is within a factor of 5 of typical N_{ss} reported [15] for thermally grown SiO_2 on Si. The calculated gate voltage dependence of the surface potential of these devices has been calculated by means of equation (2) and is shown in Figure 5. The intercept of the V_s versus V_g curves with the $V_g=0$ axis, V_{so}, is 15kT/q and is essentially independent of surface preparation; it was used to calculate the position of the pinned surface Fermi level, $E_F^* = 0.16$ eV by means of an energy level diagram such as shown in Figure 6. It was determined in terms of the fundamental band gap, $E_g=1.35$ eV, as well as the Fermi level E_F calculated in terms of the free carrier and intrinsic carrier concentrations of SI InP. This value of E_F^* is in fair agreement with those shown in Table 1 obtained from various measurements made on undoped and donor-doped n-type InP which had various free electron concentrations, different crystallographic orientations and different

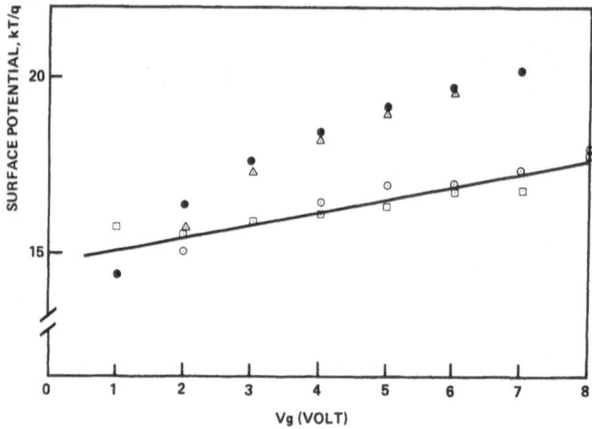

Figure 5 Surface potential as a function of gate voltage. The sample identification is the same as for Figure 2.

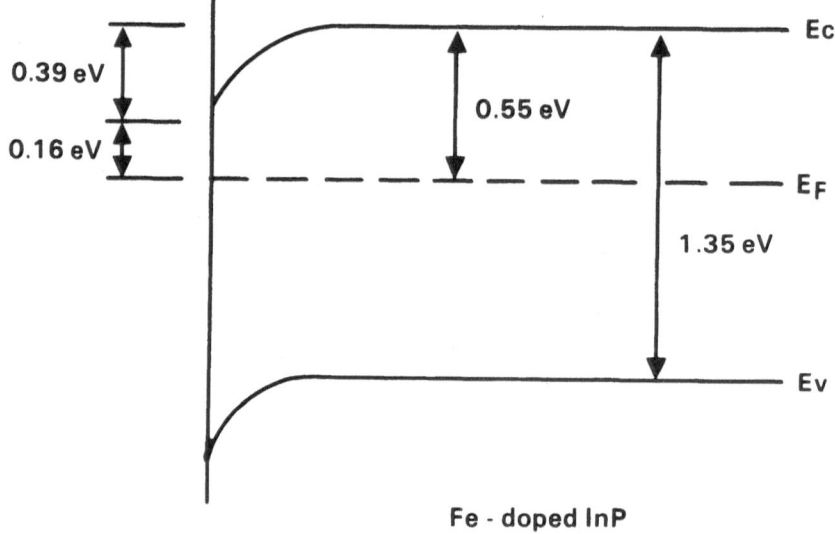

Fe - doped InP

Figure 6 Surface energy band diagram for Fe-InP.

chemical treatment of their surfaces. All of these specimens were found to be slightly depleted; in contrast, the surfaces of SI InP are accumulated. A plausible interpretation is that E_F^* is pinned at the same energy level in n-type and SI InP. Increasing compensation of the residual donors in SI InP causes the Fermi level to move away from the conduction band toward the intrinsic Fermi level. This bends the energy bands at its surfaces downward (Figure 6), leading to surface accumulation. Consequently, an enhancement type transistor [1] made on SI InP is an accumulation mode MISFET in contrast to other enhancement devices, which depend on total depletion of the channel under the gate or on inversion mode operation.

Table 1 **Undoped or donor-doped InP**

Measurement technique	Orientation	E_F^* (eV)	Reference
X-ray photoemission	<110>	0.14 ± 0.1	Spicer *et al.* [16]
Raman scattering	<111>	0.27	Pinczuk *et al.* [17]
Photovoltage	<100>	~0.15	Williams and McGovern [18]
Capacitance versus voltage	<100>	0.12	Meiners [19]

References

1. Meiners, L.G., Lile, D.L. and Collins, D.A. (1979). *Electron Lett.*, **15**, 578
2. Kawakami, T. and Okamura, M. (1979). *Electron. Lett.*, **15**, 743
3. Grant, A.J., Cameron, D.C., Irving, L.D., Greenhalgh, C.E. and Norton, P.R. (1979). *Durham Conference on Insulating Films on Semiconductors*, 2—4 July, 1979
4. Mizuno, O. and Watanabe, H. (1975). *Electron. Lett.*, **11**, 118
5. Iseler, G.W. (1978). *Inst. Phys. Conf. Ser.* **45**, 144
6. Haeusler, J. and Lippmann, H.J. (1968). *Solid St. Electron.*, **11**, 173
7. De Mey, G. (1973). *Archiv Elektron. Ubertrag.*, **27**, 309
8. De Mey, G. (1974). *Archiv Elektron. Ubertrag.*, **28**, 335
9. W. Versnel (1979). *Solid St. Electron.*, **22**, 911
10. Van der Pauw, L.J. (1958). *Philips Res. Repts*, **13**, 1
11. Pauwels, H.J. (1971). *Solid St. Electron.*, **14**, 1327
12. Hemenger, P.M. (1973). *Rev. Sci. Instrum.*, **44**, 698
13. Seiwatz, R. and Green, M. (1958). *J. Appl. Phys.*, **29**, 1034
14. Blakemore, J.S. (1962). *Semiconductor Statistics*. Oxford; Pergamon Press
15. Goetzberger, A., Klausmann, E. and Schulz, M.J. (1976). *CRC Crit. Rev. in Solid St. Sci.*, **6**, 1
16. Spicer, W.E., Chye, P.W., Seath, P.R., Su, C.Y. and Lindau, I. (1979). *J. Vac. Sci. Technol.*, **16**, 1422
17. Pinczuk, A., Ballman, A.A., Nahory, R.E., Pollack, M.A. and Worlock, J.M. (1979). *J. Vac. Sci. Technol.*, **16**, 1168
18. Williams, R.H. and McGovern, I.T. (1975). *Surface Sci.*, **51**, 14
19. Meiners, L.G. (1979). Report SF-19 Colorado State University, Fort Collins, Colorado

Cr REDISTRIBUTION IN EPITAXIAL AND IMPLANTED GaAs LAYERS

N.T. LINH, A.M. HUBER, P. ETIENNE, G. MORILLOT,
P. DUCHEMIN and M. BONNET*
Thomson-CSF Central Research Laboratory,
BP No. 10, 91401 Orsay, France

Abstract

Quantitative profiles of Cr have been determined by SIMS in cap-annealed, implant-annealed, MBE, VPE and MOCVD samples. The Cr redistribution in VPE and MOCVD layers is well explained by an out-diffusion process. The diffusion coefficients in VPE growth (750°C) and MOCVD growth (600°C) are 3×10^{-13} and 7×10^{-15} cm² s⁻¹, respectively. In cap-annealed, implant-annealed and MBE samples, a surface segregation mechanism is more appropriate to explain experimental results. A Cr background level of $2-3 \times 10^{15}$ cm⁻³ is observed in all layers grown on Cr-doped substrates. This level can be reduced if the Cr concentration in the substrate is reduced. MBE layers with Cr content as low as $1-4 \times 10^{14}$ cm⁻³ have been grown.

Introduction

The semi-insulating GaAs substrates which are commonly used in MESFET technology are Cr-doped. Although the electrical properties of Cr in GaAs have been frequently studied, this is not the case for its diffusion or, more generally, for its redistribution after annealing. Nevertheless, the epitaxial growth or ion implantation of active layers necessarily involves such a process. Therefore, Cr redistribution can modify the properties of the substrate or of the active layers.

Tuck and Adegboyega [1] have studied carefully with radiotracer, the diffusion of Cr into GaAs, the Cr and GaAs samples being in a sealed ampoule. Tuck *et al.* [2] also studied, for this Cr which is initially diffused in GaAs, the out-diffusion into an epitaxial layer or, more simply, the out-diffusion to the atmosphere of a sealed ampoule. These experiments pointed out the complicated character of the diffusion of Cr under given experimental conditions.

This paper reports the results concerning the problem of Cr redistribution in the cases encountered in the preparation of active layers for the fabrication of MESFETs: implantation, cap-anneal and epitaxy on Cr-doped substrates.

*Thomson-CSF Microwave Division.

This Cr is not introduced in the crystal by diffusion as described by Tuck *et al.* [2] but by crystallizing a GaAs melt containing Cr. The Cr-doped substrates utilized are purchased from various well-known suppliers. Several epitaxy techniques will be considered: molecular beam epitaxy (MBE), vapour phase epitaxy (VPE) and metal-alkyl decomposition in AsH_3 (MOCVD).

Experimental results

The profiles of Cr are determined by SIMS (CAMECA IMS 300 [3, 4]). The quantitative analysis is based on a calibration with samples which are preliminarily analysed by spark source mass spectrometry and by atomic absorption. The two methods were found to agree within $\pm 20\%$. By using an electrostatic analyser and a photomultiplier instead of a converter cathode as detection set-up, the detection limit has been improved to 5×10^{13} cm^{-3} or less. The experimental blank signal at this mass is 3 cps and 5×10^{13} cm^{-3} of Cr give a signal of 9 cps; this experimental value has been obtained with a slice of undoped material pulled in a PBN boat [5]. More often, undoped crystals have a level of between 5×10^{14} and 10^{15} cm^{-3}. A level of 5×10^{15} cm^{-3} gave a signal of 1000 cps, which is quite high compared to the blank signal. Therefore, the profiles which will be shown are smooth.

The statistical results of various experiments show that SIMS works within $\pm 15\%$ of precision, when the samples are doped in the range of 5×10^{15} to 10^{17} cm^{-3}. The depth resolution is about $100 - 200 \text{Å}$, but the parameter which limits the precision of the determination of the interface is the thickness homogeneity of the layers or their smoothness.

CAP-ANNEALED AND IMPLANTED LAYERS

Cr redistribution in cap-annealed and implanted layers has been studied by various authors [4, 6, 7]. The Cr profiles, which will be discussed below, are represented in Figure 1 of a previous publication [4].

MBE LAYERS

The experimental MBE set up has been described elsewhere [8]. The semi-insulating GaAs substrates of (100) orientation are purchased from various suppliers: RTC, Sumitomo, Laser Diodes, Varian and MCP. As far as the Cr distribution is concerned there is no significant difference between these substrates. Before being introduced in the vacuum chamber, they are chemically cleaned by two techniques: polish-etching with a bromine 0.75%—methanol solution or chemical etching with $H_2SO_4 : H_2O_2 : H_2O$ (5:1:1). It is shown by Auger spectroscopy that these are contaminated by oxygen, which can be removed by heating the substrate above 530°C before starting epitaxial growth. The annealing time and annealing temperature constitute important parameters which influence the Cr redistribution.

Figure 1 shows a typical profile of Cr in the case where the substrate is annealed for 30 minutes at 530°C before growing the layer at the same temperature. The Cr distribution in the substrate is quite similar to the one observed in cap-annealed layers: Cr accumulates at the interface and depletes underneath. The accumulation and depletion zones cover 2000 and 1000 Å, respectively. Note that the Cr concentration at the interface region is, once more, higher than the solubility limit of Cr in GaAs [9]. As in the case of cap-annealed and implanted layers, the actual value of Cr at the interface has to be taken with caution. The oxygen profiles have also been determined: they do show a peak at the interface which varies with the samples analysed from 5×10^{15} to 5×10^{16} cm^{-3}, but there is no correlation between the intensity of this peak and the height of the Cr peak.

The Cr concentration on the layer side decreases from the interface and reaches a level of 2×10^{15} cm^{-3} within 1000—1500 Å. Most of the layers analysed have a background level of between 10^{15} and 4×10^{15} cm^{-3}. Because of the indeterminacy in the Cr concentration near the surface [10, 11], the profile is not shown within 1000 Å of the surface.

The layer represented in Figure 1 is Sn doped with n~10^{17} cm^{-3} and $\mu(300\ K)$~3700 cm^2 V^{-1}s^{-1}.

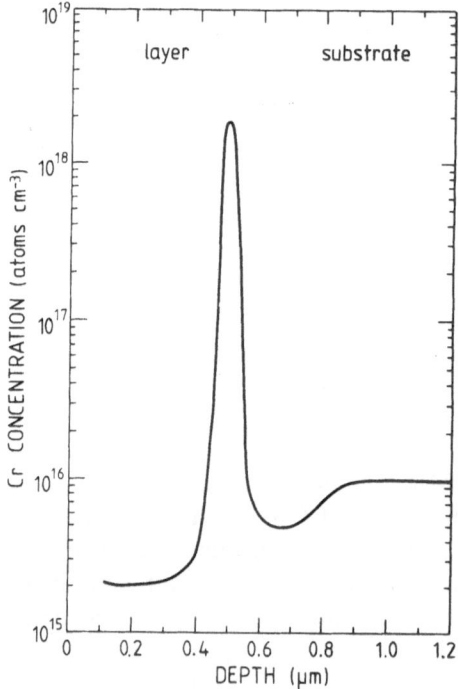

Figure 1   Cr profile in MBE layer: substrate heated to 530°C for 30 minutes before growth.

Figure 2 represents the Cr distribution of a layer grown at 580°C, the substrate being heated up from room temperature to growth temperature in 2 minutes. It should be noted, however, that once the sample is introduced into the MBE machine, the vacuum chamber is baked overnight. During this baking time, the temperature of the substrate reaches 180°C. The Cr profile shows short accumulation and depletion zones. The transition from the accumulation peak to the background level is very sharp ($\sim$500 Å). The free carrier concentration of this layer is 5×10^{14} cm^{-3} with mobility of 7100 cm^2 V^{-1} s^{-1} at 300 K and 36 000 cm^2 V^{-1}s^{-1} at 77 K.

Other procedures of heat treatment of substrates have been tried and it has been found that a Cr background concentration of 10^{14} cm^{-3} can be obtained [12].

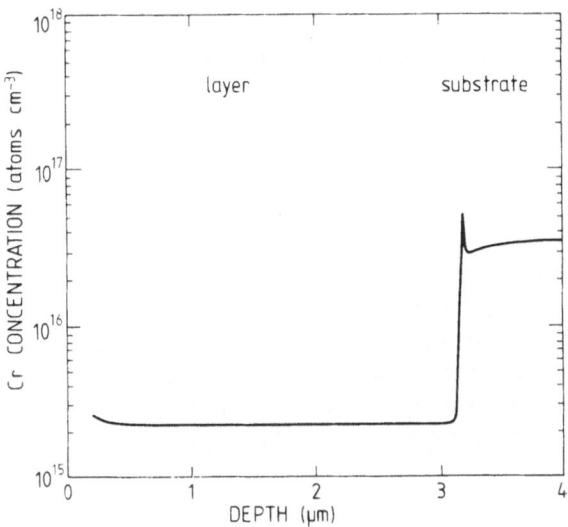

Figure 2   Cr profile in MBE layer: substrate heated to 180°C for 15 hours before growth.

VPE LAYERS

In the Ga/AsCl$_3$/H$_2$ technique, the growth temperature is 755°C and the substrate is etched *in situ* by the AsCl$_3$+H$_2$ mixture before growth. The continuous line in Figure 3 shows a typical Cr profile. The general feature of most of the curves does not depend on the substrate used. The transition region covers about 1000—1500 Å. In most of the samples, the background level in the layer is between 1 and 2×10^{15} cm^{-3}. The layer shown in Figure 3 is sulphur doped to 7×10^{16} cm^{-3} and has a mobility of 4700 cm^2 V^{-1}s^{-1}.

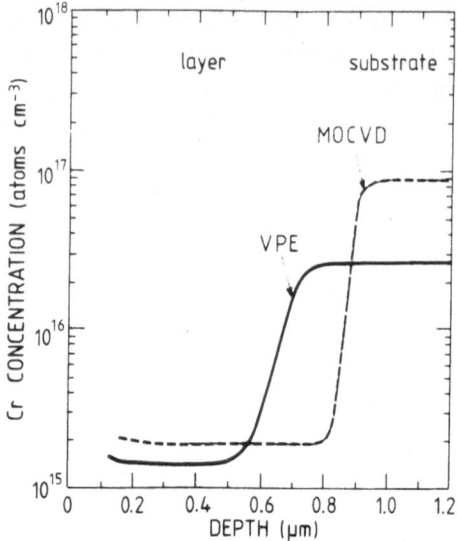

Figure 3   Cr profiles of VPE and MOCVD layers.

MOCVD LAYERS

The MOCVD layers have been grown at reduced pressure [13], at $600°C$. Before growth the substrate is heated under AsH_3. The temperature reaches its equilibrium value after 5—10 minutes. The dashed line in Figure 3 represents a typical profile which shows a very sharp transition between layer and substrate. The transition region covers $\sim 500 \text{Å}$. The background level varies from sample to sample between 1 and $4 \times 10^{15} \text{ cm}^{-3}$. The layer shown in Figure 3 is Si doped to $5 \times 10^{16} \text{ cm}^{-3}$ and has a mobility of $5000 \text{ cm}^2 \text{ V}^{-1}\text{s}^{-1}$.

Discussion

First, one can note that a level of Cr in the low range of 10^{15} cm^{-3} is observed in all the layers shown, so it may be asked whether the SIMS technique and/or apparatus are limited in sensitivity or whether they artificially alter the results when the Cr concentration is below 10^{15} cm^{-3}.

In fact, in the last few years we have analysed more than 150 GaAs samples (bulk or layers) doped or unintentionally doped with Cr. Except for about ten of them, they contain Cr at a level higher than 10^{15} cm^{-3} even in unintentionally doped samples. Among the low doped samples are bulk crystals pulled in a BN boat (supplied by MR) with $[Cr] = 5 \times 10^{13} \text{ cm}^{-3}$, MBE layers with $[Cr] = 10^{14} \text{ cm}^{-3}$ and some annealed samples $[Cr] = 7 \times 10^{13} \text{cm}^{-3}$ [12]. As low doped samples have been analysed we believe that the background level of 10^{15} cm^{-3} reported here can be taken with confidence.

Let us consider the Cr redistribution in VPE and MOCVD layers (Figure 3). These curves correspond typically to out-diffusion profiles. The diffusion coefficients determined from these curves are 3×10^{-13} and $7 \times 10^{-15} cm^2 s^{-1}$ for VPE and MOCVD, respectively. These values are close to those calculated from the diffusion law [14] $D = 4.3 \times 10^3 \exp(-3.4 \, eV \, kT^{-1}) cm^2 s^{-1}$ which gives 7.7×10^{-14} and $1.2 \times 10^{-16} cm^2 s^{-1}$ for the two temperatures of growth in VPE and MOCVD.

The background level of $\sim 10^{15} cm^{-3}$ can be attributed to an autodoping effect by the back side of the substrate. But Tuck *et al.* [2] have reported that this effect would give a background below $5 \times 10^{14} cm^{-3}$. Moreover, we have analysed some layers grown on n^+ substrates which are not intentionally doped with Cr, but in which the Cr concentration is about $10^{15} cm^{-3}$, and we found $10^{15} cm^{-3}$ of Cr in the layers. In the case of MBE, layers have been grown on undoped substrates whose Cr concentration is between 4×10^{14} and $3 \times 10^{15} cm^{-3}$. The concentration of Cr in these layers is found to be the same as in the substrate. So, it is definitely demonstrated that the background level of Cr in MBE samples is due to a fast diffusion of Cr from the substrate.

If we consider the data of segregation coefficients of Cr (acting as deep traps [15]) the limit of solubility of Cr (on a Ga site) would be about 5×10^{14} and $10^{13} cm^{-3}$ at 750 and 600°C, respectively. These values are supported by the fact that intentionally Cr-doped layers are semi-insulating only if the residual carrier concentration is lower than $\sim 10^{14} cm^3$ [15, 16]. If these values of Cr solubility are correct, the background level of epitaxial layers would not be constituted by Cr on Ga sites. Is this Cr electrically active? For FET applications, the layers studied in this paper are n-doped in the range of 5×10^{16} to $2 \times 10^{17} cm^{-3}$. So, the relatively low Cr concentration ($10^{14} - 10^{15}$ cm^{-3}) has no detectable effect on the mobility of free carriers [17]. Recently, Mitonneau, Chane and Andre [18], using optical DLTS, have shown that deep traps attributed to Cr can be detected at a concentration of about 10^{15} cm^{-3} in both their VPE and MOCVD layers. These data would explain quite well our SIMS results but these authors reported that their deep trap concentration drops to less than $5 \times 10^{14} cm^{-3}$ beyond $2 \mu m$ from the interface. Our analyses on thick layers have shown that Cr is still present at the level of $1 - 2 \times 10^{15} cm^{-3}$ at 20 μm from the interface. In conclusion, for lack of experimental evidence, the nature of the Cr detected in the epitaxial layers is not exactly known.

The Cr profiles in cap-annealed, implanted and MBE layers suggest the following mechanism of Cr redistribution in these layers: at the melting temperature of GaAs (1238°C), the Cr atoms introduced in the GaAs lattice are in thermodynamic equilibrium and occupy the Ga site. At lower temperatures, the limit of solubility being reduced [15], these atoms are in non-equilibrium. Therefore, if a Cr-doped sample is annealed, the Cr can relax towards its equilibrium value. This relaxing phenomenon can be a bulk segregation or a surface segregation, i.e. a migration of Cr atoms towards the surface where they accumulate. The surface segregation is a well-known process in the case of impurities in metals [19, 20]. This process takes place by diffusion from the bulk to the surface. The proposed mechanism can explain the profiles of cap-annealed and implant-annealed samples (Figure 1) where the total Cr atoms in the accumulation region are roughly equal to the total Cr atoms depleted underneath. This is also the case with the MBE layer in Figure 2. But in MBE samples grown on substrates annealed at 530°C, this simple mechanism is not completely satisfactory, since the accumulated Cr is in excess compared to the depleted Cr (Figure 1). The segregation mechanism in vacuum is certainly complicated by the non-congruent evaporation of GaAs at temperatures higher than 500°C. The diffusion coefficients determined in cap-annealed samples are close to those calculated from the diffusion law reported above [12], but diffusion in implant-annealed and vacuum-annealed MBE layers is faster.

In conclusion, the Cr redistribution in VPE and MOCVD layers is well explained by a Cr out-diffusion from the substrate. In cap-annealed, implant-annealed and MBE samples, a surface segregation mechanism can explain the experimental results. The Cr background level in epitaxial layers seems to be due to a fast diffusion of Cr from the substrates.

ACKNOWLEDGEMENTS

The authors wish to thank P. Delescluse for helpful discussions on diffusion, G.V. Smith and C. Whitehead (The Plessey Co. Ltd) for the analyses of Cr in standard samples by spark source mass spectrometry, P.N. Favennec (CNET Lannion) for cap-annealed and implantation experiments, D.W. Woodard (Cornell University), P.R. Jay and B.T. Debney (The Plessey Co. Ltd) for communicating unpublished results. This work was partly supported by the DRET contract No. 77—359.

References

1. Tuck, B. and Adegboyega, G.A. (1979). *J. Phys. D: Appl. Phys.*, **12**, 1895
2. Tuck, B., Adegboyega, G.A., Jay, P.R. and Cardwell, M.J. (1979). *Inst. Phys. Conf. Ser.*, **45**, 114
3. Huber, A.M., Morillot, G., Merenda, P. and Linh, N.T. (1979). *Proc. SIMS II*, Stanford, p. 91. New York; Springer Verlag; also Huber, A.M., Morillot, G., Linh, N.T., Debrun, J.L. and Valladon, M. (1978). *Nuclear Inst. Meth.*, **149**, 543
4. Huber, A.M., Morillot, G., Linh, N.T., Favennec, P.N., Deveaud, B. and Toulouse, B. (1979). *Appl. Phys. Lett.*, **34**, 858
5. We thank Mr Cajan from Metals Research France for the gift of some undoped samples
6. Favennec, P.N. and L'Haridon, H. (1979). *Appl. Phys. Lett.*, **35**, 699
7. Evans, C.A., Jr, Deline, V.R., Sigmon, T.W. and Lidow, A. (1979). *Appl. Phys. Lett.*, **35**, 291
8. Etienne, P., Massies, J. and Linh, N.T. (1977). *J. Phys. E: Scientific Instrum.*, **10**, 1153
9. Haisty, R.W. and Cronin, G.R. (1964). *Proc. 7th Int. Conf. on the Physics of Semiconductors*, p. 1161. Paris; Dunod

10. Kim, H.B., Barett, D.L., Sweeney, G.G. and Heng, T.M.S. (1977). *Inst. Phys. Conf. Ser.,* **33b,** 136

11. Slodzian, G. and Hennequin, J.F. (1968). *C.R Acad. Sci. Paris,* **263B,** 1246

12. Delescluse, P., Etienne, P., Huber, A.M. and Linh, N.T. To be published.

13. Duchemin, J.P., Bonnet, M., Koelsch, F. and Huyghe, D. (1978). *J. Cryst. Growth,* **45,** 181; also Duchemin, J.P., Bonnet, M., Koelsch, F. and Huyghe, D. (1979). *J. Electrochem. Soc.,* **126,** 1134

14. Casey, H.C. (1973). *Atomic Diffusion in Semiconductors,* (Ed. D. Shaw), p. 417. New York; Plenum

15. Woodard, D.W. (1979). Thesis, Cornell University, USA

16. Otsubo, M. and Miki, H. (1974). *Japan J. Appl. Phys.,* **13,** 1655; also Otsubo, M. and Miki, H. (1977). *J. Electrochem. Soc.,* **124,** 441

17. Debney, B.T. and Jay, P.R. To be published in *Solid St. Electron.*

18. Mitonneau, A., Chane, J.P. and Andre, J.P. (1979). *21st Electronic Materials Conf.,* Boulder, USA

19. Lagues, M. and Domange, J.L. (1975). *Surface Sci.,* **47,** 77

20. Blakely, J.M. and Shelton, J.C. (1975). In *Surface Physics of Materials,* vol. 2 (Ed. J.M. Blakely). New York; Academic Press

INVESTIGATIONS OF Cr IN GaAs USING PHONONS

P. BURY*, L.J. CHALLIS, P.J. KING, D.J. MONK, A. RAMDANE, V.W. RAMPTON and P. WISCOMBE
Department of Physics, University of Nottingham, Nottingham NG7 2RD, UK

Abstract

The paper reviews published spectroscopic work on Cr in GaAs using a variety of phonon techniques from 0.1 to 1000 GHz. All this work was done on semi-insulating material. It then describes recent Nottingham work on n-type, p-type and semi-insulating samples before and after photoexcitation using acoustic paramagnetic resonance, relaxation measurements, thermal conductivity and magnetothermal conductivity (frequency crossing). The spectra observed are thought to be largely due to Cr^{2+} but cannot be explained by the model with a static Jahn-Teller effect, which describes the EPR data at two very different frequencies. Possible explanations are advanced.

Introduction

Phonon spectroscopy provides a number of powerful techniques for studying magnetic ions in crystals. The phonons displace the surroundings of a magnetic ion and this produces electric fields at the ion which oscillate at the phonon frequencies and which can induce transitions within the ion. The form of the electric potentials produced in this way and so the phonon selection rules are quite different from those for electric dipole transitions for photons $(\Delta M_S=0, \Delta M_L=0, \pm 1)$. For example, phonon-induced transitions can occur between states of the same parity whereas these are of course 'forbidden' for photons. Such transitions can often be induced by photons using the weaker magnetic dipole transition if powerful coherent sources are available and this is the basis of EPR where the usual selection rules are $\Delta M_S=\pm 1$, $\Delta M_L=0$. The selection rules for phonon transitions are wider: $\Delta M_S=0, \Delta M_L=\pm 1, \pm 2$ etc. and this frequently results in the observation of a large range of transitions which are complementary to those observed using photons.

In this paper we first review the published phonon work on chromium in GaAs and then describe recent experiments conducted in Nottingham, with some comments on their interpretation.

*On leave from Department of Physics, Technical University of Advanced Transport Engineering, Zilina, Czechoslovakia.

Previous work

Acoustic paramagnetic resonance (APR) is the resonant absorption of an acoustic wave. The separations of the low lying energy levels are tuned to the acoustic frequency using a magnetic field or externally applied stress. The first APR measurements on Cr:GaAs were conducted at 9.4 GHz by Ganapol'skii [1] on a semi-insulating (SI) sample. Three resonances were found at 0.8 kOe, 2.1 kOe and 6.0 kOe. The linewidths and amplitudes of these resonant absorptions were found to vary with magnetic field angle, but their position was approximately isotropic in field.

Ganapol'skii attributed the upper and lower APR lines to Cr^{2+} at substitutional tetrahedral sites and the mid-field line to sites which, in addition, are tetragonally strained. Theoretical arguments were used to eliminate explanations based on interstitial sites or on Cr^{3+}. Further APR measurements at 1, 1.5 and 2 GHz have been conducted on SI samples at 1.5 K by Tokumoto and Ishiguro [2, 3] but to date no detailed explanation of these has been achieved.

Tokumoto and Ishiguro [2] have also made measurements of the non-resonant relaxation absorption of phonons. Here, the low lying levels of the ions are modulated by an acoustic wave and repopulation causes a peak in the absorption as a function of temperature. Measurements were made at frequencies from 75 MHz to 2 GHz and from 4 to 50 K. A pronounced uniaxial stress-dependent peak was found in the region of 20 K and the phonon selection rules determined. These were found to be consistent with relaxation between < 100> tetragonal distortions at a site of T_d symmetry, the phonon modes of E_g symmetry showing a peak whereas those of T_{2g} symmetry did not. The authors identify the relaxing ion as Cr^{2+} on the grounds that Krebs and Stauss ESR data [4] are consistent with Cr^{2+} in T_d site symmetry with tetragonal distortions produced by static Jahn-Teller effects.

Higher frequencies can be generated by attaching a heater to the sample. The phonons have an approximately black-body distribution with a peak at $\sim 80 T_0$ GHz, where T_0 is the heater temperature, and their attenuation down a sample may be measured by temperature sensors. If the heater power is fixed, the experiment measures the thermal resistivity (W) and the excess attenuation due to ions can be obtained at $80 T_0$ GHz from W/W_{pure}. The first measurements of this type on an SI GaAs:Cr sample were made by Chaudhuri *et al.* [5], who found an attenuation maximum at 6.5 K corresponding to a transition frequency of ~ 500 GHz. In later measurements made by Vuillermoz, Langier and Mai [6] on highly doped Cr samples, the scattering appears to be mainly by precipitates.

If the heater is pulsed, time-of-flight techniques may be used to observe longitudinal and transverse phonon modes in specific crystal directions. Such an experiment was carried out by Narayanamurti, Chin and Logan [7] on four chromium- and oxygen-doped semi-insulating samples. The attenuation was stronger in more concentrated chromium-doped samples and the ratio of transverse to longitudinal attenuation peaked at 5 ± 1 K, essentially in agreement with the result of Chaudhuri *et al.* [5]. The attenuation is highly anisotropic and the ion responsible is again much more strongly coupled to E_g modes than to T_{2g} modes. Because of this, Narayanamurti *et al.* [7] also concluded, like Tokumoto and Ishiguro [2], that the defect must be tetragonally distorted. However, this conclusion assumes that the scattering process involves the reorientation of a defect from one direction to another. If this is not the case the conclusion is certainly not valid since it is quite usual that E_g modes are more strongly coupled to electronic transitions than T_{2g} modes, even in pure T_d symmetry.

Finally, we mention the high-frequency measurements of Narayanamurti [8] using superconducting tunnel junctions. The investigation was conducted at frequencies up to $\sim$300 GHz on an SI sample, and a broad absorption band was found in the region of $\sim$200 GHz.

Phonon measurements at Nottingham

Investigations are being carried out at Nottingham to extend this work in several directions. Previously, only SI samples have been used and so we are using n- and p-type samples where sub-band gap illumination can be used to change the valence states of the chromium ions. We report briefly on our findings so far.

ACOUSTIC PARAMAGNETIC RESONANCE (APR)

APR measurements at 9.6 GHz have been made on SI samples ($\sim$1$\times$10^{17} cm^{-3} Cr) from Czechoslovakia and an n-type (GA 735-1) and a p-type (GA 781) sample from the Plessey Co. In the SI samples a large number of strong absorption lines are found together with many weaker ones. Among the stronger lines are the three found by Ganapol'skii and another which occurs at a field that has a $<$111$>$ minimum suggesting trigonal symmetry which might therefore be due to some complex like that causing the 0.84 eV photoluminescence since this also exhibits trigonal symmetry (Eaves *et al.*, this volume; Killoran *et al.*, preprint). Most of the lines vary rapidly in intensity as the magnetic field is rotated. Figure 1 shows the position of the major lines as a function of magnetic field in the (110) plane for propagation parallel to $<$001$>$. Their intensity increased by a few percent when irradiated with sub-band gap light.

In the n-type sample only two extremely weak features were found in the dark, but under illumination strong APR lines appeared which decayed when the illumination was removed in times characteristic of those observed in the photoconductivity [9]. The line positions in the n-type sample corresponded to those of the SI samples but the relative intensities of the lines were not the same and groups of lines could be identified from their intensities.

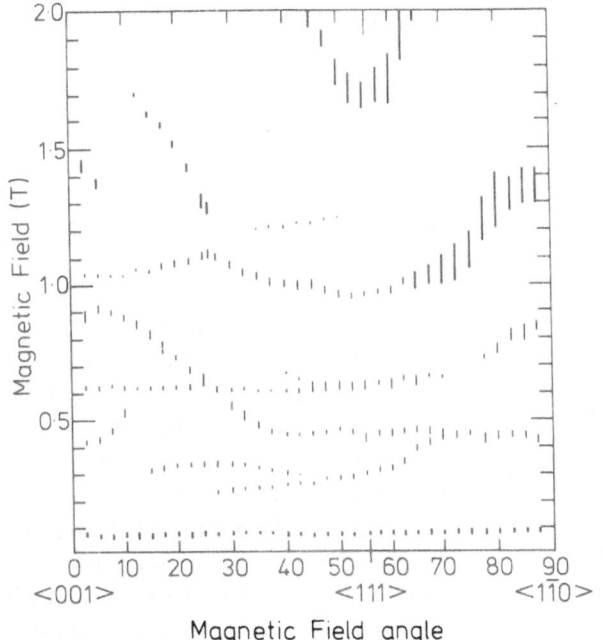

Figure 1   The positions in magnetic field of the major APR lines at 1.7 K and 9.6 GHz and as a function of field angle ($\mathbf{k}\|<001>$, field in the (110) plane). The density of the points gives an indication of the strength of the absorption.

It thus appears that the production of Cr^{2+} and Cr^{3+} by illumination carries with it the appearance of strong APR absorptions. We attribute most of our lines to Cr^{2+} since we have not found any APR absorptions in a p-type sample, although we have not ruled out Cr^{3+} as being the cause of some of the weaker features. Our results suggest that a number of differently distorted sites are responsible. Attempts are being made to fit the data to various models one of which, unlike that of Ganapol'skii, uses a dynamic Jahn-Teller effect and static lattice strains.

PHONON RELAXATION MEASUREMENTS

Acoustic attenuation has been measured at 0.58 and 1 GHz on a range of samples including the SI Czech samples, the Tokumoto and Ishiguro sample TI#5 and n-type (GA 735-064 and GA 785) and SI samples (GA 735-i) from the Plessey Co. In the GA 735 n-type and SI samples we found a large peak similar in form to that of Tokumoto and Ishiguro [2]. Sub-band gap illumination produced *no* change in peak shape or height, however. For the n-type sample we were able to monitor the large changes in photoconductivity and at the same time monitor the acoustic attenuation. We found no peak in the Czech samples, either in the dark or when illuminated. We conclude that the relaxation peak, while it has symmetries associated with

Cr^{2+} from other measurements, does not appear to be directly related to Cr^{2+} since it does not correlate with the changes in Cr^{2+} concentration associated with illumination or sample type.

THERMAL AND MAGNETOTHERMAL CONDUCTIVITY

A brief report of some of this work has already appeared [10]. From the thermal conductivity of six samples (three SI, two n-type and one p-type, from Tokumoto and Ishiguro and from the Plessey Co.) it was clear that the additional scattering due to Cr doping was very much less in the n-type samples than in the SI or p-type samples, confirming that Cr^{1+} is a weak phonon scatterer. The data are plotted in Figure 2 in the form W/W(1), which displays the additional resonant scattering attributable to Cr^{2+} or Cr^{3+} (sample 1 is the n-type sample with the highest conductivity). The predominant valence state believed to be present is indicated. The main peak varies from 7 to 9.5 K, indicating resonant scattering at ~600 GHz in all samples. The scattering is strongest in the samples believed to contain Cr^{3+} and little Cr^{2+} but there is also scattering in TI#4 which is not thought to contain Cr^{3+}, suggesting that both Cr^{2+} and Cr^{3+} have resonant frequencies of ~600 GHz. There is some evidence of structure in GA 735(e) and of scattering at ~200 GHz (~2 K) in agreement with Narayanamurti [8]. The scattering in the n-type sample GA 735(064) is appreciably increased by mid-gap radiation and this photothermal resistivity decays slowly, in times similar to that of the photoconductivity, to a persistent value ($\tau \gg 10$ h), approximately twice that in the dark. It seems likely that the photocentre that decays is Cr^{2+}.

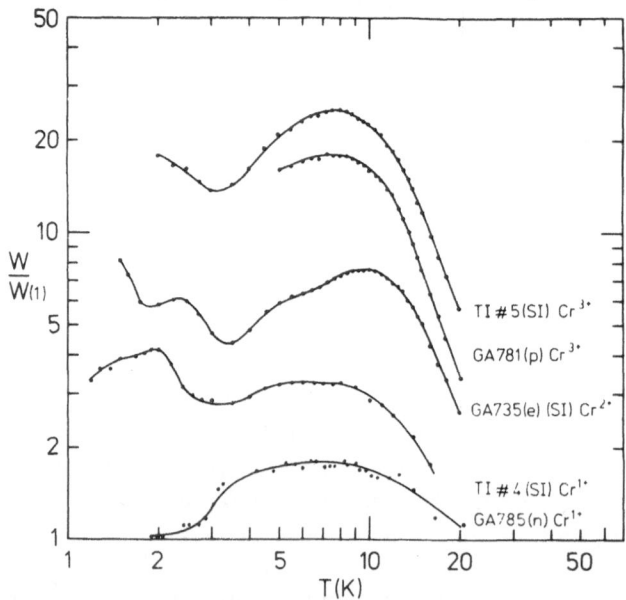

Figure 2   The additional thermal resistivity due to resonant scattering W/W(1); W(1) is the resistivity of a sample thought to contain only Cr^{1+}. The probable predominant valence state is indicated. This is estimated from the concentration of compensating impurities believed to be present.

Finally, we report briefly on the effects of magnetic field which have been measured in several samples and in fields up to 13 T. Of particular interest are the data on TI#4 which is believed to contain a small concentration of Cr^{2+}. The magnetothermal resistivity displays minima at certain fields whose positions are independent of temperature. Such effects are very well known in dielectrics and are due to frequency crossings, two transition frequencies becoming equal at a particular field, and in this case the most striking feature is that the fields at which crossings occur are all completely isotropic for all field directions in the (110) plane, which is clearly inconsistent with a strong tetragonal distortion.

Discussion

The complexity of the APR spectrum could indicate that Cr is present in several different sites, with an associated defect in several possible positions producing effects that are large compared with the spin-orbit splitting. It seems improbable, however, that most of the Cr is in such sites and it seems more likely that the majority of ions are basically in the same site, e.g. T_d but that their levels, which are clearly strongly coupled to the lattice, are shifted ≥ 10 GHz (the APR quantum) by lattice strains. In this situation, which exists for example for Cr^{2+} in CaO [11], a peak in the APR spectrum occurs when the number of ions at resonance is a maximum, i.e. when there is a turning value in $(\partial \nu / \partial \epsilon)_{\nu_0}$ where ϵ is the strain. There could, of course, be several turning values leading to several APR lines and the relative heights of these lines could depend on the strain distribution and so be sample-dependent. Evidence that strongly supports this model for Cr^{2+} comes from the highly isotropic frequency crossing data. The frequencies involved are probably very much larger than the level shifts due to lattice strain so that the T_d symmetry of the lattice point is seen.

The zero field thermal conductivity data indicate splittings ~ 600 GHz (20 cm^{-1}) in both Cr^{2+} and Cr^{3+} and this, and more convincingly the tunnel junction work, also suggests splittings ~ 200 GHz (7 cm^{-1}). The 20 cm^{-1} splitting seems too large for second order spin-orbit splitting of the $J=3$ ground state of Cr^{2+} (9.6 $\lambda^2/\Delta \sim 3$ cm^{-1} and 21.6 $\lambda^2/\Delta \sim 7$ cm^{-1} [1]) and too small for first order splitting of the $^5T_2$ (3 $\lambda \sim 150$ cm^{-1}) unless, of course, there is a strong dynamic Jahn-Teller effect with a reduction factor $\gamma \sim 0.1$. If there is, one might interpret both the 7 and 20 cm^{-1} transitions as arising from spin-orbit splitting of the $^5T_2$ manifold and we put this forward as a tentative suggestion.

How can all this be reconciled with EPR and optical data? There seems no immediate problem with the zero field optical data [13] since the data span ~ 700 GHz (20 cm^{-1}). However, the EPR data for Cr^{2+} show convincing evidence for a strong tetragonal distortion [4, 12]. Could this indicate that the low strain turning values leading to absorption peaks in the APR are forbidden to magnetic dipole transitions and that these transitions only become allowed for high values of strain? If so, the EPR data seen may be for sites with a neighbouring tetragonal defect and there is a natural explanation why the T_d sites are not detected in present material.

Certainly, as stressed by Bates, the ground states of Cr^{2+} and Cr^{3+} following a static Jahn-Teller distortion are both orbital singlets and could not therefore be responsible for the very strong phonon scattering observed in this work.

ACKNOWLEDGEMENTS

We are most grateful to M.J. Cardwell, P.R. Jay and the Plessey Co. and to T. Ishiguro and H. Tokumoto for generously providing samples, to the British Council and the Science Research Council for financial support, and to C.A. Bates and L. Eaves and those listed above for very helpful discussions.

References

1. Ganapol'skii, E.M. (1975). *Sov. Phys. Solid State,* **16,** 1868
2. Tokumoto, H. and Ishiguro, T. (1977). *Proc. 6th Int. Conf. on Internal Friction and Ultrasonic Attenuation in Solids,* p. 177, Tokyo
3. Tokumoto, H. and Ishiguro, T. (1979). *J. Phys. Soc. Japan,* **40,** 84
4. Krebs, J.J. and Stauss, G.H. (1977). *Phys. Rev. B,* **16,** 971
5. Chaudhuri, N., Wadhwa, R.S., Phoola, T. and Sreedhar, A.K. (1973). *Phys. Rev. B,* **8,** 4668
6. Vuillermoz, P.L., Langier, A. and Mai, C. (1975). *J. Appl. Phys.,* **46,** 4623
7. Narayanamurti, V., Chin, M.A. and Logan, R.A. (1978). *Appl. Phys. Lett.,* **33,** 481
8. Narayanamurti, V. (1979). *Proc. Int. Conf. on Phonon Scattering in Condensed Matter,* Brown University. Plenum Press (to be published)
9. Eaves, L. and Williams, P.J. (1979). *J. Phys. C,* **12,** L725
10. Challis, L.J. and Ramdane, A. (1979). *Proc. Int. Conf. on Phonon Scattering in Condensed Matter,* Brown University. Plenum Press (to be published)
11. Bates, C.A., Maynard, C.M., Rampton, V.W. and Shellard, I.J. (1979). *J. Phys. C,* **12,** 3561
12. Wagner, R.J. and White, A.M. (1979). *Solid. St. Commun.,* **32,** 399
13. Lightowlers, E.C. and Henry, M.O. (1978). *Inst. Phys. Conf. Ser.,* **43,** 307

CORRELATION BETWEEN EPR ASSESSED Cr²⁺ AND ELECTRICAL COMPENSATION IN SEMI-INSULATING GaAs:Cr

A. GOLTZENE, C. SCHWAB and G.M. MARTIN*
Laboratoire de Spectroscopie et d'Optique du Corps Solide,
Associé au CNRS No. 232, Université Louis Pasteur,
5 rue de l'Université, 67000 Strasbourg, France

Abstract

A set of different samples of semi-insulating GaAs:Cr has been assessed independently for their Cr content using EPR measurements, and for their electrically active centres derived from experiments combining Hall, DLTS and optical absorption measurements. In the dark, only Cr^{2+} centres were observed by EPR whereas Cr^{3+} signals were always negligible and Cr^{1+} signals could never be observed. The Cr^{2+} concentration increases with the residual concentration of shallow donors ($N_D - N_A$) and of a deep donor, probably the 'pseudo-oxygen' defect or EL2, thus indicating that compensation occurs by trapping of free electrons on the Cr^{3+} ions. The residual concentration of EL2 is found to be of the order of 10^{16} cm^{-3}, in agreement with former evaluations. Furthermore, at 4.2 K the EL2 level lies above the Cr acceptor level.

Introduction

The evident problem in the study of semi-insulating GaAs is the determination of the concentration of shallow donors and of compensating deep acceptors. The aim of the present paper is to show to what extent Cr ions may be used as a direct probe of the compensation in semi-insulating GaAs:Cr.

It is known that three charge states, namely Cr^{1+}, Cr^{2+} and Cr^{3+}, of the Cr ion have already been identified in this material from electron paramagnetic resonance (EPR) measurements [1—7]. Among the former ions, Cr^{2+} seems of particular interest since it would be directly involved in the two expected redox buffers [8] formed by the pairs Cr^{1+}—Cr^{2+} or Cr^{2+}—Cr^{3+}. Nevertheless, it should be noted that until now the Cr^{2+} direct compensating behaviour has not been clearly established. For instance, the existence of more complicated compensating defects cannot be rejected entirely, especially since it is known that semi-insulating GaAs is obtained either by oxygen doping [9] or even without any intentional doping [10].

*Laboratoire d'Electronique et de Physique Appliquée, 3 avenue Descartes, 94450 Limeil-Brevannes, France.

For the purpose of the present study, a set of semi-insulating GaAs samples, containing different dopants (Si, Te) and compensated by Cr during growth, were measured quantitatively for their Cr content by EPR. This programme was particularly appropriate since detailed electrical data were already available for this same set [11].

Experimental

EPR MEASUREMENTS

The EPR measurements were performed by means of a Thomson-CSF THN 252 X Band Spectrometer working at 9.3 GHz. All data were recorded at 4.2 K, the samples being immersed in the He bath of a finger cryostat inserted in a TE 102 rectangular cavity.

The samples were cut to bars of about $3 \times 3 \times 8$ mm^3, oriented along a $<110>$ axis.

In the whole set of samples the only strong signal was that corresponding to the Cr^{2+} signature [5]. For an external magnetic field $H_0 \| <001>$ and $H_0 \| <\bar{1}10>$, this spectrum is characterized by an isolated line around 5 kG, which allows a convenient integration for the intensity determinations. It should be noted that the intensity ratio for the two field directions is strongly strain-dependent so that, in order to obtain the correct intensities, it was necessary for both orientations to sum their two contributions, which were further weighted by the transition probabilities as determined by Stauss et al. [6]. Incidentally, we have found that in our case the ratio of the intensities for both directions was close to $\frac{1}{2}$ so that the residual strain level is certainly low.

Relative calibration of the Cr^{2+} content was done by using a secondary standard consisting of a small CuGaS$_2$ crystal doped with ^{61}Ni [12—14], whose Ni^{1+} signal does not saturate at 4.2 K if the incident microwave power is kept below 1 mW. The purpose of this secondary standard is to take account of the variations of the resonant cavity quality factor Q with the load.

Two methods have been used for evaluating the Cr^{2+} and Ni$^+$ concentrations from the recorded derivative spectra: either a graphical double integration or a comparison of the peak to peak amplitude (since the linewidths remained nearly constant for both centres) were performed. The comparison of the two methods yielded data in agreement within $\pm 10\%$.

Absolute spin number determination was performed by calibrating one of the GaAs-CuGaS$_2$ couples, actually that containing the GaAs sample No. 5 where the Ni^{1+} signal was the most intense, against an NBS SRM 2601 ruby standard. Since the latter saturates at low temperature, the room temperature value resulting from the double integration was corrected to obtain the 4.2 K value by multiplying it by the temperature ratio 291/4.2. Of course, this procedure does not account for the Q variations between 4.2 and 291 K;

nevertheless, we estimate that the overall error in the spin numbers is kept to within $\pm 25\%$.

In some instances, very weak Cr^{3+} lines near 4.5 kG could be observed. This fact is in accordance with our former deduction of a low residual strain level. On the other hand, it is known that the EPR spectrum of the Cr^{3+} centre can be noticeably enhanced by exerting a uniaxial stress on the sample, but this could not be done in our finger cryostat, so the $[Cr^{3+}]$ determination proposed recently [6] could not be applied.

Finally, the Cr^{1+} line at $g=1993$, which is not strain-dependent, was never observed above the noise level in our samples. Thus it may be deduced that, if at all present, the Cr^{1+} concentration is well below 10^{15} cm^{-3}. All experimental data are summarized in Table 1.

ELECTRICAL ASSESSMENT

As already mentioned, this set of samples has been previously assessed for its electrical properties, using the procedures described in [11].

These investigations showed clearly that even in the most insulating material, both a deep acceptor (DA) and a deep donor (DD) were present besides the shallow acceptors (SA) and donors (SD). Thus the equation for charge neutrality may be written:

$$n + N_{SA}^i + N_{DA}^i = p + N_{SD}^i + N_{DD}^i \tag{1}$$

where n and p are the concentrations of the free electrons and holes, and the N^i are the concentrations of ionized centres.

Since the material is semi-insulating, n and p may be neglected and equation (1) reduces to:

$$N_{DA} \; f(DA) \sim (N_{SD} - N_{SA}) + N_{DD}(1 - f(DD)) \tag{2}$$

where f denotes the Fermi function.

High temperature (400 K) Hall measurements allow the determination of the sum $(N_D - N_A) + N_{DD}$; the contribution of the first term is varied by doping with shallow donors Si or Te.

Assuming that the deep acceptor is related to the Cr centre and the deep donor to the 'pseudo-oxygen' centre, EL2, which are known to be the most important deep centres in bulk GaAs:Cr, equation (2) becomes, at low temperature:

$$N_{Cr} \; f(Cr) \sim (N_D - N_A) + N_{EL2} \tag{3}$$

if EL2 is located above the Cr level, or:

$$N_{Cr} \; f(Cr) \sim (N_D - N_A) \tag{4}$$

if EL2 is located below the Cr level, provided that the energy separation between EL2 and the Cr level is higher than kT.

Table 1 lists some relevant parameters for the various samples along with our $[Cr^{2+}]$ determination by EPR.

Table 1

Samples	Growth technique	Dopant	[Cr] in the melt (m Cr/m GaAs)	[Cr] (opt) (10^{16} cm^{-3})	[Cr^{2+}] (EPR) (10^{16} cm^{-3})	[N$_D$-N$_A$] (10^{16} cm^{-3})	[N$_D$-N$_A$]+[EL2] (10^{16} cm^{-3})
1	CZ	—	7×10^{-4}	8	3.2	$\leqslant 0.1$	1
2	—	—	2.5×10^{-4}	1	1.2	$\leqslant 1$	$\leqslant 2$
3	—	—	2.5×10^{-4}	2	2.0	~ 2	~ 3
4	HB	Si	—	4.5	4.1	1.5	2.5
5	—	Si	3×10^{-4}	2	6.9	2.5	3.5
6	—	Si	3×10^{-4}	3.6	6.7	2.5	3.5
7	—	Si	3×10^{-4}	12	4.5	2.5	3.5
8	HB	Si	—	15	10.3	3	4
9	CZ	Te	—	22	16.5	7.5	8.5
10	HB	Si	—	27	6.7	7.5	8.5

Abbreviations: CZ=Czochralski;
HB=horizontal Bridgman;
EPR=EPR assessment;
opt=optical absorption assessment.

The accuracy of the $(N_D - N_A) + N_{EL2}$ is estimated to be within $\pm 30\%$.

Figure 1 shows the plot of the Cr^{2+} concentration as determined by EPR as a function of the total donor concentration, $(N_D - N_A) + [EL2]$.

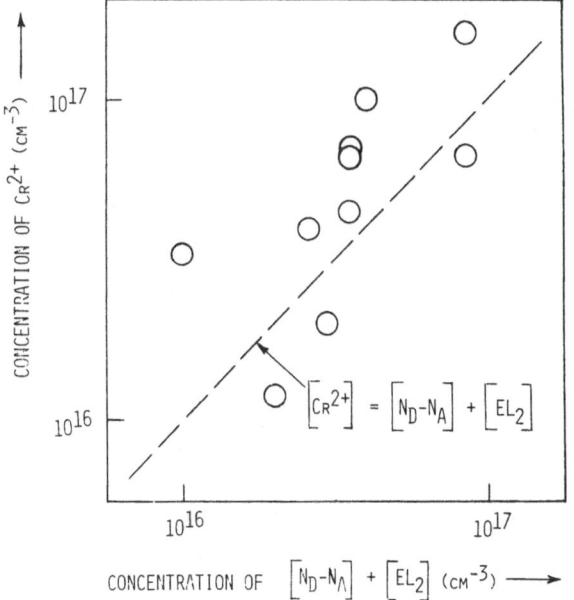

Figure 1 Cr^{2+} concentration versus $(N_D - N_A) + [EL2]$ (see text for definitions).

Discussion

First, it is to be noted that there is a direct correlation between the Cr^{2+} content and the $(N_D - N_A) + [EL2]$ values, which increase simultaneously. This is the first time that such a correlation between a Cr^{2+} state derived from EPR and the true residual concentration of donors could be established.

Second, within the limits of experimental error, the $[Cr^{2+}]$ and $(N_D - N_A) + [EL2]$ values are identical. This means that $N_{Cr}f(Cr)$ is equal to $[Cr^{2+}]$ and that the compensation mechanism proceeds by capture of free electrons on Cr^{3+} ions in these samples, thus leading to an increase of $[Cr^{2+}]$. Furthermore, this result justifies the analysis of the Hall data in terms of a single deep Cr acceptor level. The latter may thus be assigned to the $Cr^{3+} - Cr^{2+}$ states according to its electronic filling.

Third, inspection of Table 1 shows that the Cr content derived from optical measurements at 1.35 μm [17] is generally greater than the concentration of Cr^{2+} determined by EPR, since the optical measurements are essentially an evaluation of the Cr^{3+} concentration [18]. This means that the total concentration of active Cr ions is larger than the true residual concentration of donors. Thus the absence of any other EPR signal, which could be assigned to Cr states different from Cr^{3+} or Cr^{2+}, is not surprising.

Finally, in sample 1 where $(N_D - N_A)$ has been established to be very low, $[Cr^{2+}] \sim [EL2]$. Clearly, this can only be true if the deep donor lies above the Cr^{2+} level at 4 K. This is evidenced on the Shockley diagram of Figure 2 which shows that in the opposite situation (case a), i.e. the EL2 level below the Cr^{2+} level, the EL2 centre could not be ionized and therefore could not be compensated by the Cr^{3+} centres. This observation may not necessarily be in contradiction to electrical measurements, locating the Cr^{2+} level above the EL2 level [19] at higher temperatures $(T > 200 K)$. Rather, it suggests that the variation of the free ionization energy of these two centres as a function of temperature is different.

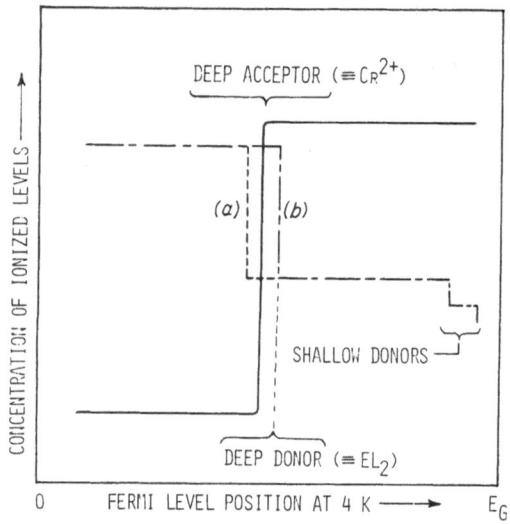

Figure 2   Schematic Shockley diagram corresponding to SI material at 4 K. (a) $E_{EL2} \lesssim E_{Cr}$. (b) $E_{EL2} > E_{Cr}$.

Also, it has been proposed recently [20, 21] that the photoluminescence lines at 0.84 eV could be related to a Cr-donor complex, instead of Cr^{2+}. The measurement of the intensity of this line in this series of samples and, more particularly, in materials like sample 1 should give some indication of a possible correlation between this line and $[Cr^{2+}]$ or $(N_D - N_A)$.

ACKNOWLEDGEMENTS

The authors gratefully acknowledge Mr G. Poiblaud (RTC, France) and Mr R. Ware (Metals Research, UK) for the supply of the GaAs wafers, and Drs A. Mircea-Roussel and S. Makram-Ebeid for fruitful discussions. They also thank Mrs M. Robino and Mr G. Schwalbach for their skilful technical assistance in the preparation of the samples.

References

1. Krebs, J.J. and Stauss, G.H. (1976). *Bull. Amer. Phys. Soc.*, **21**, 89
2. Kaufmann, U. and Schneider, J. (1976). *Solid St. Commun.*, **20**, 143
3. Stauss, G.H. and Krebs, J.J. (1977). *Inst. Phys. Conf. Ser.*, **33a**, 84
4. Krebs, J.J. and Stauss, G.H. (1977). *Phys. Rev. B*, **15**, 17
5. Krebs, J.J. and Stauss, G.H. (1977). *Phys. Rev. B*, **16**, 971
6. Stauss, G.H., Krebs, J.J., Lee, S.H. and Swiggard, E.M. (1979). *J. Appl. Phys.*, **50**, 6251
7. Goltzené, A., Poiblaud, G. and Schwab, C. (1979). *J. Appl. Phys.*, **50**, 5425
8. Goltzené, A., Poiblaud, G. and Schwab, C. (1980). *Rev. Phys. Appl.*, **15**, in press
9. Woods, J.F. and Ainslie, N.G. (1963). *J. Appl. Phys.*, **34**, 1469
10. Swiggard, E.M., Lee, S.H. and von Batchelder, F.W. (1977). *Inst. Phys. Conf. Ser.*, **33b**, 23
11. Martin, G.M., Farges, J.P., Jacob, G., Hallais, J.P. and Poiblaud, G. (1980). To be published in *J. Appl. Phys.*
12. von Bardeleben, H.J., Schwab, C. and Goltzené, A. (1975). *Phys. Lett.*, **51A**, 460
13. Kaufmann, U. (1975). *Phys. Rev. B*, **11**, 2478
14. Troeger, G.L., Rogers, R.N. and Kasper, H.M. (1975). *J. Phys. C: Solid St. Phys.*, **8**, L222
15. Mitonneau, A., Mircea, A., Martin, G.M. and Pons, D. (1979). *Rev. Phys. Appl.*, **14**, 853
16. Huber, A.M., Linh, N.T., Valadon, M., Debrun, J.C., Martin, G.M., Mitonneau, A. and Mircea, A. (1979). *J. Appl. Phys.*, **50**, 4022
17. Martin G.M., Verheijke, M.L., Jansen, J.A.J. and Poiblaud, G. (1979). *J. Appl. Phys.*, **50**, 467
18. Martin, G.M. (1980). This volume
19. Martin, G.M., Mitonneau, A., Pons, D., Mircea, A. and Woodard, D.W. (1980). To be published in *J. Phys. C: Solid St. Phys.*
20. White, A.M. (1979). *Solid St. Commun.*, **32**, 205
21. Picoli, G., Deveaud, B. and Galland, D. (1980). This volume

ACTIVATION OF Cr^{1+} (3d^5) LEVEL IN GaAs:Cr INDUCED BY HYDROSTATIC PRESSURE

A.M. HENNEL*, W. SZUSZKIEWICZ*, G. MARTINEZ,
B. CLERJAUD**, A.M. HUBER†, G. MORILLOT† and P. MERENDA†
Laboratoire de Physique des Solides associé au CNRS,
Université Pierre et Marie Curie,
4, Place Jussieu, 75230 Paris Cedex 05, France

The absorption spectrum of n-type GaAs:Cr, in which the Cr^{2+} EPR signal remains stable under illumination, was measured under hydrostatic pressure up to 10 kbar at liquid nitrogen temperature. The intracentre Cr^{2+} transition ($^5T_2 \rightarrow {}^5E$) disappears with increasing pressure; simultaneously, a rapid rise in resistivity is observed. These results can be explained by the existence of the Cr^{1+} level in the conduction band. Under hydrostatic pressure, this level lowers relative to the conduction band minimum and its population increases, i.e. conduction electrons are trapped by Cr^{2+} centres. Above 9 kbar the saturation of the observed effects appears, indicating that at this pressure the Cr^{1+} level is below the conduction band. This observation locates the Cr^{1+} level no higher than 100 meV above the conduction band minimum at 77 K, i.e. about 1.6 eV above the valence band maximum.

The existence of the Cr^{2+} charge state of the chromium impurity in GaAs is characterized optically by a broad absorption band located around 7000 cm^{-1} already reported by Bois and Pinard [1]. This absorption band is assigned to an intracentre absorption ($^5T_2 \rightarrow {}^5E$) of the Cr^{2+} levels and the corresponding zero-phonon lines have been recently observed by Clerjaud, Hennel and Martinez [2].

We report here on the behaviour of this absorption band under hydrostatic pressure at liquid nitrogen temperature.

The samples were prepared by diffusion of chromium into n-type GaAs doped with tellurium and silicon. Table 1 gives the conditions of preparation of these samples, which have been characterized by different procedures. Under diffusion some of the samples remain n-type and will be the subject of this report. The analysis of the impurity content has been performed by spark mass spectrometry and secondary ion mass spectroscopy (SIMS) described by Huber, Morillot and Merenda [3]. The results are that the donor

---

*On leave from the Institute of Experimental Physics, Warsaw University, Hoza 69, 00-681 Warszawa, Poland.
**Laboratoire de Luminescence II, Equipe de Recherche associée au CNRS, Université Pierre et Marie Curie, 4 Place Jussieu, 75230 Paris Cedex 05, France.
†Thomson-CSF-LCR, 91401 Orsay, France.

concentrations (Si and Te) are in the range of 10^{18} cm^{-3}, which corresponds to the free carrier concentration measured by Hall effect on Cr$_0$ sample. Besides these impurities, iron is found in the range of 5×10^{15} cm^{-3}, a result which correlates with the observation of Fe^{3+} in the EPR spectrum of the samples. Finally, the chromium concentration has been measured by SIMS and results are reported in Table 1. It is clear that if the chromium impurity compensates the donor impurities, it is not the only source of compensation. The annealing at high temperatures itself induces a non-negligible amount of compensation.

Table 1 Condition of diffusion of chromium-doped GaAs samples and results of their characterization. The chromium concentration of the Cr$_4$ sample has been estimated by the absorption cross-section

Sample label	Chromium diffusion		Carrier concentration at 77 K	Chromium concentration (SIMS)
	Temp (°C)	Time (days)		
Cr$_0$	—	—	n=10^{18} cm^{-3}	1.0×10^{15}
Cr$_4$	1130	1	n=1.5×10^{17} cm^{-3}	1.8×10^{17}
Cr$_6$	1100	3	n-type $\rho=10^3$ Ω–cm	9.0×10^{16}
Cr$_7$	1120	3	n-type $\rho=10^7$ Ω–cm	2.0×10^{17}

The dominant features observed in these samples by EPR measurements are that the Cr^{2+} signal does not vary under illumination with an infra-red laser line $(1.09\,\mu m)$ and, especially, never decreases within the experimental error (1%). This state is therefore stable and the main optical transition in which it is involved does not correspond to a charge transfer process. The optical absorption due to this state has been analysed in detail by Hennel *et al.* [4] and the cross-section of this absorption, σ, can be deduced when the chromium concentration is known. We found that σ for Cr$_6$ and Cr$_7$ is of the order of 1×10^{-17} cm^2 at the maximum of the absorption. This value has been used to calibrate the chromium concentration in other samples.

The pressure experiments were performed up to 10 kbar at 77 K using helium gas as a pressurizing medium and a special in-out system inside the high pressure, which allows the performance of absolute absorption measurements with a maximum uncertainty of 0.05 cm^{-1} for 2 mm thick samples. The whole assembly will be described elsewhere [5]. The pressure is purely hydrostatic and the absolute measurements of the absorption are reliable. When the pressure is applied to chromium-doped n-type samples, three types of behaviour are observed depending on the relative concentration of free carriers n and chromium centres N_{Cr}.

If n is negligible compared with N_{Cr}, the absorption band shifts towards high energies, with practically no change in the absorption strength [4].

When n is lower than N_{Cr}, but of similar order of magnitude (as in the case of Cr_4 sample) the observed behaviour is completely different (Figure 1). The strength of the intracentre absorption decreases and at the same time a very broad band extending over more than 3000 cm^{-1} increases with pressure. During the same experimental run, the resistivity of the sample was measured and increased very sharply; this is reported in Figure 2, together with the decreasing variation of the cross-section of the intracentre absorption.

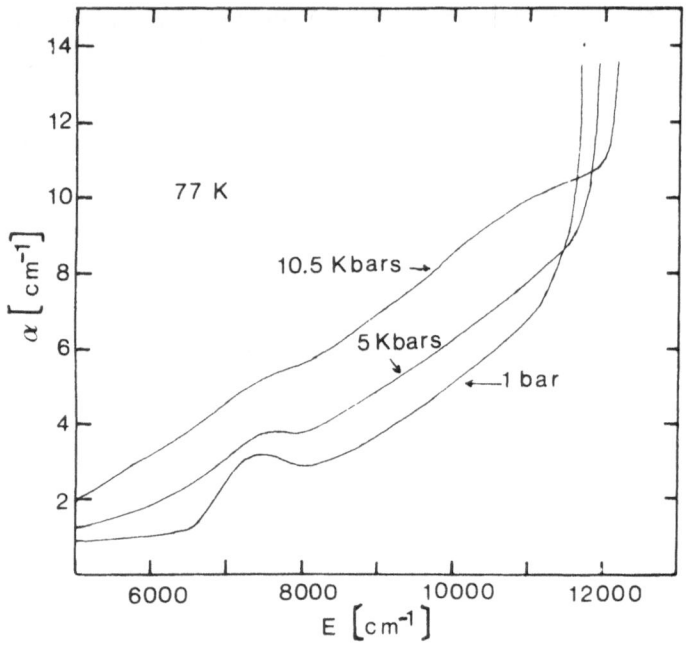

Figure 1   Absorption of the Cr_4 sample under pressure at 77 K.

These two experimental facts can be interpreted as follows: the intracentre absorption is the optical 'signature' of the Cr^{2+} charge state. Its decrease means that this charge state progressively disappears under pressure. The abrupt change in the resistivity must be, at least in part, connected to the trapping of free carriers in the sample by some deep acceptor level. Considering these two facts together leads us to believe that, under pressure, the Cr^{2+} level (in fact a one electron trap) is progressively converted into a Cr^{1+} level (a two electron trap). This transfer is possible when the Fermi level for the carriers overlaps the Cr^{1+} level, an overlap induced by pressure which increases the gap at a rate of $11 \times 10^{-6} \text{ eV bar}^{-1}$ compared to that for the impurity level $\sim 2 \times 10^{-6} \text{ eV bar}^{-1}$. So, at 1 bar and 77K the Cr^{1+} level should be well inside the conduction band.

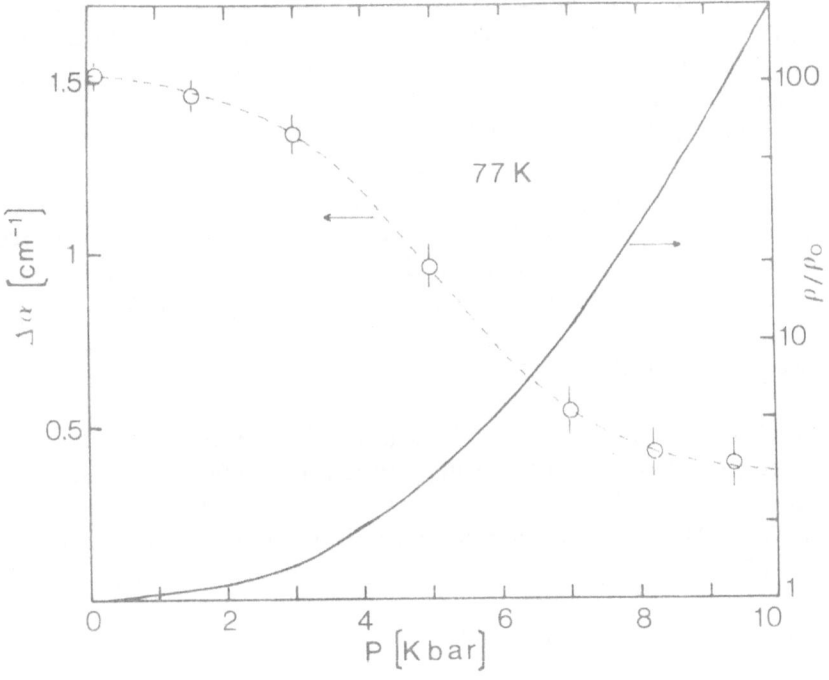

Figure 2   Variation of the intracentre absorption (Δa) and of the relative resistivity ρ/ρ_0 of the Cr$_4$ sample under pressure at 77 K.

These effects are reversible and do not depend on the way in which a given pressure is reached, i.e. by increasing or decreasing it. The fact that some residual Cr^{2+} charge state remains means that at high pressure all electrons are pratically trapped and, since n<N$_{Cr}$, some Cr^{2+} are not converted into Cr^{1+}. On the other hand, the broad absorption band which appears should be assigned to transitions involving the Cr^{1+} level. This transition is probably a photoionization process involving the p mixed part of the wavefunction of the conduction band due to its non-parabolicity.

In the third case, when n>N$_{Cr}$, the observed behaviour follows logically from the preceding one [4]. The cross-section of the intracentre absorption now decreases to zero whereas the resistivity of the sample first increases and then saturates. In this case all chromium centres have been converted into Cr^{1+} and some free carriers remain, giving the residual conductivity.

We can try, using a simple model and neglecting a change in the mobility, to locate the Cr^{1+} level inside the conduction band. This is probably questionable because a careful inspection of the experimental data shows that the change in the resistivity is significantly larger than a simple trapping process would predict. The mobility variation is therefore probably also non-linear and significant. We can, however, tentatively give for the Cr^{1+} level an energy of 70±20 meV above the conduction band at 77 K and 1 bar.

A more precise value and a detailed analysis of the effects involved by such a trapping require Hall effect measurements under pressure, which were not possible with the optical high pressure cell but which are now in progress using another apparatus.

It is to be noted that the results of these experiments are not consistent with the observation of the 'Cr^{1+}' EPR spectra [6, 7] in semi-insulating samples because a number of electrons of the order of 10^{19} cm^{-3} is necessary in order to obtain a cross-over of the Fermi level with the Cr^{1+} level. The recent attribution of the g=1.993 and 115 Gauss linewidth signal to the hole trap Cr^{4+} (3d^2)[8] remove that contradiction.

ACKNOWLEDGEMENTS

We are grateful to Mr G. Poiblaud (RTC Co., France) who has kindly provided the n-type doped GaAs ingot used for chromium diffusion, and to U. Kaufmann and J. Schneider for making their results available prior to publication. This work was supported in part by DGRST, under contract No. 78-7-0307.

References

1. Bois, D. and Pinard, P. (1974). *Phys. Rev. B*, **9**, 4171
2. Clerjaud, B., Hennel, A.M. and Martinez, G. (1980). *Solid St. Commun.*, **33**, 983
3. Huber, A.M., Morillot, G. and Merenda, P. (1979). *J. Microsc. Spectrosc. Electron.*, **4**, 493
4. Hennel, A.M., Szuszkiewicz, W., Balkanski, M., Martinez, G. and Clerjaud, B. (1980). To be published
5. Dahan, N., Barrau, B., Pinzutti, G. and Martinez, G. (1980). To be published
6. Kaufmann, U. and Schneider, J. (1976). *Solid St. Commun.*, **20**, 143
7. Krebs, J.J. and Stauss, G.H. (1977). *Phys. Rev. B*, **16**, 971
8. Kaufmann, U. and Schneider, J. (1980). *Appl. Phys. Lett.*, **36**, 747

OPTICAL AND THERMAL PROPERTIES OF Cr IN GaAs

T. JESPER, B. HAMILTON and A.R. PEAKER
Department of Electrical Engineering and Electronics,
UMIST, PO Box 88, Manchester M60 1QD, UK

Abstract

A variety of characterization techniques has been applied to the GaAs:Cr system and in particular the photoionization cross-sections, electron capture cross-section and thermal emission rates of the dominant chromium-related trap in GaAs have been measured. The temperature dependence of the photoionization thresholds can be accounted for by the temperature variation of the band gap. The hole emission rate is much greater than that for electrons and even in n-type material the DLTS signal is dominated by hole emission.

Introduction

Various workers have produced evidence which suggests that chromium may exist in three stable charge states: Cr^{3+} ($3d^3$), Cr^{2+} ($3d^4$) and Cr^{1+}($3d^5$), which implies that chromium can act as a double acceptor state, removing two electrons from the conduction band in n-type material. Brozel *et al.* [1] arrived at this conclusion through a detailed study of compensation in chromium-doped ingot material; optically excited ESR measurements, in which electrons are pumped between the bands and the various charge states, have led Kaufman and Schneider [2], Krebs and Stauss [3] and Kaufman and Koschel [4] to the same conclusion. Recent photocapacitance studies by Szawelska and Allen [5] indicated that two acceptor states were present and that the singly charged acceptor was characterized by a threshold energy for the transition $Cr^{2+} \rightarrow Cr^{3+}$ (trap to conduction band) of 0.74 eV, and for $Cr^{3+} \rightarrow Cr^{2+}$ (valence band to trap) of 0.81 eV, both measured at about liquid nitrogen temperatures. The doubly charged acceptor was characterized by a threshold energy of ~0.46 eV for the $Cr^{1+} \rightarrow Cr^{2+}$ transition.

In this work we have applied two deep level techniques, photocapacitance and DLTS, to the same set of samples in an attempt to rationalize the optical and thermal behaviour of chromium-associated defect states. We have found that while at least two chromium-associated states exist, the data do not imply that these are different charge states of the same centre. Furthermore, in most samples measured the dominant chromium-associated state was accompanied by a second state which is both close in energy and which seems to interact with the main chromium state. Analysis of samples in which this second state is below the detection limit shows that the measured optical and thermal properties, while superficially disparate, are in fact consistent when detailed properties such as capture cross-sections and the variation of binding energy with temperature are taken into account.

Survey measurements

The DLTS spectra recorded for n-type material fall into two groups (Figure 1). All spectra contained a low energy electron trap (EM3); the electron emission rate for this trap was found to be electric field dependent, and the concentration of the trap was always small, typically 1% of the concentration, determined optically, of trap EM1. Extrapolated to zero electric field, the thermal activation energy for this trap was found to be: $E_T=(E_C-0.32)$ eV. All spectra showed high temperature features. The majority of samples yielded two closely spaced peaks EM1 and EM2 (upper trace in Figure 1). The discrimination between these peaks varied from sample to sample and, in some cases, EM2 appeared only as a shoulder on EM1. In some samples only the highest temperature peak, EM1, was present. The concentration of trap EM2 varied from 4×10^{14} cm^{-3} to below the detection limit over the range of samples measured, while the concentration of EM1, calculated from the DLTS peak heights, varied between $\sim4\times10^{13}$ cm^{-3} and 4×10^{14} cm^{-3}.

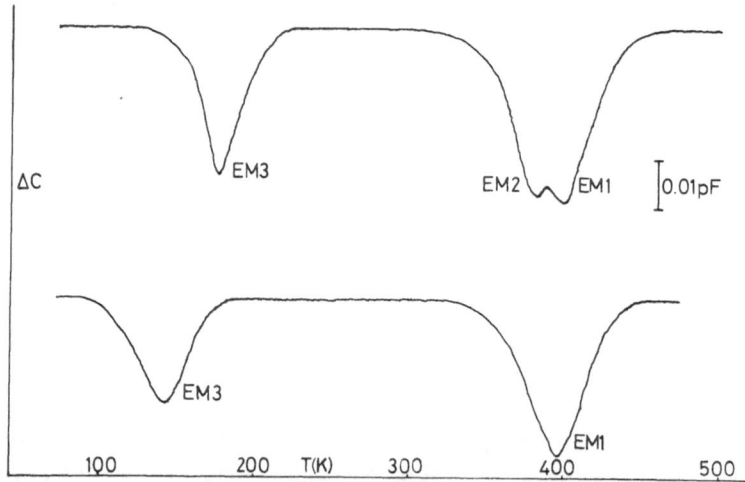

Figure 1 DLTSscans of n-type GaAs:Cr showing spectra obtained for material containing both EM1 and EM2 and material containing only EM1. The samples in the upper and lower traces had $(N_D-N_A)=1.9\times10^{16}$ cm^{-3} and 6.0×10^{16} cm^{-3}, respectively; standing capacitance $C(0)=34.9$ pF and 95.0 pF, respectively. The emission rate window is 10.2 s^{-1} and the samples were held at -1.00 V reverse bias. The shift in position for the peaks labelled EM3 is due to the electric field.

The thermal activation energies for trap EM1 derived from the DLTS data fell into the range $0.82\,\mathrm{eV} < E < 0.90\,\mathrm{eV}$.

The photocapacitance response of all the n-type chromium-doped samples measured at ~100 K showed a large feature with a threshold near 0.7 eV (Figure 2). This dominant photocapacitance effect was present only in chromium-doped samples.

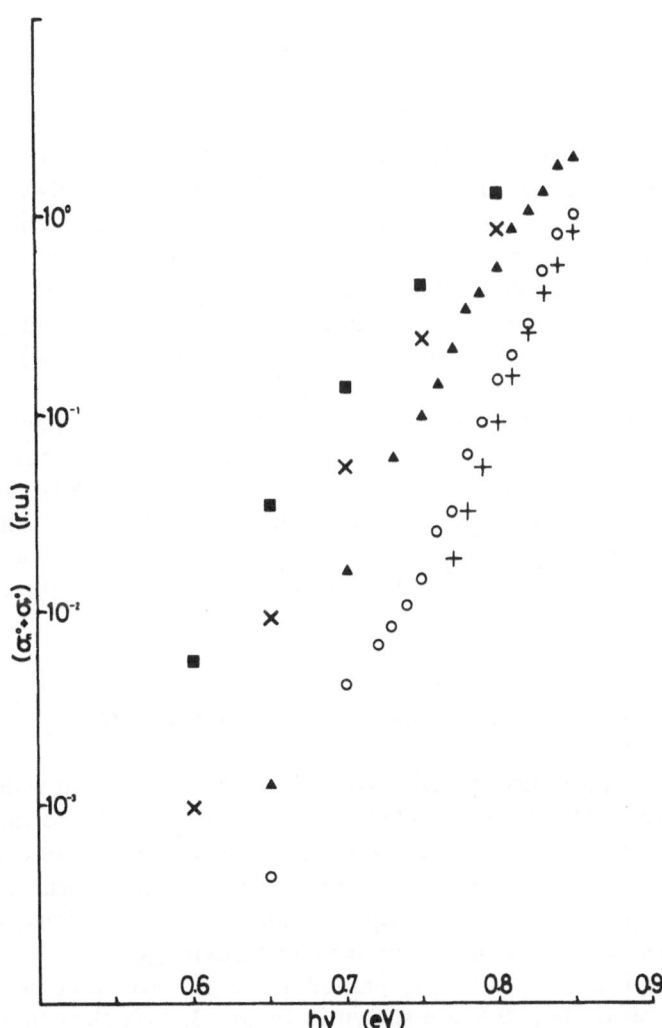

Figure 2 Photoionization cross-sections $(\sigma_n^0 + \sigma_p^0)$ for a typical n-type sample for a range of temperatures. $(+)\,T = 49$ K; open circles $= 93$ K; triangles $= 150$ K; crosses $= 210$ K; squares $= 270$ K.

The concentration of the trap EM1 in the range of n-type samples was estimated from the photocapacitance data using in each case the maximum amplitude of ΔC, the optically induced capacitance change. The final steady state electron occupancy of the trap, f_T, is obtained from the optical electron and hole emission rates, e_n^0 and e_p^0: $f_T = e_p^0 / (e_n^0 + e_p^0)$. Clearly, the maximum amplitude for ΔC gives the true trap concentration only when $e_n^0 \gg e_p^0$; in the case of trap EM1 the maximum observed ΔC occurred at $h\upsilon = 0.85$ eV, when e_p^0 cannot be zero, and the trap not completely empty of electrons. The estimated concentrations, which must then represent an underestimate, fell into the range:

$$5 \times 10^{14} \text{ cm}^{-3} < N_T < 8 \times 10^{15} \text{ cm}^{-3}$$

It is clear from these survey methods that traps EM1 and EM3 are associated with chromium; neither was observed in control samples. Trap EM2 may be a chromium-associated defect but is not always present in chromium-doped samples. The concentration of EM1, determined optically, was consistently about 100 times greater than EM3 and so it is unlikely that both are different charge states of the same defect and, of the two, EM1 is the more likely to be the chromium deep acceptor state, as a trap only 0.32 eV below E_c cannot produce the well-known semi-insulating (SI) behaviour of GaAs:Cr.

However, two major problems are apparent when an attempt is made to reconcile the photocapacitance and DLTS data for trap EM1. Firstly, the calculated binding energies are different: $E_{opt} \sim 0.7$ eV and $E_{th} \sim 0.9$ eV, both measured with respect to the conduction band. Secondly, the estimated concentrations, N_T, of trap EM1 are greatly at variance: $10^{-1} > N_{T_{DLTS}}/N_{T_{opt}} > 10^{-2}$. In the remainder of this paper, these differences are investigated.

Temperature dependence of deep trap parameters (EM1)

The photocapacitance and DLTS measurements are made at widely differing temperatures (100 K and $\sim$420 K, respectively). In order to rationalize the two, it is important to have information on the temperature variation of those deep level parameters which influence the optical and thermal transition rates, namely, the optical emission cross-sections, the thermal capture cross-sections and the electron binding energy.

The optical cross-sections, plotted on a relative scale and measured between 50 K$<$T$<$270 K are shown in Figure 3. It is clear that the overall emission rate is considerably increased at the higher temperatures. Thermal emission was insignificant for the range of measurements shown. In a purely qualitative sense, this sort of behaviour is expected when phonon interactions are strong, and such defects may exhibit multiphonon activated capture cross-sections [7].

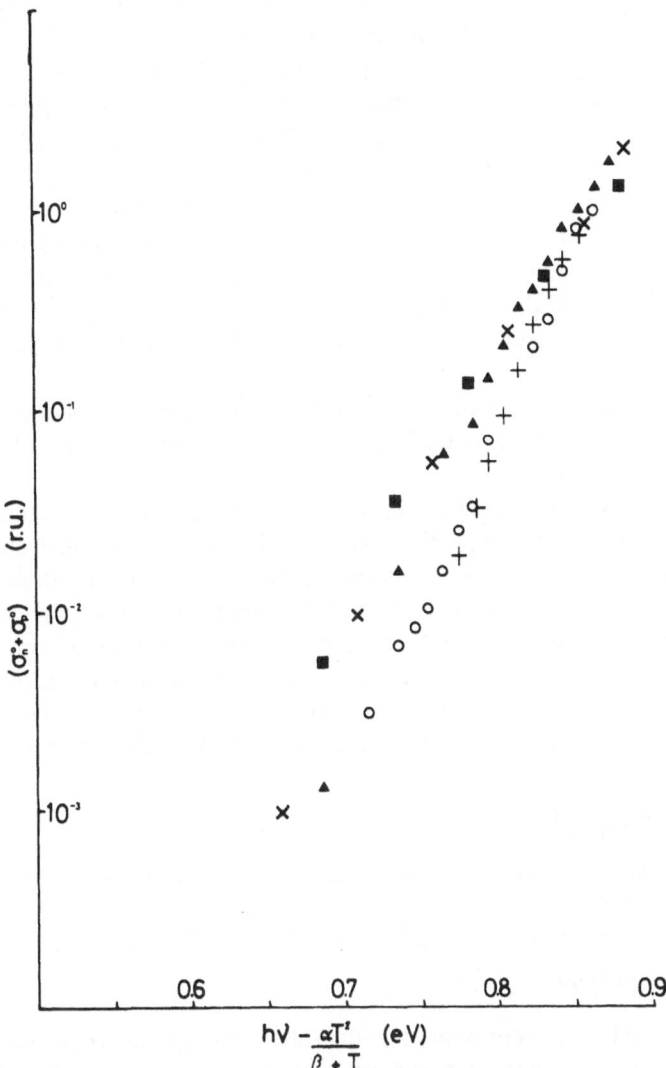

Figure 3 Photoionization cross-section $(\sigma_n^0 + \sigma_p^0)$ plotted against photon energy with a correction for the temperature dependence of the band gap included. Symbols as in Figure 2.

In order to investigate this possibility, the electron and hole capture cross-sections of trap EM1 were measured. The electron capture cross-sections were measured using two methods. In the high temperature regime $(T > 300\,K)$ the attenuation of the DLTS peak observed as the filling pulse is shortened [8] was used. At lower temperatures, when DLTS becomes impracticable, the cross-section was measured by (partially) removing electrons from the trap by photoionization, and then measuring the capacitance decay as the traps are gradually refilled by a series of short (constant width) forward bias pulses.

The DLTS derived cross-sections agree with those measured by the optically emptying and pulse filling methods. In samples which contained only EM1 in their high temperature DLTS spectra, the measured values of σ_n were 10^{-20} cm^2 < σ_n < 5×10^{-20} cm^2, for the range 450 K>T>50 K. Thus, σ_n is temperature independent within this range. The hole capture cross-section of EM1 was measured using the optical emptying and pulse filling methods in p-type material. The values obtained were: $\sigma_p \sim 3\times10^{-17}$ cm^2 with little temperature dependence. This behaviour is in close agreement with that reported by Lang and Logan [9] for a dominant hole trap in chromium-doped n-type material.

In the case of the isolated EM1 trap, it appears that strong phonon interaction is not the explanation for the strong temperature variation of the optical emission rates.

An alternative explanation may lie in the temperature variation of the energy depth of the trap. For example, if the level moved closer to the conduction band edge as T increased, then at a fixed photon energy the observed emission rate would increase with increasing T because the transition is further above threshold. If, as a first approximation, it is assumed that the level is pinned to the valence band, then the level movement would be accounted for by the variation of the band gap with temperature. In this case the electron binding energy $(E_c - E_T)$ would be replaced by the expression:

$$E_{c_0} - \frac{\alpha T^2}{\beta + T} - E_T$$

where E_{c_0} is the conduction band edge energy at absolute zero, $\alpha = 5.4\times10^{-4}$ and $\beta = 204$ [10].

By plotting emission rate against a modified photon energy:

$$(h\upsilon - \alpha T^2)/(\beta + T),$$

the energy axis then represents an equivalent energy (above threshold) for all temperatures. The data, in this form, are shown in Figure 3, which indicates that this simple approximation does bring about reasonable agreement over the whole temperature range.

DLTS analysis

The magnitude of the deep level parameters discussed above can now be used to predict the thermal release of the trap EM1. Since the trap is close to the middle of the gap, both thermal electron and hole emission should be taken into account, and the total emission rate, τ^{-1} is in the general case:

$$\tau^{-1} = e_n + e_p = \sigma_n N_c v_e \exp(-\{E_c - E_T\}/kT) + \sigma_p N_v v_h \exp(-\{E_T - E_v\}/kT)$$

where the symbols have their usual meaning.

Consider the electron emission term; by taking our value of $\sigma_n \sim 2\times10^{-20}$ cm^2, and inserting the idea that E_T is pinned to the valence band, e_n can be written:

$$e_n = 4.6T^2 \exp -\left(\frac{E_{c_0}-(\alpha T^2/(\beta+T))-E_T}{kT}\right)$$

The effect of the temperature dependent term in the hole emission rate can be expressed in a similar form, and using $\sigma_p \sim 3\times10^{-17}$ cm^2, we obtain

$$e_p = 3.7 \times 10^4 T^2 \exp -\left(\frac{E_T-E_v}{kT}\right)$$

where the energy term is independent of temperature.

In order to make a quantitative comparison of e_n and e_p some value of $E_{c_0}-E_T$ must be chosen, which then fixes E_T-E_v. The choice of $E_{c_0} - E_T$ is best made from the low temperature photoionization threshold, and a value of 0.72 eV was chosen.

The ratio of electron to hole emission rates in the middle of the DLTS temperature range (430 K) becomes:

$$e_n/e_p \text{ (at } 430\,\text{K}) = 0.08.$$

The hole emission rate is more than ten times faster than the electron emission rate. The implication is that even though EM1 is *detected* as an electron trap in n-type material by virtue of some net electron emission during the DLTS experiment, its thermal release properties are in fact dominated by the hole emission process. In essence, the large values of σ_p and N_v (compared to σ_n and N_c) and the proximity of the level to the centre of the gap produce a large hole emission rate in the temperature range used for DLTS. The energy derived from the DLTS activation plot must therefore reflect the thermal binding energy for the hole. We estimate this to be 0.8 eV but this choice is obviously subject to errors. The important point is that the activation plot should, in principle, give a larger energy than the electron binding energy since the level is in the upper half of the gap. The observed activation energies lay in the range 0.82 eV $<$ E $<$ 0.90 eV, and so our estimate of 0.8 eV is close to the lower bound of the experimental range.

The small 'thermal concentration' of EM1 now follows very simply: because $e_p \gg e_n$ the steady state occupancy f_T at the end of the thermal release capacitance transient is close to one, i.e. only a small fraction of the traps, on average, lose an electron. Taking the 'optical concentration' to be the true concentration:

$$N_{T_{DLTS}}/N_{T_{opt}} = 1 - f_T = e_n/(e_n + e_p) \simeq 7\times10^{-2} \qquad \text{(at 430 K).}$$

The range of concentration ratios observed in the whole of this work was:

$$10^{-1} > N_{T_{DLTS}}/N_{T_{opt}} > 10^{-2}$$

and the estimate of 7×10^{-2} is reasonable in view of the errors involved.

Conclusions

Three traps have been detected in chromium-doped GaAs. Of these, a trap at $E_v + 0.8$ eV (EM1) appears in all samples in the largest concentration. It is thought that this state is pinned to the valence band, that phonon interactions are not important in its optical transitions. The hole capture cross-section of EM1 is about three orders of magnitude greater than its electron capture cross-section and both are largely independent of temperature provided that a second deep trap, EM2, is not present.

The DLTS response of EM1 is dominated by hole emission. This leads to a larger activation energy than that expected for electron emission, and also to a fractional emptying of the trap and an apparently reduced concentration compared with the photocapacitance experiments.

ACKNOWLEDGEMENTS

We gratefully acknowledge materials supply from RSRE (Baldock) and the Plessey Co., and many useful discussions with Dr D.R. Wight (RSRE) and Drs J.W. Allen and H.R. Szawelska of St Andrew's University. One of us (T.J.) gratefully acknowledges financial support from an SRC Studentship. This work has been carried out with the support of Procurement Executive, Ministry of Defence.

References

1. Brozel, M.R., Butler, J., Newman, R.C., Ritson, A., Stirland, D.J. and Whitehead, C. (1978). *J. Phys. C: Solid St. Phys.*, **11**, 1857
2. Kaufman, U. and Schneider, J. (1976). *Solid St. Commun.*, **20**, 143
3. Krebs, J.J. and Stauss, G.H. (1977). *Phys. Rev. B*, **15**, 17
4. Kaufman, U. and Koschel, W.H. (1978). *Phys. Rev. B*, **17**, 2081
5. Szawelska, H.R. and Allen, J.W. (1979). *J. Phys. C: Solid St. Phys.*, **12**, 3359
6. White, A.M., Day, B. and Grant, A.J. (1979). *J. Phys. C: Solid St. Phys.*, **12**, 4833
7. Henry, C.H. and Lang, D.V. (1977). *Phys. Rev. B*, **15**, 989
8. Lang, D.V. (1974). *J. Appl. Phys.*, **45**, 3023
9. Lang, D.V. and Logan, R.A. (1975). *J. Elect. Mat.*, **4**, 1053
10. Aspnes, D.E. (1976). *Phys. Rev. B*, **14**, 5331

OBSERVATION OF VERY DEEP LEVELS BY OPTICAL DLTS

B. DEVEAUD and B. TOULOUSE*
Centre National d'Etudes des Télécommunications (LAB/ICM/MPA),
22301 Lannion, France

Abstract

Current transient spectroscopy (OTCS) is known to be able to give the electrical characteristics of the traps in semi-insulating material, as classic DLTS is in n- or p-type semiconductors. The method is limited by difficulties occurring for very deep traps. We explain here the difficulties — mainly the occurrence of negative transients — and set up an experimental procedure to interpret the data. Our first results show that, in Bridgman GaAs:Cr, the HL1 and EL2 levels are not the only levels that are present. At comparable concentration, the HL1 trap can be detected but the EL2 trap cannot.

Electrical characterization of semi-insulating semiconductors is known to be difficult. Hurtes *et al.* [1] and Martin and Bois [2] have presented a technique, called OTCS, that allows determination of the parameters of the traps present in the material. Under certain conditions, they could determine their concentration. However, although the method can easily be used for levels that are not too deep, the levels near the centre of the gap, which are in fact the most interesting ones, give rise to difficulties in the interpretation of the data (mainly due to the presence of negative peaks).

We present here some theoretical considerations that explain the occurrence of negative peaks for very deep levels. This leads us to propose a new procedure for interpreting the OTCS experimental data.

Experimental technique

The technique that we have used is almost the same as that described in [1] and [2]. First, samples are chemically polished on both sides to remove preparation damage. Then Au—Ge ohmic contacts are alloyed on the back of the samples and a semi-transparent gold contact is evaporated on the front face. This contact has some rectifying characteristics: the direct current is about twice the reverse current. The sample thickness is typically $300\,\mu$m.

---

*Present address: INSA: 35031 Rennes, France.

The semi-transparent contact is illuminated by light pulses from a GaAs laser diode. The sample is polarized at 10 V and the current is amplified by a low noise d.c. current amplifier. The sample temperature can be set between 90 and 390 K. The whole process is controlled by an HP 9821 calculator: temperature regulation, light pulse command, current sampling at t_1 and t_2, averaging and plotting of the results.

In our configuration, the photocurrent was high compared to the dark current, except at high temperatures (320—360 K, depending on the sample conductivity). The light pulses were long enough (100 ms) to reach steady state at all temperatures. The laser wavelength is long enough (0.85 μm) for a wide region of the sample to be reached by light; this reduces any possible surface effect.

Basic equations

Most of the basic equations have already been derived [1, 3]. We recall here the most important points and derive a more general equation.

We deal with semi-insulating samples; in the dark we suppose that there are no free carriers, the occupancy $n_T(\infty)$ of a given trap of concentration N_T is thus:

$$n_T(\infty) = \frac{N_T}{1+(e_n/e_p)} \tag{1}$$

Under optical excitation, at the end of the light pulse

$$n_T(0) = N_T \left(1+\frac{e_n+\delta_p v_p \sigma_p}{e_p+\delta_n v_n \sigma_n}\right)^{-1} \tag{2}$$

where $e_n(e_p)$ is the electron (hole) emission rate, $\sigma_n(\sigma_p)$ the capture cross-section for electrons (holes), $\delta_n(\delta_p)$ the concentration of photo-created electrons (holes) and $v_n(v_p)$ the thermal velocity of electrons (holes). The current $i(t)$ created by the carriers at the time t is thus:

$$i(t) = Cq\frac{AW}{2}\left\{\mu_n e_n n_T(t) + \mu_p e_p \left(N_T - n_T(t)\right)\right\} \tag{3}$$

and the occupancy of the trap varies as follows:

$$\frac{dn_T}{dt} = -n_T(e_n + e_p) + N_t e_p$$

where A is the area of the contact and W is the 'active thickness'. The equation of the current transient

$$i(t) = Cq\frac{AW}{2}N_T(\mu_n e_n - \mu_p e_p)\left[\left(1+\frac{e_n+\delta_p \sigma_p v_p}{e_p+\delta_n \sigma_n v_n}\right)^{-1}-\left(1+\frac{e_n}{e_p}\right)^{-1}\right]\exp-(e_n+e_p) \tag{4}$$

Under conditions of strong excitation e_n and e_p can be neglected as compared to $\delta_p \sigma_p v_p$ and $\delta_n \sigma_n v_n$.

Consequences

If $e_n \gg e_p$ (or $e_p \gg e_n$), equation (4) reduces to [1, 2]:

$$i(t) = Cq(AW/2)N_T \mu_n e_n \exp(-e_n t) \quad (\text{or } Cq(AW/2)N_T \mu_p e_p \exp(-e_p t)) \quad (5)$$

and the data can be interpreted as in classic DLTS except that the maximum of the OTCS peak occurs for $e_n = 1/t_1$ (as soon as $t_2 > 4t_1$). Both electron and hole traps give rise to positive peaks.

If e_n and e_p are of the same order of magnitude, the transient time constant is $\tau = 1/(e_n + e_p)$ and negative peaks can occur. For example, a trap having:

$$\mu_p e_p > \mu_n e_n \quad \text{and} \quad \sigma_p v_p / \sigma_n v_n < e_n / e_p$$

that is to say

$$\sigma_p v_p / \mu_p < \sigma_n v_n / \mu_n \quad \text{and} \quad E_p < E_n + kT \ln\left(\frac{\mu_n^2 \gamma_n v_p}{\mu_p^2 \gamma_p v_n}\right) \quad (6)$$

gives rise to a negative peak. The γ are defined by the relationship

$$e_n = \gamma_n \sigma_n T^2 \exp(-E_n/kT)$$

(see [5] for terminology). Physically, this refers to a trap located in the lower half of the band gap, having a capture cross-section for electrons greater than the capture cross-section for holes. The symmetrical case, of course, also gives a negative peak.

We have computed the influence of the dark current/photo current ratio and it appears that if the heights of the peaks are greatly affected when that ratio becomes high, their positions are only slightly shifted towards lower temperatures.

Experimental results

We have focussed our attention on the very deep levels. The data are in general very complicated due to the presence of several peaks, some of them being negative. Some samples only exhibit one peak in the range 300—400 K so we began our work on such samples. Figure 1 shows OTCS spectra obtained on FS 20344 (Sumitomo) for three time constants ($t_1 = 10$, 40 and 200 ms). Assuming, as is usually done, $e_n \gg e_p$ (or $e_p \gg e_n$) equation (5) together with an Arrhenius plot of $T^2 i$ gives $E_T = 1.3$ eV and $\sigma_n = 3 \times 10^{-8}$ or $\sigma_p = 4 \times 10^{-9}$ cm². The whole spectrum cannot be fitted with these values, which have no physical meaning, and equation (5). A better fit is obtained if we assume $e_n \sim e_p$ and if we apply equation (4). Figure 1 shows the fit obtained for $E_n = 0.75$ eV, $E_p = 0.7$ eV, $\sigma_n = 2 \times 10^{-15}$ cm² and $\sigma_p = 5 \times 10^{-16}$ cm². We are not sure that the fitting parameters are unique. However, starting from known values for some traps, it is possible, by adjusting them, to obtain a good fit of very complex spectra (see Figure 2). In order to obtain the true parameters of the involved levels, another procedure is necessary.

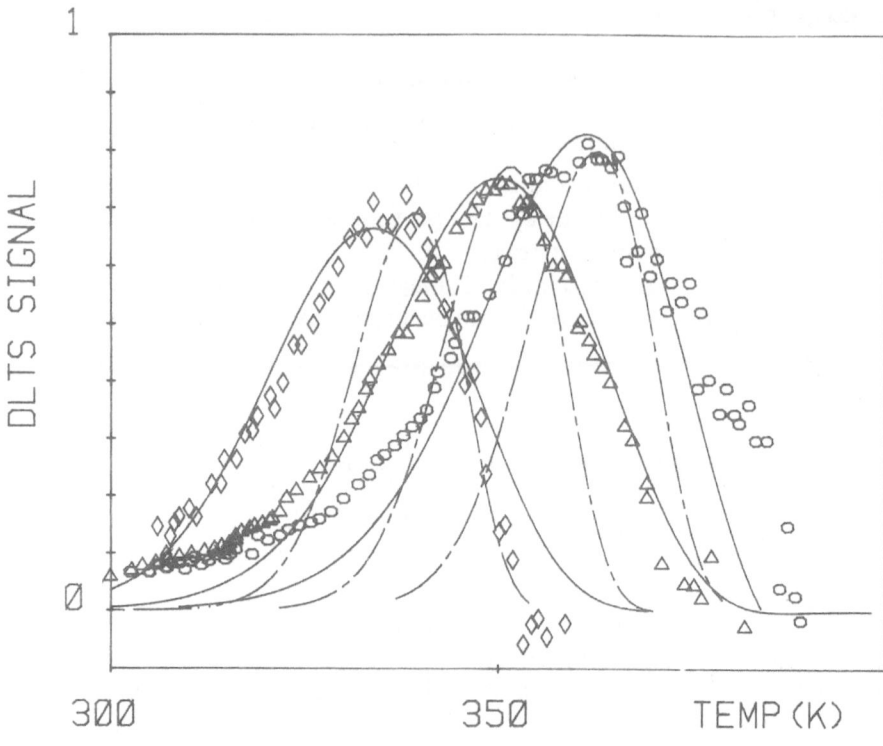

Figure 1   OTCS plot on FS 20344 for t_1=10 ms (circles), 40 ms (triangles) and 200 ms (diamonds). Dashed lines = spectra that would be obtained for a trap having E_n=1.3 eV and σ_n=3×10⁻⁸ (see text). Continuous lines = fit obtained for a trap with e_n close to e_p: E_n=0.75 eV, E_p=0.7 eV, σ_n=2×10⁻¹⁵, σ_p=5×10⁻¹⁶.

We record, at different temperatures, the current transient of the sample. This transient is then decomposed, into its different exponential components (see Figure 3), positive or negative, and each component is substracted before obtaining the next one. Such a decomposition gives a good representation of the transient (Figure 4). Let us suppose that $E_n > E_p$ + a few kT.

$$\ln(e_n + e_p) = \ln(\gamma_n \sigma_n) + \ln\left(1 + \frac{\gamma_p \sigma_p}{\gamma_n \sigma_n} \exp\left(\frac{E_n - E_p}{kT}\right)\right) - \frac{E_n}{kT} \tag{7}$$

We are dealing with centres where $e_n \sim e_p$ and so the term $\gamma_p \sigma_p / \gamma_n \sigma_n$ is very small. We thus have approximately

$$\ln(e_n + e_p) \sim \ln(\gamma_n \sigma_n) - \frac{E_n}{kT}. \tag{8}$$

An Arrhenius plot of each $1/\tau = e_n + e_p$ gives us the larger of E_n and E_p and the corresponding capture cross-section.

Starting from the values obtained by this method, it is then possible to fit the whole DLTS spectrum, two of the four values E_n, E_p, σ_n and σ_p being known. The fitting procedure allows us to distinguish clearly an electron trap from a hole trap. If the trap behaviour were complex, for example if the

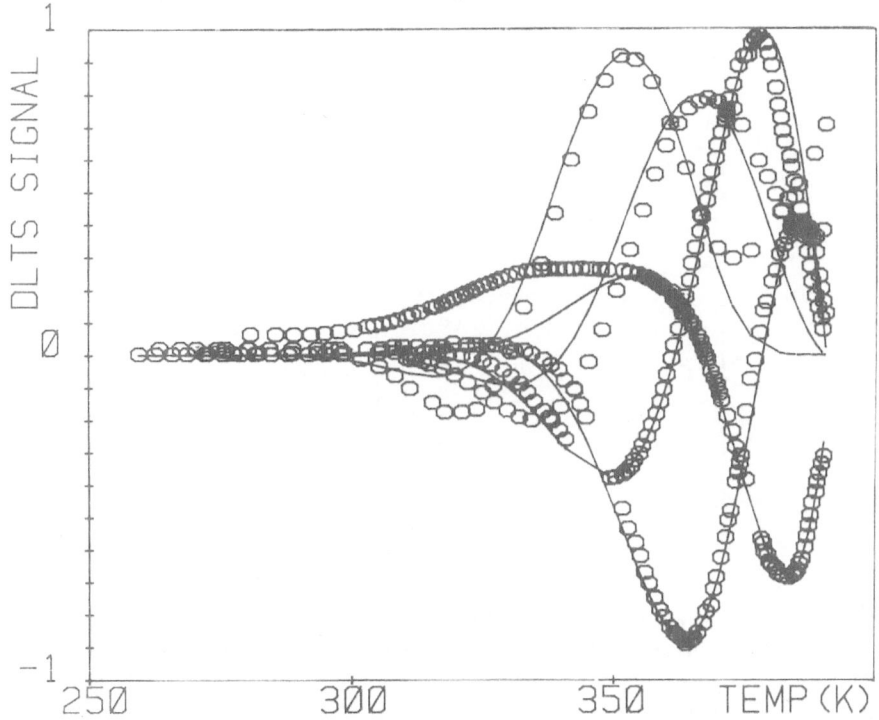

Figure 2   OTCS plot on RTC 512 for t_1=1, 10, 40, 200 and 800 ms (circles). The fit is obtained with four traps and shows that the negative peaks can be explained.

transient were negative, it would be possible to obtain the four parameters of the trap. An example of this type of fit is shown in Figure 5. The OTCS spectrum, obtained for t_1=2s is shown as well as the fit obtained, assuming the presence of two deep levels A and HL1. The presence of HL1 [4] is detected by fitting the transient. The parameters E_p and σ_p of A are given in the same way whereas E_n and σ_n are the values that fit the OTCS spectrum for t_1 varying from 10 ms to 8 s. The spectrum is the difference of the two peaks shown in Figure 5. It is clear that the simple Arrhenius plot procedure will not allow one to obtain the true parameters of HL1 for example.

The trap EL2, which should appear on the same temperature range, is not seen. Taking the values in the literature [5, 6] for EL2 and assuming a concentration equivalent to that of HL1, the peak has almost the same position but its intensity is 20 times lower. So, unless a very high concentration of EL2 were present in the material, this trap would not be detected by OTCS, using above gap excitation. Martin and Bois [2] have used a Nd:YAG laser and so, their $n_T(o)$ being different, they are able to see EL2.

A proper knowledge of the electrical parameters of all deep levels in semi-insulating material would therefore need several excitation procedures at different wavelengths, as well as a precise deconvolution of the current transients and of the OTCS spectra. However, we have shown that levels other than HL1 and EL2 were present, in comparable concentrations, near the centre of the gap of GaAs:Cr.

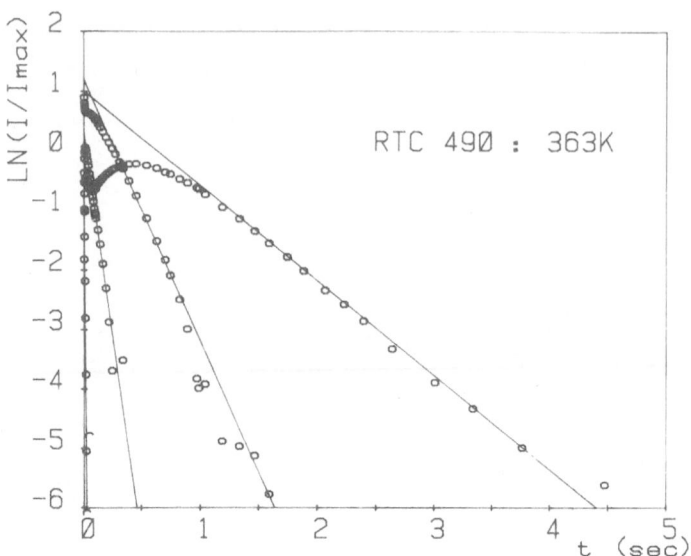

Figure 3 Current transient at 363 K for RTC 490 (circles). The long time part of the transient is fitted by an exponential decay, and then subtracted to give the second transient, which is then fitted . . . The fit, followed by an Arrhenius plot, gives four levels at: 1.1, 0.8, 0.8 and 0.65 eV.

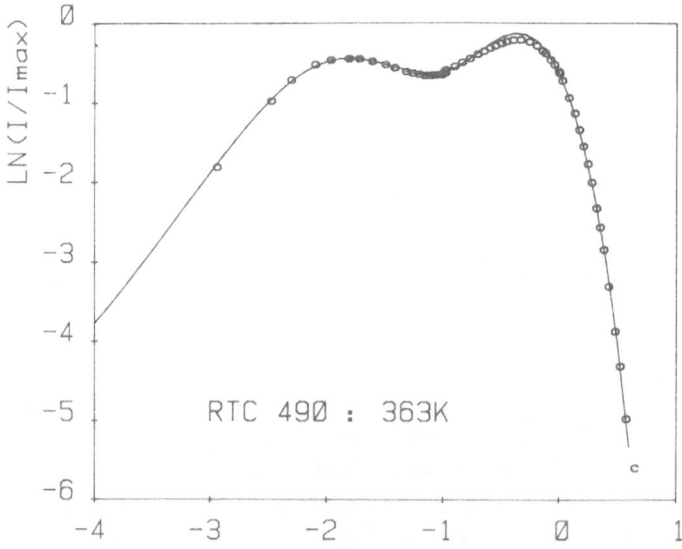

Figure 4 Current transient of RTC 490 at 363 K: experimental (circles) and theoretical sum (line) of the exponential components obtained following procedure described in Figure 3.

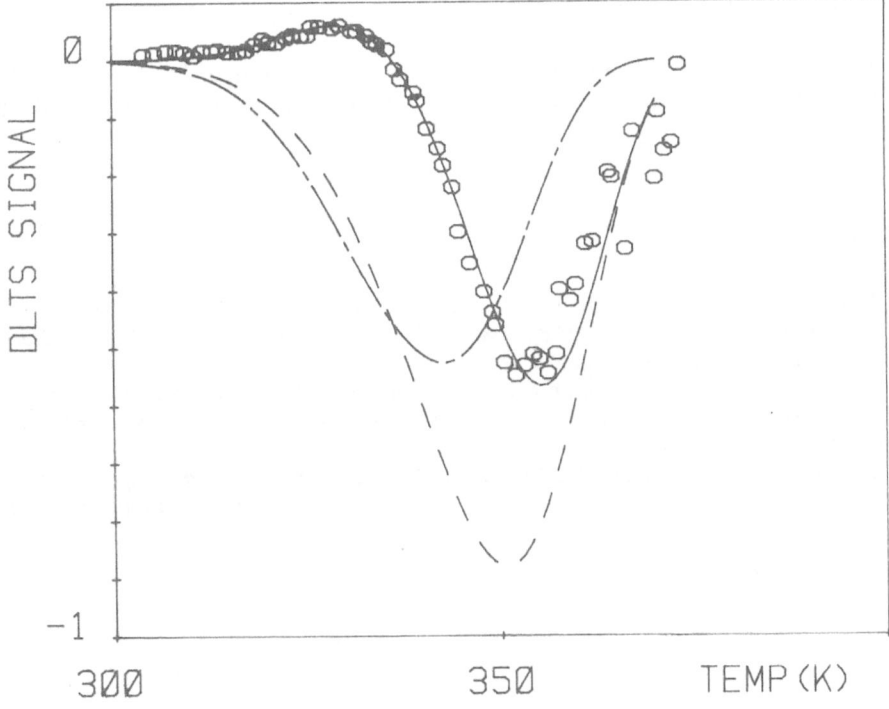

Figure 5 Fit of the OTCS spectrum for RTC 490, $t_1=2s$. Circles = experimental points; continuous line = best fit obtained by taking the values deduced from the transient deconvolution.

Trap A: $E_n=0.5eV$, $E_p=1.1eV$, $\sigma_n=3\times10^{-19}cm^2$, $\sigma_p=10^{-11}cm^2$ (—).
Trap HL1: $E_n=0.87eV$, $E_p=0.81eV$, $\sigma_n=1.3\times10^{-17}cm^2$, $\sigma_p=2\times10^{-15}cm^2$ (—·—).

Conclusion

Starting from some considerations on the transient current equations, we have set up a new procedure that allows interpretation of the OTCS spectra. The principle of the method is to fit first the current transients and then the OTCS spectra with the obtained values. We are able in this way to determine the four parameters, E_n, E_p, σ_n and σ_p for some of the levels. While the method, used with excitation close to the gap energy, cannot detect EL2, we have shown that several levels, besides HL1, are present near the middle of the band gap, the most surprising being a level with $E_n=0.5$ eV, $E_p=1.1$ eV, $\sigma_n = 3\times 10^{-19}$ cm^2, $\sigma_p = 10^{-11}$ cm^2. To summarize this work, the main characteristic of this type of sample seems to be the great number of levels, shallow as well as deep, in the band gap.

ACKNOWLEDGEMENTS

We wish to thank M. Pelous for his technical assistance, Dr A. Mircea for helpful discussions and M. Poiblaud (RTC) for providing most of the samples we have used.

References

1. Hurtes, C. Boulou, A., Mitonneau, A. and Bois, D. (1978). *Appl. Phys. Lett.*, **32**, 821
2. Martin, G.M. and Bois, D. (1978). *Conf. on Characterization Techniques for Semiconductor Materials and Devices,* Seattle, **78**, 32
3. Sah, C.T., Forbes, L., Rosier, L.L. and Tasch, A.F. (1970). *Solid St. Electron.*, **13**, 759
4. Martin, G.M., Mitonneau, A., Pons, D., Mircea, A. and Woodard, D.W. (1980). To be published in *J. Phys. C.: Solid St. Phys.*
5. Mitonneau, A., Mircea, A., Martin, G.M. and Pons, D. (1979). *Rev. Phys. Appl.*, **14**, 853
6. Bois, D., Chantre, A., Vincent, G. and Nouaillhat, A. (1978). *Inst. Phys. Conf. Ser.*, **43**, 295

PHOTOCONDUCTIVITY OF EPITAXIAL GaAs:Cr

M.A. AMATO, M.C. ARIKAN and B.K. RIDLEY
Department of Physics, University of Essex, Colchester CO4 3SQ, UK

Abstract

An investigation of photoconductivity in epitaxial GaAs:Cr under different illumination levels shows evidence of two levels with optical thresholds at 0.75 and 0.79 eV, and of a broadened trap distribution. Analysis of the spectral dependence of the photoresponse could be carried out, satisfactorily, with our simple model rather than with the Lucovsky model. For the 0.75 eV level we obtained a Huang-Rhys factor of 3, corresponding to a Franck-Condon shift of 0.09 eV and a thermal level at 0.66 eV.

The photoconductivity of an epitaxial sample of semi-insulating GaAs:Cr, grown by the alkyl method at RSRE, Baldock, UK, was measured at 296 K between photon energies of 0.6 and 1.4 eV. The sample was 14.2 μm thick and mounted on a glass substrate. The nominal concentration of chromium was 1×10^{16} cm^{-3}, and the conductivity in the dark, σ_0, at 296 K was 6.6×10^{-9} $(\Omega - \text{cm})^{-1}$.

Steady state measurements of the intensity dependence showed that the photoconductivity, $\Delta\sigma$, obeyed a fractional power law, namely, $\Delta\sigma \alpha I^\gamma$ (I=intensity) with $\gamma = 0.9$ and independent of wavelength. In terms of an exponentially broadened trap distribution, such an exponent implies a broadening of 225 meV, which is rather large [5].

Spectral photoconductivity curves were obtained by conventional techniques involving chopped monochromatic light at 13 Hz, but with the precaution that $\Delta\sigma/\sigma \ll 1$ for all but the weakest illuminated cases, but in no case did $\Delta\sigma/\sigma$ exceed 0.5. The response per photon varied with the intensity of steady background illumination. The latter was varied by using either a Ge or an Si filter together with neutral density filters. The result in the dark and the results with background illumination through an Si filter are shown in Figure 1. Those for the Ge filter lie between the dark curve and the weakest Si filtered illumination and, for clarity of presentation, are not shown. These latter curves all show a shoulder just below 0.8 eV, but this shoulder weakens at stronger background illumination levels and becomes invisible. This suggests that two levels are operative in the dark, one at about 0.75 eV and the other at about 0.79 eV, and that the latter ceases to be operative at high background illumination levels. There is also some structure apparent near 1.2 eV.

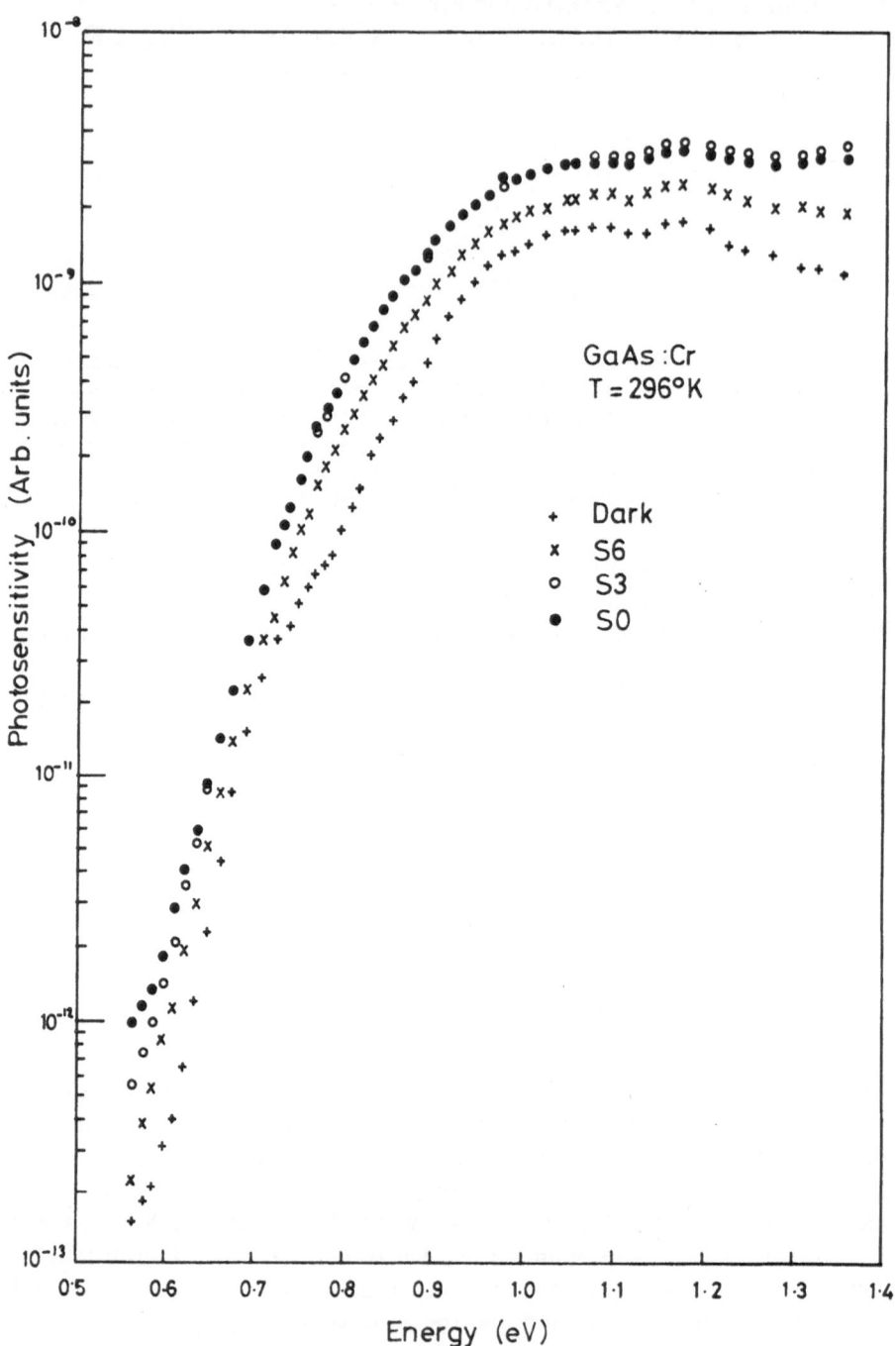

Figure 1 Photoconductivity per photon for various background illumination levels. (Continued on next page)

Figure 1 (contd)

	σ $(\Omega-cm)^{-1}$			$\dfrac{\Delta\sigma \text{ (at max.)}}{\sigma}$		
	I	II	III	I	II	III
Dark	6.50×10^{-9}	6.61×10^{-9}	7.43×10^{-9}	0.16	0.49	0.06
S6	2.70×10^{-8}	2.84×10^{-8}	3.79×10^{-8}	0.06	0.15	0.03
S3	1.63×10^{-7}	1.73×10^{-7}	2.45×10^{-7}	0.01	0.06	0.01
S0	1.23×10^{-6}	1.31×10^{-6}	1.86×10^{-6}	0.01	0.001	0.01

Region I: (0.50—0.70 eV), II: (0.70—1.05 eV) and III: (1.05—1.40 eV)

A plot of photosensitivity (change of conductivity per photon), P, versus conductivity at three photon energies is shown in Figure 2 where is shown, also, a tentative fit with a theory [1] involving a single type of carrier being captured by the midgap level, or levels, in the presence of a trap distribution of breadth 225 meV. The large value of $\Delta\sigma/\sigma$ at the lowest conductivities almost certainly accounts for the deviations between theory and experiment in that regime. Otherwise the fit is plausible. If the carriers are principally electrons, the deep level volume capture rate is about 4×10^{-6} cm^3 s^{-1} corresponding to a capture cross-section of about $2-10^{-13}$ cm^2, which is very large.

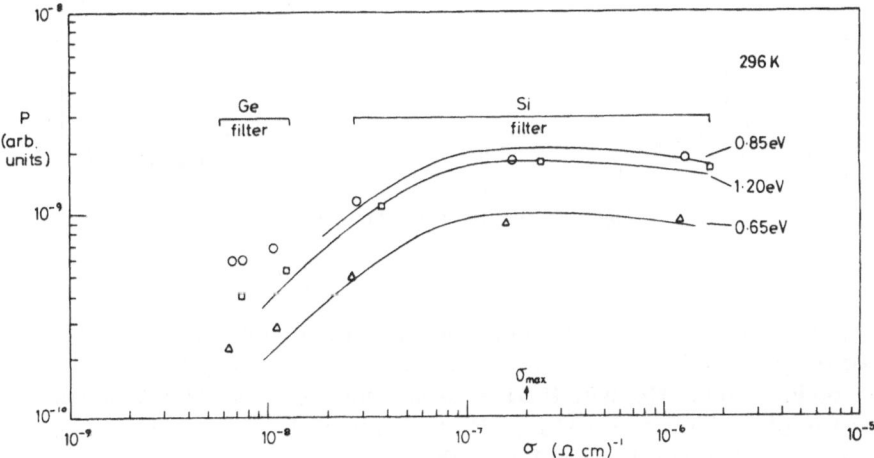

Figure 2 Photosensitivity versus conductivity. Theoretical curves are of the form $P=B^{8/9} (\sigma^2+\sigma^2_{max}/8)^{-1/2}$ corresponding to a trap broadening of 225 meV.

The spectral shape of the photosensitivity curve agreed better with a model of photoionization involving the excitation of the electron from a |s>—like, tight orbit to a |s>—like band [2, 3], rather than with a Lucovsky model [4]. The intercepts at low and high background illumination levels are at 0.79 and 0.75 eV, respectively (Figure 3).

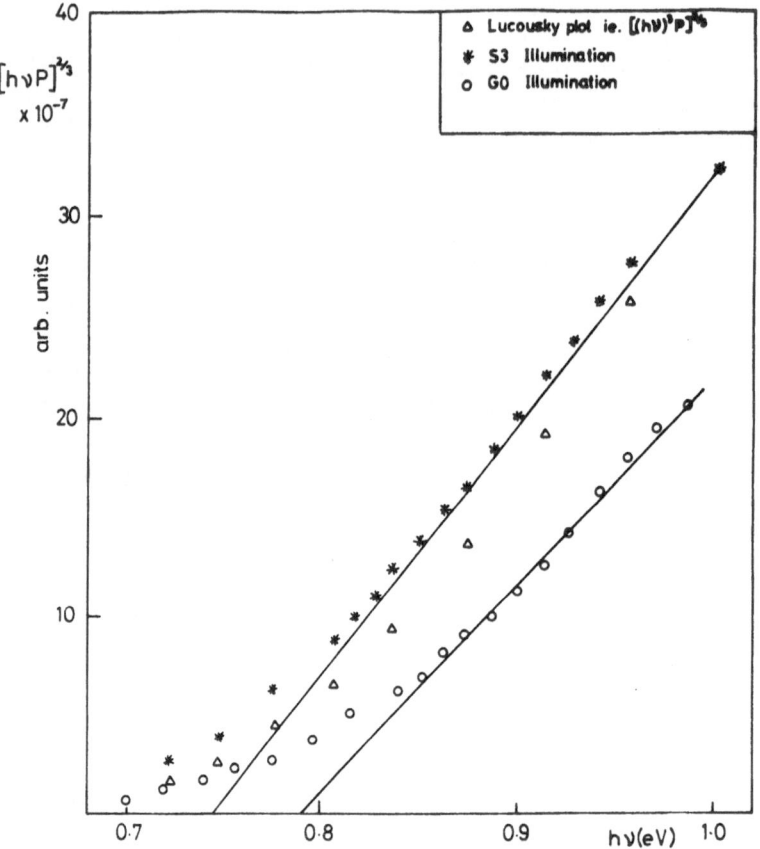

Figure 3   Spectral shape of photosensitivity.

Finally, an excellent fit of photoionization theory and experiment is shown for the 0.75 eV level in Figure 4. The theory assumes a |s>—|s> transition involving a tight orbit with thermal broadening described by a single phonon energy of 0.03 eV and a Huang-Rhys factor of 3, and taking into account non-parabolicity in the band by $k \cdot p$ theory. The thermal ionization energy of the 0.75 eV level is therefore 0.66 eV. The choice of phonon energy was based on group theoretic arguments which predict that only LA and LO phonons couple significantly to the electron in this case.

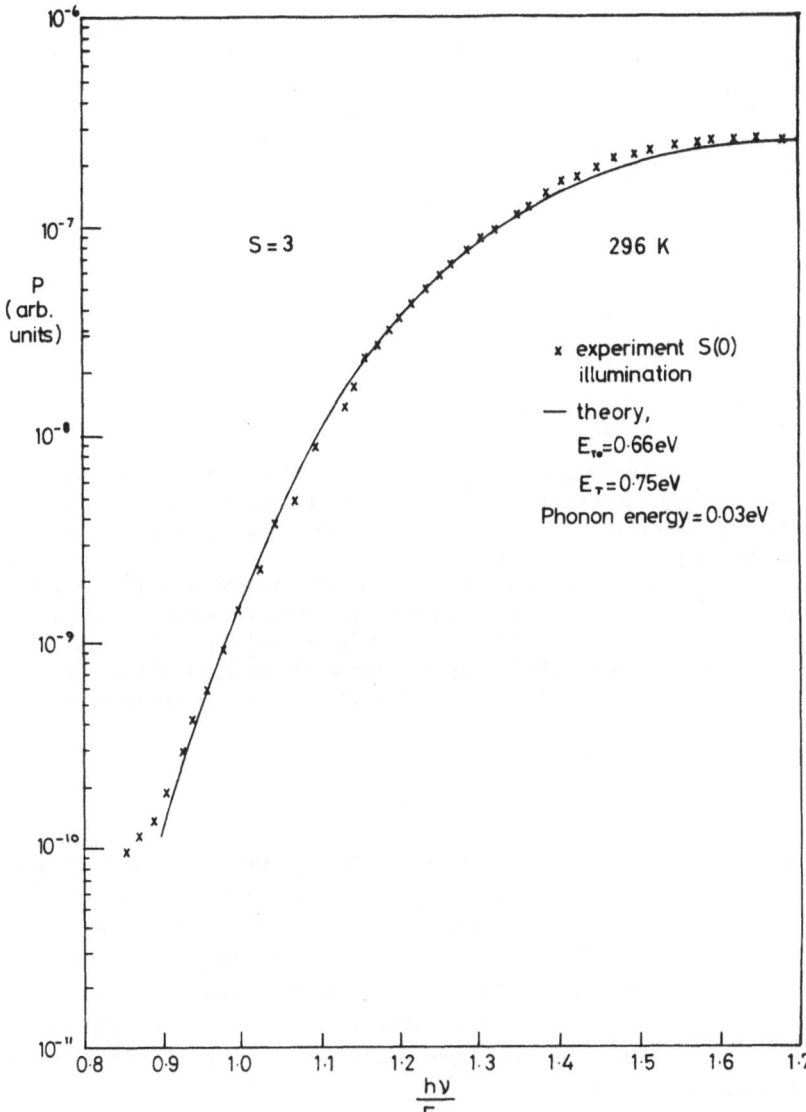

Figure 4 Fit of theoretical thermally broadened photoionization curve to experiment.

ACKNOWLEDGEMENT

The authors are indebted to MOD (CVD) for financial support.

References

1. Arikan, M.C., Hatch, C.B. and Ridley, B.K. (1980). *J. Phys. C: Solid St. Phys.*, in press
2. Ridley, B.K. (1980). *J. Phys. C: Solid St. Phys.*, in press
3. Amato, M.A. and Ridley, B.K. (1980). *J. Phys. C: Solid St. Phys.*, in press
4. Lucovsky, G. (1965). *Solid St. Commun.*, **3**, 299
5. Rose, A. (1963). *Concepts in Photoconductivity and Allied Problems.* Chichester; Wiley

A POSSIBLE INTERPRETATION OF THE 0.839 eV LINE IN GaAs:Cr

G. PICOLI, B. DEVEAUD and D. GALLAND*
Centre National d'Etudes des Télécommunications (LB/ICM/MPA),
22301 Lannion, France

Abstract

The often observed luminescence band at 0.839 eV GaAs:Cr is generally interpreted as an internal transition (^5E—$^5T_2$) of the Cr^{2+} ion. This interpretation has already been excluded and a new interpretation has been proposed by White. We present here another possible explanation.

The number of levels can be explained by a ^5E—^5E transition of a ^5D state in C_{3v} symmetry. Cr^{2+} (on a Ga site) coupled to a donor on As site or Cr^0 (interstitial) coupled to an acceptor on an arsenic site are more probable defects to explain qualitatively the 0.839 eV line and the associated 0.75 eV hump. The interstitial model gives rise to a possible interpretation for the 0.574 eV line and its satellite at 0.535 eV that we have resolved.

0.839 eV line

Lightowlers and Henry [1] describe the 0.839 eV line as being an internal transition between a set of seven levels (excited state) and a set of five levels (ground state). It has now been proved [2, 3, 9] that this line was not an internal transition (^5E—$^5T_2$) in Cr^{2+}. The two main reasons are (a) that the ground state, as detected by EPR, for Cr^{2+} [4] does not correspond to the ground state of the 0.839 eV line; and (b) that an absorption line, in agreement with EPR data, has been observed at 0.825 eV by Clerjaud, Hennel and Martinez [3].

A new interpretation was first proposed by White [2], in which the 0.839 eV line is an excitonic recombination at a deep isoelectronic trap. He presents two models, either a neutral donor-neutral chromium acceptor or an isolated neutral chromium centre. We propose here an alternative explanation in which the 0.839 eV line is an internal transition of a chromium atom in C_{3v} symmetry.

The 0.839 eV line is clearly related to chromium doping and the number of levels suggests symmetry other than T_d with some orbital degeneracy (no static Jahn-Teller effect) and high spin. The simplest way of obtaining another symmetry is to couple the chromium atom to a first neighbour

---

* Section de Résonance Magnétique, DRF-CENG, 85 X 38041 Cedex, France.

impurity, as proposed by White [2]. The electronic structure of the ion in the crystal is described, as usual, by the successive action of the crystal field (V_c) and of the spin-orbit coupling (λLS):

$$H = H_0 + V_c + \lambda LS,$$

H_0 being the free ion Hamiltonian (without spin-orbit coupling). The crystal field lifts the degeneracy of this level, giving states that belong to the irreducible representations of the C_{3v} point group. A sufficient number of levels can be obtained only for $3d^4 - {}^5D(Cr^{2+})$ or $3d^6 - {}^5D(Cr^0)$ configurations. In both cases the levels obtained are labelled for orbital states:

$$D_2^+ \rightarrow \Gamma_1 + 2\Gamma_3 \qquad (A_1 + 2E \text{ following Muppiken's notation}).$$

The spin-orbit term couples a spin S=2 with orbital states belonging respectively to the Γ_1 and Γ_3 representations, and gives the following splitting:

orbital state: $\Gamma_1(A) \rightarrow \Gamma_1 + 2\Gamma_3$ (three levels)
orbital state: $\Gamma_3(E) \rightarrow 2\Gamma_1 + 3\Gamma_3$ (seven levels)

A $3d^4$ ion would be Cr^{2+}, that is to say a substitutional chromium on gallium site having accepted one electron. A $3d^6$ ion would be a Cr^0 ion, that is to say an interstitial ion. In both cases a first neighbour impurity gives a C_{3v} symmetry. If the C_{3v} potential is only a perturbation on the T_d potential, we obtain the level structure shown in Figure 1.

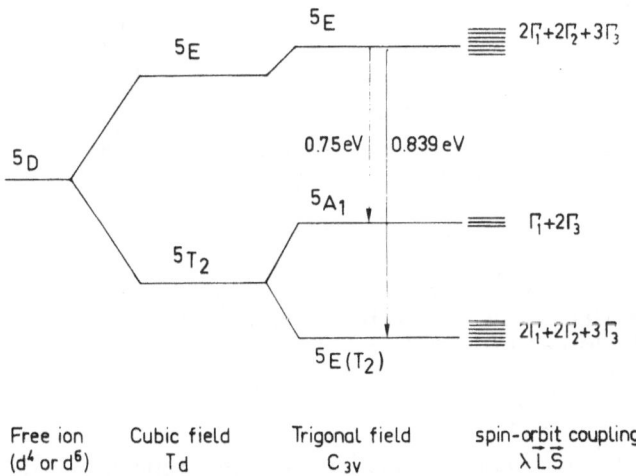

Figure 1   Electronic structure of a Cr^0 or Cr^{2+} ion in C_{3v} symmetry. The sign of the T_d crystal field and the order of the levels is as discussed in the text.

This model explains qualitatively the high degeneracies observed in luminescence and absorption. Furthermore, the 0.75 eV hump, always present with the same amplitude ratio to the 0.839 eV line, can be explained by a $^5E \rightarrow ^5A_1$ transition (Figure 2). The order of the three levels ($^5E, ^5A_1, ^5E$) that allows such an interpretation is consistent, for Cr_{Ga}^{2+} with coupling to a positive charge on an arsenic site (oxygen or any other donor substitutional for arsenic). For interstitial Cr^0, we believe that there are no bonds (made with sp³ hybridization). This should give an inversion between ^5E and $^5T_2$ with respect to the classic situation for a 3d⁶ ion. Such an explanation has already been used for the EPR spectra of some transition metals on interstitial sites in silicon [5]. The $^5E, ^5A_1, ^5E$ order is thus possible if the interstitial chromium is coupled to a negative charge on an arsenic site (silicon, oxygen, vacancy).

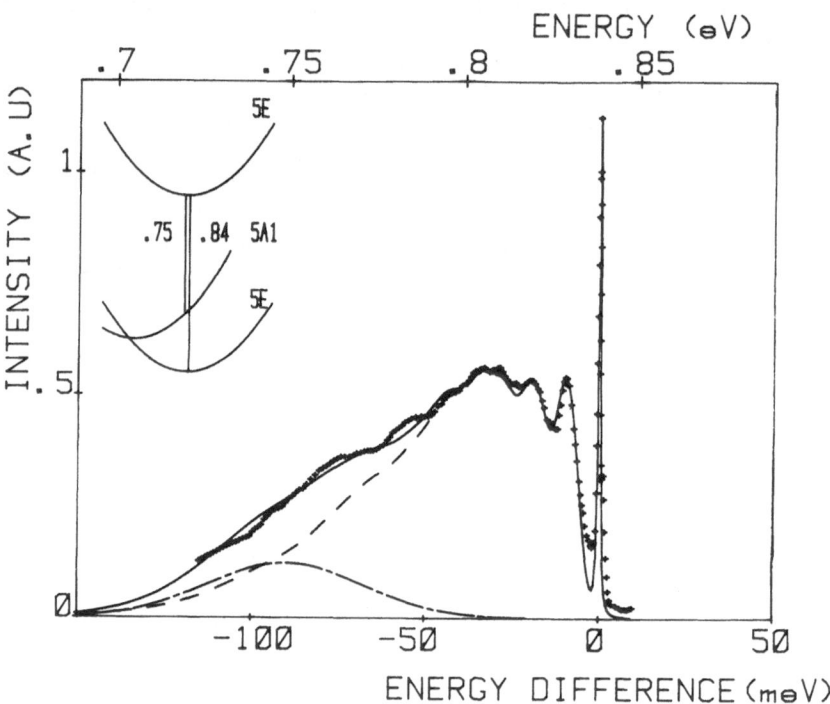

Figure 2   Luminescence band at 0.839 eV (experiment). The 0.839 eV line corresponds to the $^5E—^5E(T_2)$ transition. A simple Huang Rhys fit of the phonon replicas (---) cannot explain the 0.75 eV hump (—·—·—). This band could be due to the $^5E—^5A_1(T_2)$ transition.

We have tentatively described the ion using the static crystal field theory. The description is qualitatively satisfactory except that using a λ value of 50 cm⁻¹ (free ion) the ground state splitting is too large. A vibronic description of the system provides a good quantitative agreement with the experimental splitting [10].

0.574 eV and 0.535 eV lines

Lightowlers and Penchina [6] have observed a zero-phonon line at 0.574 eV. We have noticed that this line is present when the 0.839 eV line is very strong and the 0.64 eV (oxygen) band almost absent. We have also resolved the replica at 0.535 eV, which has the same structure as the 0.574 eV line and does not seem to be a phonon replica because of its temperature dependence.

The fine structure of the excited state of the 0.574 eV line is similar to the structure of the ground state of the 0.839 eV line. These remarks lead to the interpretation of the 0.574—0.535 eV system as the complementary transition to the 0.839 eV one, that is to say:

$$Cr^{(2+ \text{ or } 0)}(C_{3v}) + hole_{VB} \rightarrow Cr^{(3+ \text{ or } 1+)}(C_{3v}).$$

The fine structure of the final state should then be that of a d^3 or d^5 ion in C_{3v} symmetry. This agrees with the structure observed only for the d^5 case, for which the spin Hamiltonian is:

$$H = D[S_z^2 - 1/3 S(S+1)] + \text{other terms.}$$

Experimental results correspond to a value of $D \sim 0.15$ meV. The 0.535 eV transition could be interpreted as:

$$Cr^0 (C_{3v}) + hole \text{ (on the coupled acceptor)} \rightarrow Cr^+ (C_{3v}).$$

Interstitial chromium?

A characteristic of some transition metal ions in silicon is that they exist on the interstitial site, and that they can be coupled with almost any acceptor. We therefore believe that our hypothesis of an interstitial chromium should be considered very carefully because of its possible consequences:

1. Explanation of the 0.575 eV/0.535 eV transition.
2. Possibility of the existence of Cr^+ (isolated interstitial: T_d) which could give an isotropic EPR spectrum.
3. Possible explanation of the EPR photoexcitation properties of GaAs:Cr.
4. The intensity of the 0.839 eV line increases with silicon incorporation (in SI material) but not with oxygen addition in the melt [7, 8].
5. Some diffusion properties of Cr.

ACKNOWLEDGEMENTS

We wish to thank M. Clerjaud and M. Pelous for many helpful discussions and M. Poiblaud (RTC) for providing most of the samples.

References

1. Lightowlers, E. and Henry, M.O. (1978). *Inst. Phys. Conf. Ser.,* **43,** 307
2. White, A.M. (1979). *Solid St. Commun.,* **32,** 205
3. Clerjaud, B., Hennel, A.M. and Martinez, G. (1980). *Solid St. Commun.,* **33,** in press
4. Krebs, J.J. and Stauss, G.H. (1977). *Phys. Rev. B,* **16,** 971
5. Ludwig, G.W. and Woodbury, H.H. (1962). *Solid St. Physics,* Vol. 13. Eds F. Seitz and D. Turnbull
6. Lightowlers, E.C. and Penchina, C.M. (1978). *J. Phys. C.,* **11,** L405
7. Deveaud, B. and Favennec, P.N. (1977). *Solid St. Commun.,* **24,** 473
8. Deveaud, B. To be published
9. Deveaud B., Hennel, A.M., Szuszkiewicz, W., Picoli, G. and Martinez, G. (1979). *Second 'Lund' Deep Level Conference,* May 1979. St-Maxime, France
10. Picoli, G., Deveaud, B. and Galland, D. To be published

Section 4:
Theoretical consideration of deep level semi-insulating systems

PROCEDURES FOR COMPARING THE THEORETICAL AND EXPERIMENTAL POSITIONS OF ENERGY LEVELS OF MULTI-ELECTRON IMPURITIES IN SEMICONDUCTORS

J.W. ALLEN
Wolfson Institute of Luminescence, School of Physical Sciences,
University of St Andrews, St Andrews, Fife, UK

Abstract

Calculations of the energy bands of semiconductors are usually based on three main approximations. The first is the independent particle approximation. The second is the reduction, by the use of local exchange, of the Hartree-Fock coupled equations to a single effective Schrödinger equation giving single-particle eigenenergies. The third is inherent in Koopmans' theorem. Energy level diagrams of the type used in semiconductor physics also contain these approximations implicitly.

It is necessary to go a stage further when dealing with multi-electron impurities such as transition metal impurities. A method was proposed earlier which allowed one to use a few empirical parameters to give a tractable treatment of the energy levels of $3d^n$ impurities. The method when applied to GaAs, for example, gives a good account of the experimental data and allows predictions about energy levels to be made.

Recently, there have been several attempts to calculate energy levels of impurities using modifications of the methods used for energy bands, and containing the approximations listed above. The single-particle energies obtained cannot be directly compared with experiment. It is possible to resolve this difficulty either by using the calculated wavefunctions as basis functions in diagonalizing a correlated Hamiltonian or, equivalently, by comparing the single-particle energies with those occurring in the semi-empirical method mentioned above.

Introduction

An energy level diagram for a deep impurity in a semiconductor, shown in its simplest form in Figure 1, means different things in different contexts. In particular, many theories of deep levels lead to energy level schemes which have a different meaning from those produced by experiment. Neglect of this has caused confusion in the past and occasionally claims have been made that a particular theory is in good agreement with a particular experiment even when the things being compared are quite different in nature. Here we give a brief outline of the relevant concepts and suggest a tentative basis for comparison of theory and experiment for multi-electron impurities.

Definitions of the energies

Most calculations of band structures of solids use the approximations that the total wavefunction can be expressed as products of single-particle wavefunctions and that the potential in the single-particle Hamiltonian is a function of the local electron density. The latter approximation, in effect, is the use of a Hartree potential together with a local-density correction to take exchange into account. One then has the eigenvalue equation

$$\mathcal{H}\psi_i = \left(-\frac{\hbar^2}{2m}\nabla^2 + V(r)\right)\psi_i = E_i\psi_i. \tag{1}$$

There is a correspondence between eigenenergies and eigenfunctions. Several impurity calculations take the same approach, e.g. those based on the Koster-Slater theory or those using scattered-wave theory. An excellent survey has been written by Pantelides [1]. One obtains a set of single-particle eigenenergies, of which Figure 1 is a diagrammatic representation. Each level can be labelled by, for example, quantum numbers or irreducible representations appropriate to a single-particle state or a few degenerate states. The levels are filled at absolute zero from the lowest energy, within the restriction of the Pauli principle. Often, the potential is made self-consistent with the filling of the levels.

Figure 1   Diagrammatic representation of a deep level in a semiconductor.

The procedure can, in principle, give a good description of the ground state, i.e. of the filled levels. The empty levels are not necessarily good approximations to possible excitations or ionizations of the system, except in the case when the potential for the excited or ionized system is adequately approximated by that of the ground state (one then has a theorem analogous to Koopmans' theorem). For band structure calculations of simple semiconductors or metals this approximation may, in practice, be good enough. One reason is that the correlation is shared between many electrons so a change in the orbital of one electron may make only a small change in that of each of the others, i.e. the electron-electron interaction is screened. For multi-electron impurities the approximation is unlikely to be good because there is less screening between the electrons on one particular atom.

Equation (1) is an approximation to the Hartree-Fock equations, which are a set of coupled integrodifferential equations. Koopmans' theorem states that the energy required to remove an electron from the system is equal to the negative of the Hartree-Fock single-particle energy of that electron, in the approximation that the wavefunctions of all the other electrons are unchanged by the ionization.

One therefore has a whole hierarchy of approximations. The total wavefunction is approximated by a determinant of single-particle functions. The Hartree-Fock equations then give the best estimates of the energies and wavefunctions. However, the Hartree-Fock potential is usually approximated by a local-density potential, sometimes with the inclusion of an empirically adjusted parameter, as in the $X\alpha$ method. The energies of the system are then equated to the single-particle eigenenergies. A problem which is immediately apparent is that in the local-density method, if the empty levels are taken to be possible levels of the system, no distinction is made between excitations and ionization. When there is a small number of electrons in a partially filled shell in an impurity atom, it is clear that the potentials in the ground, excited and ionized states may be quite different, and the approximations inherent in Koopmans' theorem break down. Some attempts have been made to cope with the problem by using a potential intermediate between that of the initial and final states in calculations of ionization energies, usually at the expense of introducing another disposable parameter.

The experimentalist means something different by Figure 1. Single-particle energies are not accessible to experiment. Instead, what is measured is the difference between total energies of the system, i.e.

$$\Delta E = E_f - E_i \qquad (2)$$

where E_f and E_i are total energies of the final and initial states. Figure 1 is now a representation of possible energy differences. The band gap is the minimum energy required to remove an electron from the valence band and place it at infinity in the conduction band, and the impurity level is defined analogously. There is no longer a correspondence between levels and configurations of the system: instead, an impurity level is characterized by two configurations, i.e. the initial and final many-electron states [2]. For impurities with a single electron or hole this is trivial but for, say, nickel in GaP one cannot talk of the d^9 or d^8 level; instead, one must use a notation which indicates that the level corresponds with a transition, say from d^9 to d^8, plus an electron at the bottom of the conduction band. A corollary is that there are one fewer levels in the energy gap than there are stable charge states.

Multi-electron impurities

The problem of calculating total energies for use in equation (2) is formidable. One could use single-particle wavefunctions to construct sums of Slater determinants for a given configuration and then find the expectation value of the full Hamiltonian

$$\mathcal{H} = \sum_i \left(-\frac{\hbar^2}{2m}\nabla_i^2\right) + \sum_i V_n(r_i) + \sum_{i>j} \frac{e^2}{r_{ij}} \tag{3}$$

where V_n is the potential arising from the nuclei. (For simplicity, spin-orbit coupling is omitted here.) For an infinite solid this is impossible, and for a cluster of adequate size it is at the moment impracticable unless somewhat arbitrary approximations to most integrals are made. It is tempting to perform the calculation using only a few wavefunctions. For example, if the impurity contains a partly-filled $3d^n$ shell one might calculate the interactions of only the localized functions with d-character falling within the energy gap. The calculations of Hemstreet and Dimmock [3] on GaAs:Cr are of this type. However, many calculations suggest that these functions are not strongly localized but can be regarded as consisting partly of d-functions and partly of host-crystal functions, while simultaneously there are levels deep in the valence band which have a large d-component. Neglect of all but a few interactions could lead to gross inaccuracies in the correlation terms arising from the non-spherical symmetry of the wavefunctions.

An alternative approach is based on a semi-empirical procedure suggested in 1964 [4] for the calculation of the energy levels of $3d^n$ impurities in semiconductors. The energy of a configuration is written down in terms of the parameters of crystal field theory. (This does not, of course, require the assumption of a point-charge model for the crystal field.) For example, if the crystal field Δ is small, the ground state energies of d^7 and d^6 configurations are

$$\begin{aligned} E(d^6) &= -6U + 15A - 35B + 7C - \tfrac{3}{5}\Delta \\ E(d^7) &= -7U + 21A - 43B + 14C - \tfrac{6}{5}\Delta. \end{aligned} \tag{4}$$

Here, terms in U come from the central core and from the spherically symmetric part of the crystal potential, and terms in B and C come from the angular variation of the d-d interaction. The values of B and C (which in general are different for different configurations) can be found experimentally from the spectra of excited states. For free atoms it is found that the difference in $E(d^{n-1}) - E(d^n)$ arising from terms in U and A varies linearly through the transition metal series. Hence, if this variation is parameterized as $\phi + (n-1)\eta$, expressions can be written down for $E(d^{n-1}) - E(d^n)$ for values of n from 1 to 10, i.e. for all members of the series. If ϕ and η are fixed empirically from known energy levels of two of the transition metal series in a semiconductor, the energy levels of the remainder of the series can be calculated.

Unfortunately, even 16 years later our knowledge of the excited state energies of $3d^n$ impurities in different charge states in any one semiconductor is too sparse for this programme to be carried out in full. One has to make the approximation that the values of B and C are the same for different charge states and different impurities in a given semiconductor. The resultant expressions for the energy differences are tabulated in [4]. Another problem is immediately met. Although extensive data have been published on the experimental values of deep levels in several semiconductors, it is only rarely that the two configurations defining the level have been established. If, however, we assume that the reported energy levels for vanadium and cobalt in GaAs [5, 6] are for transitions from the divalent to the trivalent state, we obtain the calculated energy level scheme shown in Figure 2.

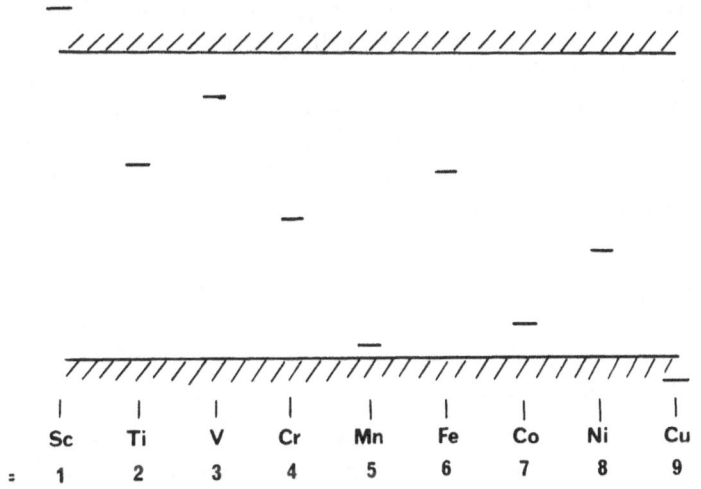

Figure 2 Energy levels calculated semi-empirically for $3d^n$ substitutional impurities in GaAs. The transitions are from the divalent to the trivalent state, and the initial occupancy of the d-states is shown.

It is now suggested that the procedure be reversed. Experimental values of the $3d^n$ impurity level positions should be corrected using information about parameters B and C from excited state spectra, in order to obtain the energy levels which would exist in the absence of a non-spherically symmetric d-d interaction. The resultant levels should be the appropriate ones with which to compare theoretical calculations which neglect this interaction, for example those using a muffin-tin potential. Ideally, one would still like a theoretical calculation of the total-energy differences of equation (2), whereas what we have are single-particle energies. One must therefore rely on the conditions required for the validity of Koopmans' theorem to be met sufficiently adequately for the spherically symmetric part of the potential, or one must use an intermediate-state potential. A minor problem is that the single-particle energies contain the crystal field splitting, and this is removed from the experimental data in the process of finding $\phi + (n-1)\eta$. Fortunately, the value of Δ is small so it is not a bad approximation to use the barycentre of the e and t_2 levels. This is not accurate because in general the e and t_2 orbitals have different radial functions and are therefore not simply the result of splitting a set of five-fold degenerate levels, but the error involved is negligible given the present uncertainty in both theoretical and experimental values of energy levels.

At this time there is no semiconductor for which we have, simultaneously, calculated and experimental values of the energy levels of several $3d^n$ impurities associated with known charge states. Because the field is rapidly expanding, this situation is likely to change soon. The object of this paper is to provide a foundation for comparison of theory and experiment which is better than the present unsatisfactory one in which unlike quantities are compared. As a first step, one might expect calculations of the increase in ionization energy as one goes through the transition metal series to be more reliable than for the absolute positions of the levels with respect to the conduction and valence bands. In our notation, the quantity η, which is the change of ionization energy in going from one member of the series to the next in the absence of d-d angular correlation, should be more accurately calculable than ϕ. Values of $|\eta|$ for ionization from a particular charge state are:

divalent impurity in GaAs (experimental)	0.45 eV [7]
monovalent impurity in ZnSe (experimental)	0.65 eV [8]
divalent impurity in ZnS (calculated)	0.80 eV [9]
divalent free ion (experimental)	1.85 eV [10].

A decrease of $|\eta|$ with increasing covalency is seen. It is hoped that the availability of these data will encourage theoreticians to explore GaAs and ZnSe where the experimental data are available, and experimentalists to explore ZnS where a calculated value is available.

References

1. Pantelides, S.T. (1978). *Rev. Mod. Phys.*, **50**, 797
2. Shockley, W. and Last, J.T. (1957). *Phys. Rev.*, **107**, 392
3. Hemstreet, L.A. and Dimmock, J.O. (1979). *Phys. Rev. B*, **20**, 1527
4. Allen J.W. (1964). *Proc. 7th International Conf. Physics of Semiconductors*, p. 781. Paris; Dunod
5. Haisty, R.W. and Cronin, G.R. (1964). *Proc. 7th International Conf. Physics of Semiconductors*, p. 1161. Paris; Dunod
6. Adrianov, D.G., Sovelev, A.S., Suchkova, N.I., Rashevskaya, E.P. and Filippov, M.A. (1977). *Sov. Phys. Semicond.*, **11**, 858
7. Allen, J.W. and Pearson, G.L. (1967). Stanford Electronics Laboratories Technical Report No. 5115-1.
8. Szawelska, H.R. Private communication.
9. de Siqueira, M.L. and Larsson, S. (1975). *Chem. Phys. Lett.*, **32**, 359
10. Griffith, J.S. (1964). *The Theory of Transition-metal Ions.* Cambridge; Cambridge University Press

JAHN-TELLER MODEL OF Cr^{2+}:GaAs

C.A. BATES, A.S. ABHVANI and S.P. AUSTEN
Department of Physics, University of Nottingham,
Nottingham NG7 2RD, UK

Abstract

A dynamic Jahn-Teller model of Cr^{2+} centres in GaAs is developed as a basis for understanding the acoustic paramagnetic resonance (APR) spectrum. It is shown that first and second order spin-orbit coupling together produce two triplets and a singlet as the ground states of the system separate by a few cm^{-1}. APR transitions are allowed between many of the Zeeman-split levels. The possibility of trigonal crystal fields is also considered and is consistent with some of the experimental data. The centres responsible for the APR spectrum appear to be very different from those seen by electron paramagnetic resonance.

Introduction

Electron paramagnetic resonance (EPR) together with optical illumination has been used to identify the three charge states Cr^{3+} ($3d^3$), Cr^{2+} ($3d^4$) and Cr^{1+} ($3d^5$) of substitutional chromium in GaAs [1]. However, despite the large number of publications and high degree of accuracy of the results, there are considerable difficulties in correlating the various optical properties of chromium with the ground states determined by EPR. In particular, the 0.839 eV luminescence and absorption line of GaAs:Cr has been ascribed to Cr^{2+}, but the ground state is not that determined by EPR [2—4].

An alternative technique which can be used to obtain detailed information on the ground states of paramagnetic centres is acoustic paramagnetic resonance (APR). The technique is complementary to EPR as the selection rules involve changes in the orbital quantum number ($\Delta M_L = \pm 1, \pm 2, ...$) with $\Delta M_S = 0$. Also, centres detectable by one technique are often undetectable by the other technique. APR was first applied to GaAs:Cr crystals to study Cr^{2+} by Ganapol'skii; longitudinal acoustic waves were used at 9.4 GHz and 1.7 K [5]. In fact, the APR measurements were carried out before Cr^{2+} was studied by EPR. Four years later, additional APR measurements were performed by Tokumoto and Ishiguro [6] at frequencies between 1 and 2 GHz, and pronounced resonance peaks were detected. It is apparent that the APR results cannot be explained in terms of the model (KS) proposed by Krebs and Stauss [1] for their EPR data as the APR transitions are forbidden in the KS model. On the other hand, the KS model for Cr^{2+}:GaAs is adequate to account for the pulsed far infra-red measurements of Wagner and White [7] and the zero phonon structure of the 0.9 eV absorption line [2], for example.

Comprehensive APR experiments at 9.6 GHz are currently being undertaken in our laboratories at Nottingham [8]. Results have been obtained on several samples to date (both n-type and semi-insulating) and many APR absorption peaks have been found. Optical treatment (in which light changes the valency of Cr) suggests that the centre is of the Cr^{2+}-type.

This paper sets up and analyses possible models for the Cr^{2+} centres detected by APR. A critical summary of the data available to date from phonon studies on GaAs:Cr is given by Bury et al. in this volume.

The model

Any paramagnetic centre giving rise to an APR absorption spectrum must inevitably show pronounced Jahn-Teller effects (JTE). In this paper, therefore, the consequences of the JTE on the magnetic properties of the low lying states of Cr^{2+} will be examined. Isomorphism procedures will be used throughout [9].

We assume, initially, that the Cr^{2+} ion replaces a Ga atom in the GaAs lattice and is thus at a site of T_d point symmetry. The orbital degeneracy of the ^5D ground term is then split into a lower triplet ($^5T_2$) and an upper doublet (^5E). It is convenient to concentrate our attention initially on the $^5T_2$ ($\ell=1$) ground state. The Hamiltonian of the Cr^{2+} ion and the surrounding lattice can be written in the form:

$$H = H_{ion} + H_{lattice} + H_{ion\,lattice} \tag{1}$$

where H_{ion} includes the spin-orbit coupling H_{so}, and any strain, $H_{lattice}$ includes the elastic and vibrational energies of the lattice, and $H_{ion\text{-}lattice}$ is the interaction between the ion and its surroundings.

In any dynamic JT system, it is usual to include only those lattice displacements of E-symmetry so that

$$H_{ion\text{-}lattice} = V_E (Q_\theta E_\theta + Q_\epsilon E_\epsilon) \tag{2}$$

where V_E is the ion-lattice coupling constant, Q_θ, Q_ϵ are E_θ - and E_ϵ - type displacements in the near-neighbour tetrahedron, and

$$E_\theta = \tfrac{1}{2}(3\ell_z^2 - \ell(\ell + 1))$$
$$E_\epsilon = \tfrac{1}{4}\sqrt{3}(\ell_\uparrow^2 + \ell_-^2). \tag{3}$$

Oz is a two-fold axis of the tetrahedron. (Full details of the isomorphism procedures, the $Q_{\theta,\epsilon}$ collective displacements of a tetrahedron and the approximations involved are given in a review by Bates [9].)

$H_{\text{ion-lattice}}$ may be diagonalized by a suitable unitary transformation [9]. If an average is taken over the ground state of the lattice, the magnetic properties of the system in a strain-free site can be expressed in terms of an effective Hamiltonian in ℓ and S only, namely

$$H_{\text{eff}} = k_1^{T_1} \lambda \gamma \ell \cdot S + \lambda \left[c(E_\theta E_\theta^S + E_\epsilon E_\epsilon^S) + b(\ell \cdot S)^2 \right] \tag{4}$$

where

$$\begin{aligned} c &= -\tfrac{2}{3}(k_1^{T_1})^2 \, \lambda(F_b - F_a) \\ b &= -(k_1^{T_1})^2 \, \lambda F_a \end{aligned} \tag{5}$$

and $k_1^{T_1}$ is the isomorphic constant ($k_1^{T_1} = -1$) for the $^5D(T_2)$ orbital states of Cr^{2+}. $E_\theta^S = E_\theta$ with the ℓ operators in (3) replaced by S operators. H_{eff} contains the spin-orbit coupling only; the first term is that obtained in first order but quenched by the reduction factor γ, and the second term is the total of the second order contributions. F_a, F_b are second order JT reduction factors. Full details of the underlying mathematical analysis, the approximations involved and a discussion of the relative magnitudes of the reduction factors are given in Bates [9].

The eigenstates, energies and g-values

The first order spin-orbit coupling term in equation (4) splits the $^5T_2$ into a lower set $J=3(A_2+T_2+T_1)$, a middle set $J=2(E+T_2)$ and an upper set $J=1(T_1)$. The second order spin-orbit coupling terms remove some of the degeneracies. The resulting energy level diagram is shown in Figure 1, and the energies of the various states are as follows:

$$\begin{aligned}
|J = 1, T_1\rangle \quad & 3\lambda(\gamma - 3\lambda F_a) - \tfrac{7}{5}\lambda^2 (F_b - F_a) \\
|J = 2, E\rangle \quad & \lambda(\gamma - \lambda F_a) + 2\lambda^2 (F_b - F_a) \\
|J = 2, T_2\rangle \quad & \lambda(\gamma - \lambda F_a) + \lambda^2 (F_b - F_a) \\
|J = 3, A_2\rangle \quad & -2\lambda(\gamma + 2\lambda F_a) + 2\lambda^2 (F_b - F_a) \\
|J = 3, T_2\rangle \quad & -2\lambda(\gamma + 2\lambda F_a) \\
|J = 3, T_1\rangle \quad & -2\lambda(\gamma + 2\lambda F_a) - \tfrac{8}{5}\lambda^2 (F_b - F_a).
\end{aligned}$$

λ typically has a value ~ 50 cm^{-1} while F_b and F_a $\sim 10^{-4}$. Thus, the splittings between the three $J=3$ states are ~ 1 cm^{-1}.

If coupling to the ^5E states is taken into account, additional terms must be added to (4). However, these additional terms have an identical form to those derived from the second order JTE and thus the general features of Figure 1 are unaltered. However, the splittings have an additional component. In the theory of Ganapol'skii [5], coupling to ^5E only was considered and the JT contributions were ignored.

The effects of a magnetic field **B** can be described by the Hamiltonian

$$H_B = \mathbf{B} \cdot (k_1^{T_1} \gamma \ell + 2S) \tag{6}$$

and this gives g-values of $\frac{1}{2}(4-\gamma)$ and $\frac{1}{8}(4-\gamma)$ for the $|J=3, T_1>$ and $|J=3, T_2>$ states, respectively. (Small second order contributions should also be included in H_B, but we neglect them here.) The effects of H_B are also shown schematically in Figure 1.

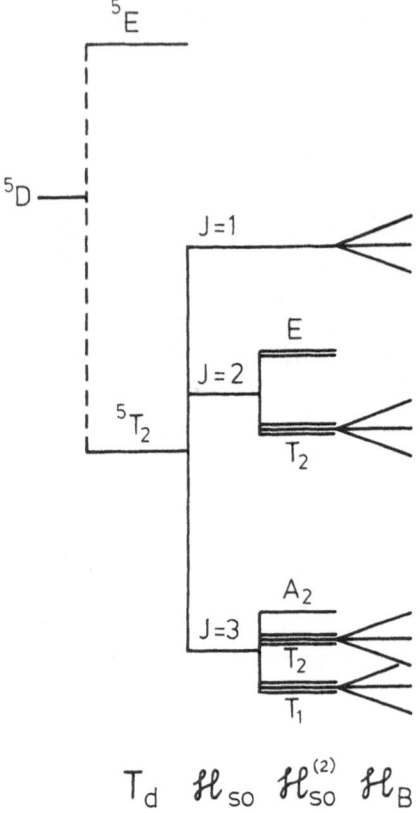

Figure 1   Schematic energy level diagram, not to scale, showing the splittings caused by the tetrahedral crystal field, T_d, first order spin-orbit coupling, H_{SO}, second order spin-orbit coupling, H_{SO}^2 and the effects of increasing B on H_B.

The acoustic transitions

The Hamiltonian describing the ultrasound may be written in the form:

$$H_{APR} = V_E(\bar{Q}_\theta E_\theta + \bar{Q}_\epsilon E_\epsilon) + \gamma V_{T_2}(\bar{Q}_4 E_4 + \bar{Q}_5 E_5 + \bar{Q}_6 E_6) \tag{7}$$

where V_{T_2} is the coupling constant of the dominant T_2-type displacements of the tetrahedron and the bars denote the instantaneous magnitudes of the relevant collective co-ordinates. To determine which transitions are allowed in Figure 1, we express the $|J=3, T_1>$ and $|J=3, T_2>$ states in terms of $|m_\rho m_s>$ as follows:

$$T_2 \begin{cases} |T_2,1> = -\sqrt{\tfrac{3}{8}}|1,2> + \sqrt{\tfrac{5}{8}}[\sqrt{\tfrac{2}{3}}|-1,0> + \sqrt{\tfrac{8}{15}}|0,-1> + \sqrt{\tfrac{1}{15}}|1,-2>] \\ |T_2,0> = \sqrt{\tfrac{1}{2}}[\sqrt{\tfrac{1}{3}}|0,2> + \sqrt{\tfrac{2}{3}}|1,1>] + \sqrt{\tfrac{1}{2}}[\sqrt{\tfrac{1}{3}}|0,-2> + \sqrt{\tfrac{2}{3}}|-1,-1>] \\ |T_2,-1> = -\sqrt{\tfrac{3}{8}}|-1,-2> + \sqrt{\tfrac{5}{8}}[\sqrt{\tfrac{2}{3}}|1,0> + \sqrt{\tfrac{8}{15}}|0,1> + \sqrt{\tfrac{1}{15}}|-1,2>] \end{cases}$$

$$T_1 \begin{cases} |T_1,-1> = -\sqrt{\tfrac{5}{8}}|1,2> - \sqrt{\tfrac{3}{8}}[\sqrt{\tfrac{2}{3}}|-1,0> + \sqrt{\tfrac{8}{15}}|0,-1> + \sqrt{\tfrac{1}{15}}|1,-2>] \\ |T_1,0> = \sqrt{\tfrac{1}{5}}|-1,1> + \sqrt{\tfrac{3}{5}}|0,0> + \sqrt{\tfrac{1}{5}}|1,-1> \\ |T_1,1> = -\sqrt{\tfrac{5}{8}}|-1,-2> - \sqrt{\tfrac{3}{8}}[\sqrt{\tfrac{2}{3}}|1,0> + \sqrt{\tfrac{8}{15}}|0,1> + \sqrt{\tfrac{1}{15}}|-1,2>]. \end{cases}$$

Detailed calculations show that

$$\langle T_2,1|H_{APR}|T_2,-1\rangle, \quad \langle T_2,1|H_{APR}|T_2,0\rangle, \quad \langle T_2,-1|H_{APR}|T_1,-1\rangle$$

are zero if V_{T_2} terms in equation (7) are dropped. On the other hand, inclusion of the V_{T_2} terms will allow all the transitions to take place. In principle, therefore, acoustic transitions between the various Zeeman-split levels derived from the $|J=3,T_1>$ and $|J=3,T_2>$ states are allowed, but their intensities depend on the relative magnitudes of the coupling coefficients and on the particular distortions $\overline{Q}_\theta, \overline{Q}_4$, etc. induced by the phonons.

The effects of random internal strain

The effects of random internal strain can be represented by a Hamiltonian H_{strain} identical in form to H_{APR}, but with $\overline{Q}_\theta$, etc. equal to the strain-induced static displacements in the tetrahedron. H_{strain} has two effects. Firstly, it changes the energies by a small amount so that each of the levels of Figure 1 is broadened. Secondly, it admixes the states, especially in regions of the cross-over so that the various APR transition probabilities are modified.

Ganapol'skii [5] observed three APR lines and explained his results in terms of precise non-zero values of $\overline{Q}_\theta$ and $\overline{Q}_\epsilon$ in H_{strain}. In our model, we prefer to regard $\overline{Q}_\theta, \overline{Q}_\epsilon$, etc. as having a distribution about $\overline{Q}_\theta=\overline{Q}_\epsilon=0$. The APR peaks then arise from various turning points of the energy level separation with strain magnitude and angle.

A trigonal crystal field

While the foregoing model forms a basis for the understanding of some of the APR lines and also other transitions observed by other phonon spectroscopic techniques (Bury *et al.*, this volume), it is not entirely satisfactory for all lines because the transitions within T_1 and T_2 will occur at values B_{res} of B which are in simple multiples of each other (e.g. 1:2:3). Transitions between T_1 and T_2, however, will not follow this pattern. The APR spectra of Ganapol'skii [5], Tokumoto and Ishiguro [6] and the more detailed results obtained in Nottingham, fail to reveal any simple relationship between the B_{res} values. (The new measurements [8] have been carried out at approximately the same frequency as those by Ganapol'skii [5]; they reproduce those originally found, plus many other additional features.)

We suggest that some of the chromium may be in sites in which an additional trigonal crystal field H_{trig} exists for the following reasons:

1. The new APR spectrum reveals a very strong absorption peak having a pronounced trigonal symmetry.
2. The Zeeman dependence and anisotropy of the luminescence spectrum recently found by Eaves et al. (this volume) can be fitted to an effective Hamiltonian H' which contains a significant trigonal field term. It is suggested that the line arises from transitions between excited states of a 'Cr^{3+}-donor pair plus a loosely bound trapped electron' complex to a 'Cr^{2+}-donor pair' ground state. The terms in H' are thought to reflect the structure of the excited state rather than that of the ground state.

A suitable interpretation of the latter model is that the donor would effectively generate a trigonal field H_{trig} at the Cr^{2+} site. With reference to an applied B, there are four possible directions for the trigonal axes which are not necessarily all equivalent. If $H_{trig} \ll H_{so}$, the T_1 and T_2 triplets in Figure 1 are each split into a singlet $|M'_L = 0\rangle$ and doublet $|M'_L = \pm 1\rangle$. When $B \neq 0$, the $|M'_L = 0\rangle$ state will cross either the $|M'_L = +1\rangle$ or the $|M'_L = -1\rangle$ state. Alternatively, if $H_{trig} \gg H_{so}$, the $^5T_2$ state will be split into $^5A_1$ and ^5E. The energy level pattern is then completely different from that shown in Figure 1; if ^5E is lower in energy than $^5A_1$, inclusion of second order JT terms involving the trigonal field and spin-orbit coupling again gives many low-lying levels. When H_B is added, there are many energy levels which cross and between which APR transitions could occur.

The APR, thermal conductivity and frequency crossing experiments are still in progress while the implications of the theoretical models detailed above with and without H_{trig} have still to be investigated thoroughly. However, there is a sufficient number of attractive features of the models to suggest that a satisfactory explanation of the experimental data could be forthcoming.

Conclusions

The possible models we have suggested to account for the APR spectrum are completely different from the KS model. This implies that at least two types of Cr^{2+} sites are present in GaAs. It explains why the EPR and APR spectra are so dissimilar and why some of the luminescence features cannot be accounted for on the KS model. Furthermore, the incorporation of the dynamic JTE is consistent with that normally found in other systems which contain Cr^{2+} [10] or Fe^{2+} [11] as impurities.

ACKNOWLEDGEMENTS

The authors wish to thank Drs P. Bury and P.J. King for making available to them all their APR data, and Dr L. Eaves for the information concerning possible trigonal centres. Our thanks are also due to Professor L.J. Challis and Drs A.-M. Vasson and A.-M. de Goer for many helpful discussions on the Cr:GaAs system. Two of us (A.S.A. and S.P.A.) gratefully acknowledge the Science Research Council for the award of research studentships.

References

1. Krebs, J.J. and Stauss, G.H. (1977). *Phys. Rev. B,* **16,** 971
2. Clerjaud, B., Hennel, A.M. and Martinez, G. (1980). *Solid St. Commun.,* in press
3. Lightowlers, E.C., Henry, M.O. and Pechina, C.M. (1979). *Inst. Phys. Conf. Ser.,* **43,** 307
4. Koschel, W.H., Bishop, S.G. and McCombe, B.D. (1976). *Solid St. Commun.,* **19,** 521
5. Ganapol'skii, A.M. (1975). *Sov. Phys. Solid St.,* **16,** 1868
6. Tokumoto, H. and Ishiguro, T. (1978). *Internal Friction and Ultrasonic Attenuation in Solids* (Eds R.R. Hasiguti and N. Mikooshiba), p. 177. Univ. of Tokyo Press
7. Wagner, R.J. and White, A.M. (1979). *Solid St. Commun.,* **32,** 399
8. Bury, P., King, P.J. and Wiscombe. P. (1980). To be published
9. Bates, C.A. (1978). *Phys. Rep.,* **35C,** 187
10. Bates, C.A. (1978). *J. Phys. C, Solid St. Phys.,* **11,** 3447
11. Johnstone, I.J., Lockwood, D.J. and Mischler, G. (1978). *J. Phys. C, Solid St. Phys.,* **11,** 2147

THE APPLICATION OF EVANESCENT STATES TO DEEP IMPURITIES

K.J. BLOW and J.C. INKSON
Cavendish Laboratory, Madingley Road,
Cambridge CB3 0HE, UK

Abstract

The complex band structure of a semiconductor can be calculated using basis functions appropriate to the impurity problem. These functions diverge at the origin and decay exponentially at large distances. The complex band dispersion relationship relates the decay length of the impurity state to the energy of the impurity and connects to the real band structure at the extremal points such as the Γ, X or L minima. The symmetry of the impurity state plays a central role in determining its properties. The wavefunctions used have been appropriately symmetrized. We show that for states of certain symmetries the relevant complex band connects to the higher minima rather than the Γ minimum. The evanescent states are matched to an impurity wavefunction within a core region. We show that for deep traps the core can have a significant effect on the properties of the impurity. We discuss some of the experimental data which can be understood simply in terms of states connecting to higher minima.

1 Introduction

In this paper we discuss the application of evanescent states to the study of the electronic structure of impurities. This approach is based on the work of Inkson [1] and is designed both to give a simple picture of the impurity state beyond the effective mass approximation and also to bring in specific properties of the defect, such as its point symmetry.

For many deep levels the major change in properties is restricted to a small core region, outside which the semiconductor is essentially undisturbed. It follows that in the outer region the wavefunction for the defect is a semiconductor one, in this case an evanescent (or localized) state. The evanescent nature of the state forces it to be in the band gap of the semiconductor; in addition, it must obey the following restrictions: it must decay in all directions and match smoothly on to the core wavefunction to form the complete state.

In considering how an impurity state will be formed, symmetry plays an important role. The core wavefunction at any energy may be decomposed into a few symmetry components (S, P, D, etc.). When the evanescent states are matched on to these core wavefunctions the symmetry of the evanescent states will govern which are present. In addition to this there are considerations of logarithmic derivatives to govern the overall contribution to the complete state but if an evanescent state has a symmetry which is not present in the core it will certainly not contribute to the final state.

From previous work [1] we have found that the evanescent states suitable for impurities are related closely to the band extrema. In this paper we discuss how this, together with a simplified version of the development of the evanescent states, can be used to highlight the symmetry considerations and hence connect the properties of the impurity to the conduction and valence band properties.

2 Evanescent solutions for the impurity problem

When developing solutions for the conduction or valence bands, all allowed solutions must obey Bloch's theorem. In an impurity problem this criterion is removed and one apparently has a much freer choice. All solutions, if they are to reproduce a localized state, must however go to zero at large distances (though they are allowed to go to infinity within the core region since they do not apply there) and must also be solutions of the crystal Schrödinger equation. In a previous paper we showed how this could be done in terms of a simple extension of $\mathbf{k} \cdot \mathbf{p}$ theory. The restriction that the energy be real limited the solutions to the form of sums of terms such as:

$$|E_A, l, m\rangle = k_l(\gamma r) Y_l^m(\theta, \phi) |E_A\rangle$$

where $|E_A\rangle$ is the Bloch wavefunction at an extremum (e.g. A may be L, X or Γ). These sets of solutions can be thought of in terms of the Bloch function part ensuring the solution to the Schrödinger equation while the Hankel function and spherical harmonics produce the evanescent envelope. This is equivalent to the way that in $\mathbf{k} \cdot \mathbf{p}$ theory one can produce wavefunctions in the band states by combination of $e^{i\mathbf{k}\cdot\mathbf{r}} |E_A\rangle$.

The necessity for modelling the core and matching to the evanescent states can be seen by considering the Lucovsky model. The Lucovsky wavefunction, $\exp(-qr)/r$, is precisely the evanescent state that would be obtained for the free electron Hamiltonian. When this is matched to a simple wavefunction in the core region, which we have modelled with a square well potential, the effect of the core can be studied. In Figure 1 we show the photoionization cross-section, σ , both with and without the core.

Note that the peak has been reduced in energy quite substantially. The core has such a marked effect because it removes the unphysical divergence in the wavefunction at the origin. This divergence causes the wavefunction to have too much weight near the impurity. The matrix elements which determine the optical cross-section correspondingly are dominated by the 'k=q criterion' [3, 4]. Once this divergence is removed, the longer wavelength terms in the description of the wavefunction are left and the peak correspondingly moves down to lower energies (longer wavelengths).

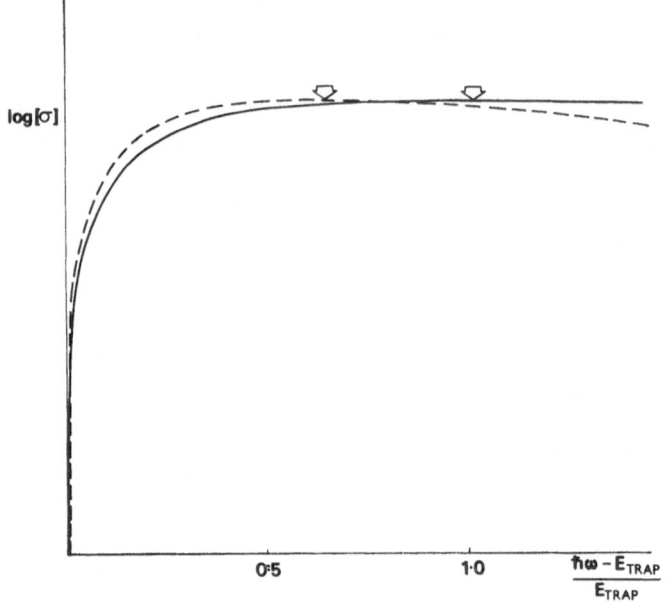

Figure 1   Effect of the core region on the photoionization cross-section. —Lucovsky model; ————
Lucovsky without core. The arrows indicate the peak in the spectrum.

The spherical harmonics will also transform according to certain representations of Γ and when these are combined with the symmetry of the Bloch functions we obtain the appropriate basis states for the computation of the evanescent 'spherical' band structure. In general, the complex band wavefunctions will be a linear combination of these basis functions with the coefficients determined by the secular determinant, as discussed in [1].

The complex band wavefunctions must be matched to the core wavefunction at some radius, r_0 say. This is done by matching the logarithmic derivatives of the core and evanescent wavefunctions. Each complex band wavefunction will match to a particular component of the impurity wavefunction as determined by the symmetry.

The logarithmic derivative of the basis functions is

$$\frac{1}{k_l(\gamma r)} \frac{\partial k_l(\gamma r)}{\partial r} + \frac{1}{u_{E_A}(r)} \frac{\partial}{\partial r} u_{E_A}(r)$$

The second term is a constant for a given extremum. The first term increases with l for a given γ so that the logarithmic derivative of the basis function increases with l. In surface calculations [2, 3] it is found that evanescent states with a high logarithmic derivative are not important in the matching process.

This can be understood in the following way. If the logarithmic derivative is high at the core radius, then either the derivative of the wavefunction is high or its amplitude is small. In the former case, the core wavefunctions would need a high curvature and hence a high kinetic energy. In the latter case, the contribution to the properties of the impurity would be negligible. Thus in any practical calculation it will be possible to terminate the basis set at some reasonable value of l.

Here, for illustrative purposes, we make the fairly crude approximation of including only the l=0 basis functions. Within this approximation Table 1 shows which extrema can contribute to the impurity wavefunctions of a particular symmetry. Note that only a few extrema are available to most symmetries, although the number of extrema which can contribute to a particular wavefunction will increase with l.

Table 1 **States are ordered vertically corresponding to their energies as in gallium arsenide**

Band edge	State symmetry				
	Γ_1	Γ_2	Γ_{12}	Γ_{15}	Γ_{25}
L_3				X	X
Γ_{15}				X	
X_3					X
X_1	X		X		
L_1	X			X	
Γ_1	X				
Γ_{15}				X	
L_3				X	X
X_5				X	X

An entry in the table indicates a possible decomposition, e.g. L_3 can be decomposed into Γ_{15} and Γ_{25}.

3 The core region and matching

Once the evanescent states have been obtained by solution of the secular determinant they must be matched to the impurity wavefunction in the core region. We begin by making a spherical harmonic decomposition of the core wavefunction, which we assume transforms according to a particular irreducible representation of Γ.

$$\psi_{core}(T) = \sum_{n,m} R_{n,m}(r)\, Y_n^m(\theta,\phi).$$

The complex band wavefunctions are:

$$\psi(\gamma r) = \sum_{k_0,l,m} \alpha_{k_0,l,m}\, k_l(\gamma r)\, Y_l^m(\theta,\phi)\, u_{k_0}(r)$$

where $\mathbf{k}_0$ is defined by

$$\nabla_k E(k)|_{k=k_0} = 0$$

and we expect the range of l and m to be small.
Define

$$\phi_{l,m}(\gamma r) = \sum_{k_0} \alpha_{k_0,l,m}\, k_l(\gamma r)\, u_{k_0}(r);$$

then

$$\psi(\gamma r) = \sum_{l,m} \phi_{l,m}(\gamma r)\, Y_l^m(\theta,\phi).$$

Thus we write the impurity wavefunction outside the core as

$$\psi_{out} = \sum_{\gamma}\sum_{l,m} A_\gamma \phi_{l,m}(\gamma r)\, Y_l^m(\theta,\phi).$$

The values of γ occurring in this sum are determined by the complex band dispersion relationships and are such that all the evanescent components have the same energy. We can now use the orthogonality properties of the spherical harmonics to write down the matching equations at the surface of the core:

$$R_{l,m}(r_0) = \sum_\gamma A_\gamma\, \phi_{l,m}(\gamma r_0)$$

$$\frac{\partial}{\partial r} R_{l,m}(r_0) = \sum_\gamma A_\gamma \frac{\partial}{\partial r} \phi_{l,m}(\gamma r_0).$$

The core region will make an increasing contribution to the properties of an impurity as its wavefunction becomes more localized.

These basis functions can be used to obtain the complete 'complex band structure' that is the relationship between the energy of the evanescent state and the decay parameter, γ, and the various l, m's involved. For the present purposes, however, let us restrict ourselves to the components of the evanescent state, i.e. of the form given above for ψ_{out} and consider the effect of symmetry. The appearance of an impurity will remove the translational group characteristic of the lattice and replace it with the point group. For example, an impurity of a cubic site will require $T_d(\bar{4}\,3m)$ — the point group of Γ.

It is most convenient to treat the Bloch functions $|E>$ separately. For instance, if the extremal point is the set of L minima then we can form linear combinations of the wavefunctions which transform according to the irreducible representations of Γ. In Table 1 we present the results of this procedure.

4 Non-Γ centres

In Section 2 we showed that some states may connect better to the higher minima than to the nearest conduction band because of symmetry considerations. These states have been referred to in the literature as non-Γ centres. There are two methods for determining which minima dominate the impurity wavefunctions.

1. Measurements on alloys such as $Al_xGa_{1-x}As$ as a function of x [5].
2. Measurements of impurity levels as a function of applied hydrostatic pressure [6].

The first method is the simpler experimentally but it is slightly more difficult to justify symmetry arguments when the local environment may change from site to site. Only within a virtual crystal approximation should the results of Section 2 apply. A study of Te-doped $Al_xGa_{1-x}As$ [5] has revealed an impurity level at about 0.1 eV below the X-band, independent of x. We would expect such a state to have Γ_{12} symmetry, since this symmetry can only connect to X_1 in a first approximation (see Table 1). Lang, Logan and Jaros [5] have proposed a model of this centre which involves a Te-V_{As} complex. Theoretical work [7] indicates that such a centre would be dominated by the properties of the arsenic vacancy and consequently have Γ_{15} symmetry. Such a state would not connect in any simple way to the extrema. This difference may be due to a number of factors. Lang *et al.* propose that the DX centre follows the X-point because of the high density of states at this point. However, the impurity seems to be unaffected by the closer high density of states L minimum for x⩽0.38. The approximation we have made may not be valid in this case, i.e. it may be that the combination of l=1 with the X minima giving the Γ_{15} symmetry required by Lang *et al.* still dominates the extrema contributions over the L minima l=0 term, which would connect the impurity to the L minima. We now estimate the logarithmic derivatives of the L and X points using the work of Cardona and Pollak [8]. Although they apply the **k·p** procedure to silicon, we would not expect the results to differ greatly between silicon and gallium arsenide. We assume the core radius to be equal to half the nearest neighbour distance and γ^{-1} to be about five core radii. This value of the core radius will be most appropriate for a deep neutral centre. As the impurity level moves further from the edge of the complex band, this approximation will improve. Using these values, we obtain the following results for the logarithmic derivatives:

$$D_X = \left|\frac{1}{u_X}\frac{\partial}{\partial r}u_X\right| = 0.8\ \text{Å}^{-1} \qquad D_L = \left|\frac{1}{u_L}\frac{\partial}{\partial r}u_L\right| = 0.5\ \text{Å}^{-1}$$

$$D_0 = \left|\frac{1}{k_0}\frac{\partial}{\partial r}k_0\right| = 0.9\ \text{Å}^{-1} \qquad D_1 = \left|\frac{1}{k_1}\frac{\partial}{\partial r}k_1\right| = 1.6\ \text{Å}^{-1}$$

These values indicate that the restriction of the complex band wavefunctions to $k_0 Y_0^0 |E_A\rangle$ will be a reasonably good approximation for a state connecting to the L-point since $D_l \gtrsim 3D_L$. However, for a state connecting to the X-point it may be necessary to include the k_1 functions to ensure that the logarithmic derivative is dominated by the envelope. This is due to the fact that $D_0 \sim D_X$ and $D_l \sim 2D_X$.

The second method is more difficult experimentally, principally due to the difficulty in determining the position of the higher minima. Iseler et al. [6] have measured the pressure dependence of (Cl, Br) and (Ga, In) in CdTe. These two pairs have a different (linear) pressure dependence which suggests that they connect to different minima. Unfortunately, the lack of experimental data on the pressure dependence of the higher minima makes it impossible to say which minima are involved.

Hoo and Becker [9] have explained the pressure dependence of the Hall effect and resistivity in Se-doped GaSb in terms of the two resonant impurity levels. These levels have an energy which is fixed below the L_1 and X_1 band edges, respectively. The state associated with L_1 is an effective mass impurity and lies about 50 meV from the band edge. The other level is deep and lies about 0.23 eV below X_1. These states move rigidly with their respective bands over the pressure range 0–10 K bar.

5 Conclusions

The wavefunction of a deep impurity state has contributions from large parts of k-space, in contrast to the simple effective mass impurity which is localized about a minimum in k-space. In this paper we have shown that the impurity level may be correlated with a few points in k-space. This is not incompatible with the large k-space view of the wavefunction since evanescent states of any band near the centre of a gap will have this property, i.e. they are localized in real space and delocalized in k-space. We have indicated, in Section 4, some of the known centres which can be understood simply in these terms.

References

1. Inkson, J.C. (1980). J. Phys. C., 13, 369
2. Burt, M.G. and Inkson, J.C. (1976). J. Phys. D, 9, 43
3. Burt, M.G. (1980). To be published
4. Blow, K.J. and Inkson, J.C. (1980). J. Phys. C, 13, 359
5. Lang, D.V., Logan, R.A. and Jaros, M. (1979). Phys. Rev. B, 19, 1015
6. Iseler, G.W., Kafalas, J.A., Strauss, A.J., MacMillan, H.F. and Bube, R.H. (1972). Solid State Commun., 10, 619
7. Jaros, M. and Brand, S. (1979). J. Phys. C, 12, 525
8. Cardona, M. and Pollak, F.H. (1966). Phys. Rev., 142, 530
9. Hoo, K. and Becker, W.A. (1976). Phys. Rev. B, 14, 5372

COMPLEXES AND THEIR EFFECTS ON III—V COMPOUNDS

MAX N. YODER
Office of Naval Research,
Arlington, Virginia 22217, USA

Abstract

The degree and type of activation of any given impurity in III-V compounds is noted to be significantly affected by the presence of other selected impurities within the host material. Certain impurity combinations are believed to favour the formation of electrically active complexes; other complexes can be formed which are electrically neutral. Selected impurities are thought to be capable of breaking previously formed complexes and substituting new complexes. Other impurities, while not capable of breaking previously formed complexes, are noted to prevent the further formation of these same complexes. *Neutral complexes in insulating material may be electrically activated by impurities capable of breaking these complexes;* thus, insulating III-V materials must be carefully qualified by screening for these delitescent conditions. Heuristic examples are given.

Background

Several factors relating to GaAs crystal properties have not been satisfactorily explained. Among these are: (a) greater than 100% activation of low doses of implanted Se [1, 2]; (b) the rules governing the amphoteric nature of the group IV impurities; (c) the rules governing the diffusion tails of chalcogen (group VI) impurities; and (d) the rules governing the activation of deep level impurities in SI material.

Impurity redistribution in GaAs

The diffusion of Zn into GaAs crystals grown with a Mn background is known to 'getter' the Mn to the surface and to leave a deep depletion of Mn just behind the Zn diffusion front [3]. Cr is particularly susceptible to being 'gettered' by stress [4, 5] created by damaged regions within or on the surface of the crystal. Damage, stress, and impurity concentration gradients created by large dose ion implantation of S into GaAs are noted to induce the background Cr to move towards the front surface during a post-implant anneal at 840°C for 20 minutes [6]. Figure 1 illustrates this action. The interesting findings are (*a*) that a significant portion of the Cr is trapped in a region beneath the surface and under the peak of the S implant profile; and (*b*) that the Cr concentration in this 'trapped' region is several times that of the background impurity level.

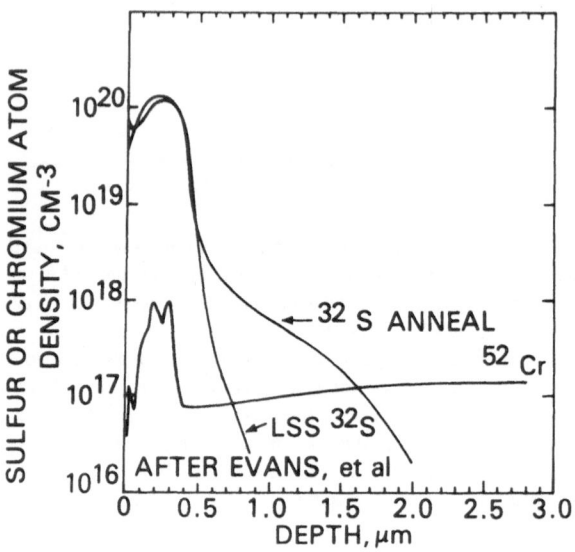

Figure 1   Cr gettered and trapped in Cr:S complex.

Group IV impurities exhibit a unique ability to resist a tendency to diffuse or be gettered. Ge, for instance, will not diffuse into high purity GaAs unless another impurity is present [7]. The incorporation of Sn into GaAs during molecular beam epitaxy (MBE) is significantly enhanced by the presence of a group VI impurity [8]. Si-Si cation-anion complexes have been identified with certainty in GaAs [9]. A most probable mechanism explaining the difficulty of incorporating Ge and Sn into GaAs (in the absence of other impurities) as described in [7] and [8] is that Ge and Sn tend to form nearest neighbour covalent complexes which are too large to diffuse or be readily incorporated into the GaAs lattice. Other impurities, judiciously chosen, must be added to break these IV—IV bonds.

Group IV and group VI implanted impurities 'activate' in quite dissimilar ways when annealed in a surface-stress-free molten GaAs capless anneal process [10]. The striking effect here is that elemental Si atoms apparently migrate to find other Si atoms (with which to form nearest-neighbour, self-compensating complexes) at temperatures >825°C. In contrast, Se has no such tendency to form a self-compensating covalent-like bond and thus its electrical activation continues to improve with temperature.

Implanted S is well known for its characteristic diffusion tail upon anneal/activation. The co-implantation of both Si and S, however, behaves quite differently, as illustrated by Figure 2. As long as the concentration of the implanted S does not exceed the concentration of the implanted Si, there appears to be no diffusion tail; moreover, activation efficiency and mobility are dramatically improved [11]. It appears that the Si is complexing with the S in preference to complexing with itself. The nearest neighbour, Si-S, cation-anion complex is then too large to diffuse. In a similar manner the work of Evans, Deline and Sigmon [6] can be explained in terms of the formation of a Cr-Se complex too large to diffuse.

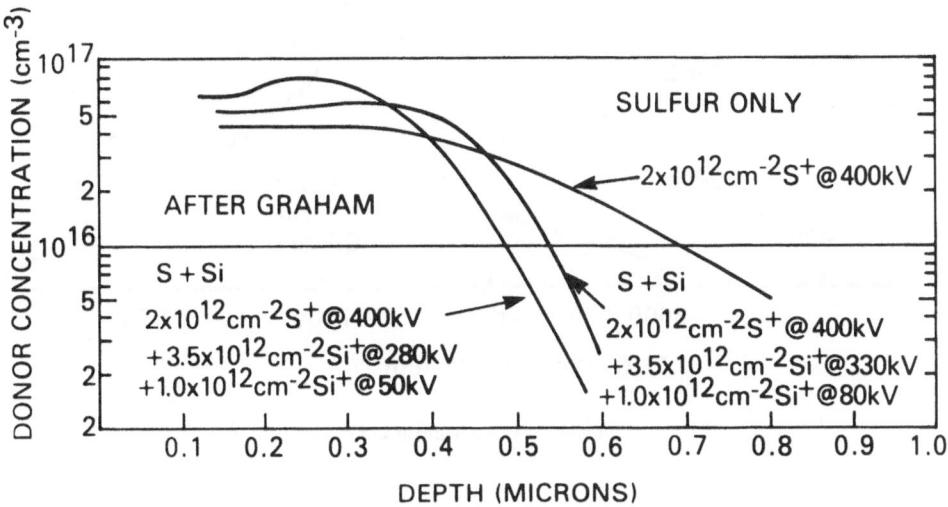

Figure 2   Probable Si:S complexes formed during GaAs anneal.

Complexes in semi-insulating GaAs

A well-recognized method of growing GaAs SI substrates which will not type convert at elevated temperatures is deliberately to add a donor impurity in a concentration sufficient to ensure that it is the dominant donor (e.g. 4×10^{15} cm^{-3}) and then to add Cr just in excess of this amount. This was done at the Naval Research Laboratory (unpublished) using the standard boric-oxide-liquid encapsulated (LEC) melts. One such melt used the Cr:Te system while the other used the Cr:Sn system. Crystals from both melts were semi-insulating and did not type convert. The photoluminescence spectra of the two crystals, however, differed remarkably. Figure 3 illustrates one example of the two cases in the region of 0.93 eV. This energy level is thought to derive from a background group IV impurity or an As vacancy complexing with Te on an As site. A logical explanation is that the Te suppressed the formation of self-compensating, nearest neighbour IV—IV background impurity complexes*. An equally interesting result of this experiment concerns the 0.63 eV spectral line traditionally attributed to oxygen. It is nearly six times higher in the Cr:Te system than in the Cr:Sn system. In fact, the 0.63 eV line is higher in the Cr:Te system than in a LEC SI Te-free GaAs crystal purposely doped with oxygen! Thus it appears that Te forms at least two complexes with residual background impurities and/or Cr.

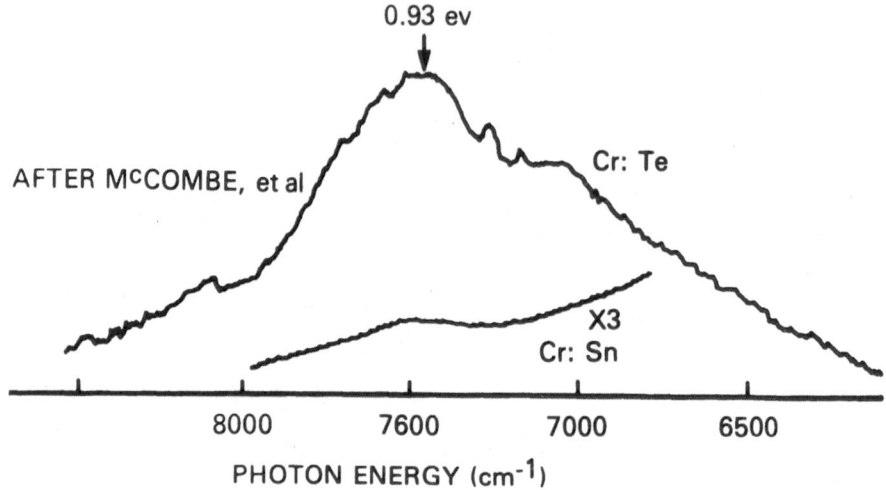

Figure 3   Te activations in GaAs.

The implantation of B into GaAs has been found to introduce As vacancies in GaAs upon activation/anneal [12]. These vacancies and the complexes of these vacancies with other impurities on anion sites were noted to cause acceptor-like action and compensation of Si-doped material [12]. Although the reason for B inducing As vacancies was not given, a logical explanation may be found by resorting to a literature search of boron arsenide. Although there are numerous articles relating theoretical and/or calculated electrical properties for this compound, there appear to be no measured properties published. In fact, there is a dearth of information regarding the definitive physical existence of this material! A plausible explanation derives from the fact that the (revised Pauling) electronegativity of boron and arsenic are equal. As such, the ionic bond between the two is very weak and there is no evidence that a true covalent bond exists. Thus, a B atom on a Ga site leaves the four adjacent As atoms with but three of their normal four bonds. The

*Si-Si, Ge-Ge and Sn-Sn bonds within GaAs are apparently not seen with photoluminescence spectroscopy.

situation, however, is probably even more pronounced. Recent evidence seems to indicate that impurities in GaAs are not randomly distributed, but tend to be drawn to each other in clusters after traversing hundreds of lattice parameters [13]! Thus, these As atoms may well be surrounded by four nearest neighbour B cations leaving them with virtually no bonding whatever. In crystals pulled from LEC melts in quartz crucibles, B concentration has been found to be 0.1 ppm; in LEC melts in BN crucibles the B concentration in the crystal was 4 ppm [14]. Thus the magnitude of probable As vacancies and their acceptor/compensating action [12] provides a plausible explanation for the SI nature of undoped LEC material. These As vacancies also provide an explanation for the previously mentioned difference in photoluminescence response between Cr:Sn and Cr:Te material; undoubtedly the Sn first fills the As vacancies and then begins to occupy Ga sites (compensation), unlike the Te which occupies first the As vacancies and then replaces additional As. These B-induced As vacancies may also explain the better activation efficiency and mobility of low level Si+S implants over Si only implants into LEC material [11] by the presence of S precluding Si from filling As vacancies.

Driving forces for impurities

Stress and concentration gradients are universally recognized impurity redistribution driving forces. Not as intuitive a force is the Fermi level [15]. It may play a significant role in Cr compensation in GaAs. Here the concentration of *electrically active* Cr is noted to track closely the Si background level [16]. The question arises, however, as to why Cr appears to be unique in this respect. A plausible explanation again relates to crystal ionicity. A pure GaAs crystal has an effective ionicity of about 33%. An impurity within the lattice 'distorts' this ionicity factor. This distortion can be avoided if both the electronegativity and the polarizability (i.e. the major components of ionicity) of the impurity resemble those of the replaced host atom. Both Ga and Cr exhibit an electronegativity of 1.6. Their ionic radii are nearly identical; thus their ionic polarization is similar. Among all the impurity acceptors, Cr appears unique in this respect. Thus Cr may interchange freely with Ga in the lattice and not distort the ionicity factor of the host lattice. With this free interchange, the Fermi level is left as the dominant driving factor to determine how much Cr shall become electrically active. In contrast, reconsider the case of Zn diffusion and its displacement of background Mn [3]. Here it is noted that both impurities act as acceptors on Ga sites and therefore the Fermi driving force difference is negligible. From an ionicity viewpoint, however, the electronegativity of Zn matches that of Ga and their ionic radii (polarizability) are also much closer than that of Mn and Ga. Thus Zn 'distorts' the ionicity of the GaAs host much less than does Mn and may account for its very fast diffusion and Mn displacement capability in GaAs.

As an example of ionicity affecting impurities on anion sites, consider first the examples of low level Se implants achieving activation efficiencies >100% [1, 2]. Since the ionicity of either the Ga-Se or the Si-Se cation-anion bond is much closer to that of the host lattice ionicity than is the ionicity of the Si-Si bond, the former bonds distort less and are to be preferred. Thus it is quite plausible that the introduction of Se into a GaAs crystal containing self-compensating and spectroscopically invisible Si-Si nearest neighbour complexes breaks some of these complexes and substitutes electrically active Si-Se complexes instead. This action is similar to the previously discussed Cr:Te case wherein Te may complex with residual group IV background impurities. Presumably S, when co-implanted with Si, prevents the formation of Si-Si bonds [11] but (because of the larger ionicity of the Si-S bond) may not be able to break Si-Si bonds already formed.

Conclusions

Both electrically active and electrically neutral complexes have been shown to exist in GaAs. The electrically neutral complex can be broken and electrically activated by a judicious choice of impurity. The ionicity of the host lattice and the ability of impurities to 'distort' the ionicity factor have been shown as plausible driving forces affecting the electrical nature of the resultant crystal. The order of factors affecting impurity distribution/diffusion are thought to be stress, ionicity, Fermi level and concentration gradient. These factors, when known, can be used in a self-consistent manner to describe and predict the action of any given impurity in the GaAs lattice. Although the ionicities of various complexes within the GaAs lattice are largely unpublished and must be interpolated, an empirical determination is suggested by a systematic study of the relative displacements and/or electrical activations of any given impurity as changed by the introduction of another impurity. Of even greater immediate significance is that one can no longer naively expect to analyse the contribution of any single impurity without taking into account other included impurities, their concentrations and locations.

ACKNOWLEDGEMENTS

The author expresses appreciation to E. Swiggard for the growth of the Cr:Te and Cr:Sn boules, and to B. McCombe for their photoluminescence spectral analysis.

References

1. Donnelly, J.P., Bozler, C.O. and Lindley, W.T (1977). *Solid St. Electron.,* **20,** 273
2. Favennec, P.N., Henry, L. and L'Haridon, H. (1978). *Solid St. Electron.,* **21,** 705
3. Peart, R.F., Weiser, K., Woodall, J. and Fern, R. (1966). *Appl. Phys. Lett.,* **9,** 200
4. Magee, T.J., Peng, J., Hong, J.D., Evans, C.A. Jr, Deline, V.R and Malbon, R.M. (1979). *Appl. Phys. Lett.,* **35,** 277
5. Magee, T.J., Peng, J., Hong, J.D., Deline, V.R. and Evans, C.R. Jr (1979). *Appl. Phys. Lett.,* **35,** 615
6. Evans, C.A., Deline, V.R. and Sigmon, T.W. (1979). *GaAs IC Symposium,* Lake Tahoe, Nevada. Research Abstracts Paper No. 15
7. Anderson, W.T., Christou, A. and Davey, J. (1978). *IEEE J. Solid State Circuits,* **SC-13,** 430
8. Collins, D.M. (1979). *Appl Phys. Lett.,* **35,** 67
9. Spitzer, W.G. and Allred, W.P. (1968). *J. Appl. Phys.,* **39,** 4999
10. Vaidyanathan, K.V., Anderson, C.L., Dunlap, H.L., Kamath, G.S. and Krumm, C.F. (1979). *GaAs IC Symposium,* Lake Tahoe, Nevada. Research Abstracts Paper No. 17
11. Oakes, J.G., Degenford, J.E. and Eldridge, G. (1980). *GaAs Monolithic Microwave Subsystem Technology,* Base Report N00014-78-C-0268, Westinghouse Electric, 7.1
12. Rao, E.V.K, Duhamel, N., Favennec, P.N. and L'Haridon, H. (1976). *Ion Implantation in Semiconductors,* p. 77. New York; Plenum Press
13. Berman, L.V., Solov'ena, E.V., Mil'vidskii, M.G., Nazhivina, L.N. and Sabanova, L.D. (1979). *Sov. Phys. Semicond.,* **13,** 388
14. Evans, C.A. Private communication
15. Kung, J.K. and Spitzer, W.G. (1974). *J. Appl. Phys.,* **45,** 2254
16. Stolte, C.A. (1975). *Technical Digest,* IEEE IEDM, p. 585

ACCEPTOR STATES AND CORE SHIFTS IN Al$_x$Ga$_{1-x}$As

A. BALDERESCHI, K. MASCHKE and F. MELONI*
Laboratoire de Physique Appliquée, EPF-Lausanne, Switzerland

Abstract

The compositional dependence of the binding energy of shallow acceptors and of the anion and cation core shifts in Al$_x$Ga$_{1-x}$As are studied theoretically and compared with experiment.

Recent experiments have shown that some electronic properties of Al$_x$Ga$_{1-x}$As have an unexpected dependence on x. Zukotynski *et al.* [1] and Dingle [2] found that the ionization energy of the Ge acceptor increases strongly with Al content and is ~4.5 times larger in AlAs than in GaAs. Ludeke, Ley and Ploog [3] investigated the core shifts in Al$_x$Ga$_{1-x}$As and found that, relative to the top of the valence band, the position of the cation states is nearly independent of composition whereas that of the anion states is ~0.6 eV deeper in GaAs than in AlAs. In this contribution we analyse and explain these compositional dependences in terms of (a) the energy band structure of Al$_x$Ga$_{1-x}$As [4], and (b) the difference between the Al and Ga atomic potentials.

We first consider the acceptor ionization energy in Al$_x$Ga$_{1-x}$As. The Luttinger valence-band parameters, $\gamma_1(x)$, $\gamma_2(x)$ and $\gamma_3(x)$, are not known experimentally for all x and were therefore obtained from recent pseudopotential energy bands [4]. Since the band-structure values of the γ_i for GaAs differ by ~30% from the accepted values [5], the calculated $\mathbf{k} \cdot \mathbf{p}$ matrix elements which enter the definition of the γ_i were scaled in order to fit the GaAs data. The same scaling parameters were then used for all x values since no experimental data are available for x≠0. The spin-orbit splitting $\Delta_{so}(x)$ was taken from [6] and the low-frequency dielectric constant $\epsilon_0(x)$ was linearly interpolated between the experimental values 12.6 for GaAs [7] and 9.7 for AlAs [8]. We used the dielectric function

$$\epsilon(q,x)=\epsilon_0(x)(q^2+\alpha^2)/(\epsilon_0(x)q^2+\alpha^2)$$

where $\alpha=1.2$ a.u. is independent of x and is a value which fits the calculated dielectric functions for III-V compounds [9]. The parameters used in the calculations are summarized in Table 1.

*Also GNSM-CNR, Istituto di Fisica, Università Cagliari, Italy.

Table 1   **Values of the parameters $\gamma_1(x)$, $\gamma_2(x)$, $\gamma_3(x)$ and $\Delta_{SO}(x)$ used in the calculations. The corresponding values of the electron effective mass at Γ are also given**

	$x = 0$	$x = 0.2$	$x = 0.4$	$x = 0.6$	$x = 0.8$	$x = 1$
$\gamma_1(x)$	6.85	6.00	5.39	4.95	4.60	4.34
$\gamma_2(x)$	2.10	1.67	1.37	1.14	0.97	0.82
$\gamma_3(x)$	2.90	2.51	2.22	2.01	1.84	1.72
$\Delta_{SO}(x)$ (meV)	340	318	300	286	280	280
$m_c{}^*(x)/m_0$	0.0665	0.0785	0.0902	0.1012	0.112	0.123

The acceptor ionization energy, $E_i(x)$, was calculated with the method described in [10] for Si and Ge and the results are given in Figure 1. The experimentally observed strong increase of $E_i(x)$ with Al concentration is well reproduced by our calculation but the dependence on x is not linear as found by Dingle [2], who proposes $E_i(x)=(40+140x)$ meV. We find that the large value of $E_i(x=1)$ is very sensitive to all parameters, in particular to α and to the γ_i. In Figure 1 we also give the values of $E_i(x)$ calculated by neglecting the contribution from the split-off valence band $(\Delta_{SO}\rightarrow\infty)$ and/or that from the dielectric dispersion $(\alpha\rightarrow\infty)$. The results clearly indicate that *both* effects are responsible for the strong increase of $E_i(x)$ with x.

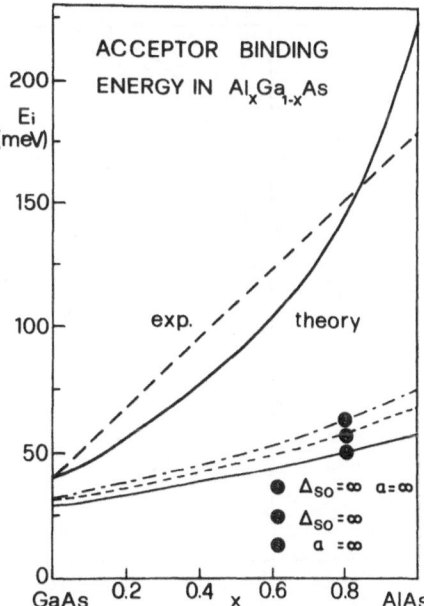

Figure 1   Theoretical and experimental ionization energy $E_i(x)$, of shallow acceptors in $Al_xGa_{1-x}As$. The curves labelled 1—3 give the values of $E_i(x)$ without the contribution from the split-off valence band $(\Delta_{SO}\rightarrow\infty)$ and/or that from the dielectric dispersion $(\alpha\rightarrow\infty)$.

We now consider the core shifts which are due to the crystal field produced by all ions and all valence electrons over the volume occupied by core electrons. Since the latter are very localized around their nucleus, the smooth crystal pseudopotential used in the energy-band calculation [4] is useless. We have therefore calculated hard-core ionic pseudopotentials for Al and Ga obtained from Hartree-Fock calculations for the Al^{3+} and Ga^{3+} ions [11]. The angular-momentum dependent ionic pseudopotential is

$$V_\ell(r) = -\frac{Ze^2}{r} + \delta V_\ell(r)$$

where $Z=3$ is the atomic valence and $\delta V_\varrho(r)$ is the deviation from Coulomb behaviour represented in Figure 2 for $\ell=0$ (the results for $\ell=1$ are similar). In Figure 2 we also give the function $r\Psi(r)$ for the outer core electron of each atom in order to show that the ionic pseudopotential of Ga is more attractive than that of Al for $r>1$ a.u. due to the incomplete screening of the Ga nucleus by the 3d electrons (these electrons are absent in the Al core). This observation allows us to understand the different behaviour of the anion and cation core shifts.

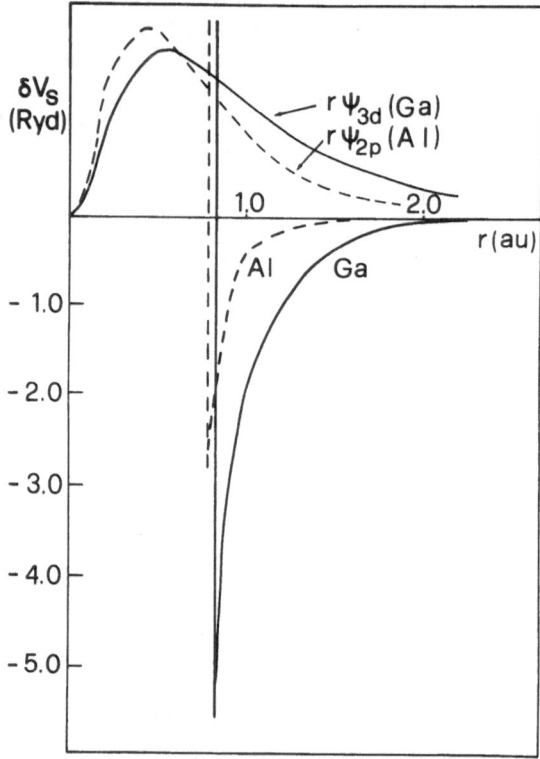

Figure 2   Deviation from Coulomb behaviour of the ionic pseudopotentials of Al^{3+} and Ga^{3+} for electrons with angular momentum $\ell=0$. The functions $r\Psi(r)$ for the Ga 3d and the Al 2p electrons are also represented in arbitrary units.

As the Ga potential is more attractive than that of Al at intermediate distances, the valence electronic charge, which is rather concentrated around As for all x values, will expand somewhat towards the cations with increasing Ga concentration and thus the As core states will shift to lower energy as observed experimentally. On the contrary, the valence charge around the cations will not depend appreciably on composition and therefore the cation core shifts will be rather small.

References

1. Zukotynski, S., Sumski, S., Panish, M.B. and Casey, H.C. Jr (1979). *J. Appl. Phys.*, **50**, 5795
2. Dingle, R. (1979). Private communication as cited in [1]
3. Ludeke, R., Ley, L. and Ploog, K. (1978). *Solid St. Commun.*, **28**, 57
4. Baldereschi, A., Hess, E., Maschke, K., Neumann, H., Schulze, K.R. and Unger, K. (1977). *J. Phys. C : Solid St. Phys.*, **10**, 4709
5. Hess, K., Bimberg, D., Lipari, N.O., Fischbach, J.V. and Altarelli, M. (1976). *Proc. XIII Int. Conf. Phys. Semiconductors* (Marves Rome), p. 142
6. Berolo, O. and Woolley, J.C. (1972). *Proc. XI Int. Conf. Phys. Semiconductors* (PWN, Warsaw), p. 1420
7. Stillman, G.E., Larsen, D.M., Wolfe, C.M. and Brandt, R.C. (1971). *Solid St. Commun.*, **9**, 2245
8. Monemar, B. (1970). *Solid St. Commun.*, **8**, 2121. This paper gives $\epsilon_\infty = 8.4$, ϵ_0 is obtained by using $\epsilon_0/\epsilon_\infty = 1.156$ as for GaAs
9. Walter, J.P. and Cohen, M.L. (1970). *Phys. Rev. B*, **2**, 1821
10. Lipari, N.O. and Baldereschi, A. (1978). *Solid St. Commun.*, **25**, 665
11. Biémont, E. Private communication

PHOTO-IONIZATION WITH PHONON PARTICIPATION IN SEMICONDUCTORS

J.M. NORAS
Wolfson Institute of Luminescence, School of Physical Sciences,
University of St Andrews, St Andrews, Fife, UK

Abstract

When the photo-ionization spectrum of an impurity in a semiconductor is measured, it is sometimes possible to find the ionization energy by suitable interpretation of the data. In many cases this has been done by fitting to a simple power law, in the derivation of which it is assumed that phonon participation may be neglected. When there is strong electron-phonon coupling it is not clear that this approach is valid, and the interpretation of the threshold energy obtained in this way is ambiguous.

In this paper we consider a general model of electron-phonon coupling and obtain an expression for the photo-ionization cross-section as a function of photon energy. With sufficient experimental information it is possible to fit the data using this expression. Where less is known about the electronic or vibrational factors, a knowledge of the general cases allows one to decide whether a straightforward power law approximation may be used. If it is valid to interpret the data by fitting to a power law, the threshold obtained is the optical ionization energy.

The theory of impurity photo-ionization has been discussed by many authors, but only a few models have been proposed for particular cases where phonon participation is included [1—6]. The present paper outlines a general approach to this problem, and gives a specific case to illustrate the results. The primary intention is to present a model that is readily comparable with experiment.

It is often assumed that electron-phonon coupling is negligible in photo-ionization, an empirical justification being that in many cases data can be fitted to a purely electronic model. Naturally, discrepancies are found especially at energies near and below photo-ionization thresholds where phonon structure and thermal broadening may occur. A more serious difficulty is the interpretation of threshold energies found in this way, since it is not clear whether these should correspond to thermal trap depths or to other quantities.

Suppose the purely electronic form for the absorption cross-section $\sigma(h\nu)$ is

$$\sigma(h\nu) = \nu^{-1} \sigma_{el}(E), \qquad E = h\nu - E_I \geqslant 0,$$
$$= 0, \qquad\qquad E < 0.$$

E_I is the ionization threshold energy. The purely electronic case is equivalent to a static lattice model. Since no vibrational kinetic energy is involved, conservation of energy ensures that a photon of energy $h\nu$ places a free hole or electron into a state E above the relevant band-edge.

If phonon participation is allowed, then in the Born-Oppenheimer and Condon approximations the electronic matrix elements remain unaffected.

Since energy may be exchanged with the lattice, energy conservation allows the ionized electron or hole to have any energy E' such that

$$E' + \epsilon_j = h\nu - E_I.$$

The ϵ_j are changes in vibrational energy which may involve the creation or absorption of several phonons of different frequencies. Thus

$$\sigma(h\nu) = \nu^{-1} \int \sum_j \sigma_{el}(E')P_j \, \delta(h\nu - E_I - E' - \epsilon_j)dE', \qquad E' \geqslant 0.$$

P_j represents the probability of a change in vibrational energy by ϵ_j. As a simple example, the single frequency configuration co-ordinate model gives:

$$\epsilon_j = j\hbar\omega, \qquad j = \ldots, -1, 0, 1, \ldots$$

and

$$P_j = \exp[-S_0(2\bar{n} + 1)] \left(\frac{\bar{n}+1}{\bar{n}}\right)^{j/2} I_j[2S_0(\bar{n}^2 + \bar{n})^{1/2}].$$

Returning to the general case, it is most convenient to develop the theory initially for zero temperature, since we may assume the ordering $\epsilon_i > \epsilon_j$ if $i > j$. Then ϵ_1 corresponds to the zero phonon transition and

$$\sigma(h\nu) = \nu^{-1} \sum_{j=1} P_j \sigma_{el}(E - \epsilon_j)$$

where $E - \epsilon_j \geqslant 0$ and $E = h\nu - E_I$.

Define L such that $E - \epsilon_j < 0$ for all $j > L$ and consider

$$\sigma(h\nu) = \nu^{-1} \sum_{j=1}^{L} P_j \sigma_{el}[(E - \theta) - (\epsilon_j - \theta)].$$

Suppose that E is sufficiently large so that $\sum_{j=1}^{L} P_j \cong 1$, and we may neglect the high energy vibrational processes. Then a Taylor expansion about $x = E - \theta$ gives

$$\sigma(h\nu) = \nu^{-1} \sigma_{el}(x) \left(\sum_{j=1}^{L} P_j\right) - \frac{d\sigma_{el}(x)}{dx} \sum_{j=1}^{L} P_j(\epsilon_j - \theta) + \ldots.$$

If we choose $\theta = \sum_{j=1}^{L} P_j \epsilon_j$ then

$$\sigma(h\nu) \cong \nu^{-1} \sigma_{el}\left(E - \sum_{j=1}^{L} P_j \epsilon_j\right) + \ldots.$$

If the correction terms are sufficiently small, then fitting the experimental data to the purely electronic form gives the threshold at $E_I + \theta$, where

$$\theta \simeq \overline{\epsilon_j} = \sum_{j=1}^{N} P_j \epsilon_j.$$

Thus the apparent threshold is different from the zero phonon threshold E_I.

It is easy to show that this result is true for all temperatures, the only change being that as thermal broadening makes the high-energy vibrational processes more significant, with rising temperature there will be an increase in the width of the region near threshold where deviations from the purely electronic behaviour are found.

Any temperature dependence of E_I or $\overline{\epsilon_j}$ will produce corresponding changes in the threshold.

The above result is useful if the photo-ionization bandwidth depends mainly on electronic factors and if the correction terms, which depend on higher moments of the phonon coupling, become small as x increases. In many systems these conditions are quite realistic and could be verified by substitution of the particular factors. It is satisfactory that the model does not require fine details of the electron-phonon coupling and that it applies in the region above threshold where data are most accessible because of the increasingly strong absorption.

To illustrate the behaviour above threshold and to confirm that the assumptions might hold in a particular case of experimental interest, take a power law $\sigma_{el}(E) = E^{s-1}$, together with a Gaussian form for the phonon spectrum. The result

$$\sigma(h\nu) = \nu^{-1}\frac{k}{\sqrt{\pi}}\exp(-k^2 x^2/2)(2k^2)^{-s/2}\Gamma(s)D_{-s}(-\sqrt{2}kx)$$

may be derived, where x is the energy above the shifted threshold. D_{-s} is a parabolic cylinder function and k is inversely related to the width of the coupled mode spectrum. When x is greater than zero by a few multiples of k^{-1}, $\sigma(h\nu)$ tends quickly to the form $\sigma(h\nu) = \nu^{-1}x^{s-1}$. Thus, plotting $(\nu\sigma(h\nu))^{2/3}$ against energy permits a linear extrapolation to an intercept on the energy axis at the position of the optical ionization threshold.

For example, if the purely electronic form contains a three-halves dependence on the energy of the free carrier produced, plotting $(\nu\sigma(h\nu))^{2/3}$ produces a linear graph away from the threshold region [8]. The intercept gives the energy threshold which corresponds to a vertical transition on a configuration co-ordinate diagram.

References

1. Kukimoto, H., Henry, C.H. and Merritt, F.R. (1973). *Phys. Rev. B*, **7**, 2486
2. Kopylov, A.A. and Pikhtin, A.N. (1976). *Sov. Phys. Semicond.*, **10**, 7
3. Jaros, M. (1977). *Phys. Rev. B*, **16**, 3694

4. Monemar, B. and Samuelson, L. (1977). *Phys. Rev. B,* **18,** 809
5. Piekara, U., Langer, J.M. and Krukowska-Fulde, B. (1977). *Solid St. Commun.,* **23,** 583
6. Stoneham, A.M. (1979). *J. Phys. C,* **12,** 891
7. Fitchen, D.B. (1968). In *Physics of Colour Centers.* New York; Academic Press
8. Szawelska, H.R. and Allen, J.W. (1979). *J. Phys. C,* **12,** 3359

INFLUENCE OF DEEP LEVELS ON SCHOTTKY BARRIER FORMATION

G.P. SRIVASTAVA
Physics Department, New University of Ulster,
Coleraine, Northern Ireland

Abstract

It is shown that Schottky barrier formation at metal-semiconductor interfaces can be influenced by deep bulk defects. We consider, in particular, the effect of a phosphorus vacancy and also an oxygen substituted at a phosphorus site on the Schottky barrier formation at InP/Al and InP/Ag interfaces. The theoretical Bassani-Iadonisi-Preziosi-Jaros method for dealing with such deep impurities predicts localized and resonant energy levels which are helpful in understanding some recent measurements of the metal-InP interface.

Within the past three decades or more many theories have been proposed for Schottky barrier (SB) formation at interfaces between metals and semiconductors. However, there is still no general agreement as to what are the most important processes involved. Recently, Williams et al. [1, 2] have measured SB heights for a range of metals deposited at room temperature on the (110) surface of InP. In their experiments, both reactive and unreactive metals were considered. For reactive metals like Al they measure low apparent barrier heights at room temperature which increase somewhat when the sample temperature is lowered to 77 K. On the contrary, for unreactive metals such as Ag, Au or Cu they measure an SB height of around 0.5 eV, even at room temperature. When clean cleaved InP surfaces are exposed to oxygen or chlorine before deposition of these unreactive metals, the contacts formed are nearly ohmic (SB height less than 0.2 eV). These authors thus find that the metal work function has little influence on the contact behaviour and suggest that chemical effects are of crucial importance (see Table 1). However, no exact mechanism has been put forward for these observations.

In this paper we attempt to explain the room temperature experimental results of Williams et al. by using the 'defect model' proposed by Lindau et al. [3]. First, we notice that there is no surface state pinning of the Fermi level [4] in the gap of the atomically clean, cleaved, relaxed (110) surface of InP. Deposition of unreactive metals gives rise to 'metal-induced states' [5] which would pin the Fermi level in the band gap of InP. This should explain the observations of Williams et al. for unreactive metals on InP. At room temperature reactive metals on InP would react, resulting in the depletion of phosphorus. Also, at room temperature phosphorus may be replaced by oxygen. Thus it is of relevance to consider the nature of defects such as phosphorus vacancies and oxygen on substitutional sites for SB formation in the above experiments.

Table 1 **SB heights for various metals on InP as measured by Williams** *et al.* [1]

System	SB height (eV) at room temperature
InP + Al	<0.2
InP + Ag	0.5
Oxygen exposed InP + Ag	<0.2

With the above arguments we have calculated, using the Bassani-Iadonisi-Preziosi-Jaros method [6], the electronic states of a phosphorus vacancy and an oxygen substituting for a phosphorus atom in the band gap of InP. The results are presented in Figure 1. Essentially, a single phosphorus vacancy forms a deep level a_1, containing two electrons, about 0.12 eV above the valence band maximum and a level t_2 containing a single electron and overlapping the conduction band. Oxygen at a phosphorus site again forms a single donor level approximately 0.1 to 0.2 eV below the conduction band minimum. (The traps V_p and O in InP are neutral when all the levels in Figure 1 are occupied.) These defects therefore pin the Fermi level close to the conduction band minimum. This interpretation, however, demands a high density of such defects [2]. We conclude, therefore, that the room temperature SB height measurements of Williams *et al.* for reactive metals on InP can be explained on the basis of the defect model.

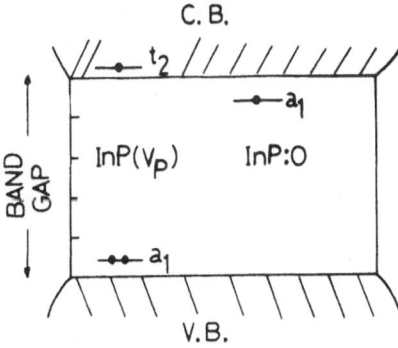

Figure 1 Calculated electronic energy levels in the band gap of InP for a single phosphorus vacancy (V_p) and for an oxygen substitution for a phosphorus (InP:O). Solid circles show the number of electrons occupying the levels. The traps InP:V_p and InP:O are neutral when all the levels are occupied.

It might be argued that here we have considered bulk defects whereas surface defects are of more concern to us. However, there is some evidence (on the basis of lattice relaxation around the defect, see also [7]) that as these defects move from the bulk towards the surface, the a_1 states stay more or less at the same energy while the t_2 state is brought down (towards the conduction band edge in our calculation) by only a few tenths of an eV. Therefore our bulk defect calculations give at least a guide to the understanding of SB formation at metal-InP interfaces.

References

1. Williams, R.H., Montgomery, V. and Varma, R.R. (1978). *J. Phys. C,* **11,** L735
2. Montgomery, V., McKinley, A. and Williams, R.H. (1979). *Surf. Sci.,* **89,** 635
3. Lindau, I., Chye, P.W., Garner, C.M., Pianetta, P., Su, C.Y. and Spicer, W.E. (1978). *J. Vac. Sci. Tech.,* **15,** 1332
4. McKinley, A., Srivastava, G.P. and Williams, R.H. (1980). *J. Phys. C,* in press
5. Zhang, H.I. and Schlüter, M. (1978). *Phys. Rev. B,* **18,** 1923
6. Srivastava, G.P. (1979). *Phys. Stat. Sol. (b),* **93,** 761
7. Daw, M.S. and Smith, D.L. (1979). *Phys. Rev. B,* **20,** 5150

IMPACT IONIZATION AND AUGER EFFECTS INVOLVING TRAPS

D.J. ROBBINS and P.T. LANDSBERG
Department of Mathematics, The University of Southampton,
Southampton SO9 5NH, UK

In this paper four theoretical models are considered for the discussion of low-field impact ionization and, its inverse process, Auger recombination, involving a single localized level. Four such processes are shown schematically in Table 1, where the corresponding Auger recombination and impact ionization rates are given for non-degenerate statistics. Here, n and p are electron and hole concentrations, n_t and p_t are concentrations of full and empty traps. The various possibilities are cut down (a) by considering only the process specified by X_1 in Table 1, although the present approach can be used for the other processes shown; and (b) by considering the band to be parabolic for energies above the minimum which are of the order of the trap depth, E_t. Here the probability $P(2')$ per unit time per unit volume, that a given energetic electron in a band state $2'$ will impact-ionize a bound state of energy $-E_t$, is discussed. The threshold for such a process lies at the electronic kinetic energy $E_{2'}=E_t$. For any band having the probability of finding a vacancy at states 1 and 2 (see Table 1) of order unity, and in particular for a non-degenerate band, $P(2')$ is independent of the statistical electron distribution in the band. One can then obtain the impact ionization coefficient X_1, and the Auger recombination coefficient T_1 by an integration over the states $2'$.

Table 1   **Impact ionization from traps**

Energy diagram for direct process				
Impact ionization rate	$X_1 nn_t$	$X_2 pn_t$	$X_3 np_t$	$X_4 pp_t$
Auger recombination rate (for which arrows must be reversed)	$T_1 n^2 p_t$	$T_2 npp_t$	$T_3 npn_t$	$T_4 p^2 n_t$

Earlier work [1] is improved both in the way it dealt with non-orthogonality effects in the matrix element, and also in deriving new estimates of P(2') and T_1 (and thus for X_1 also) for deep levels. In particular, four models are developed, two based on hydrogenic effective mass theory, labelled (a_1) and (b_1), and two based on a trap potential of δ-function type, the Lucovsky model, (a_2) and (b_2), the corresponding band state being a Bloch function for models (a_1) and (a_2), and a continuum Coulomb function for models (b_1) and (b_2). The work of [1] is based on model (a_1). The uncertainties in the theory are largely due to the lack of orthogonality between the wavefunctions employed to which particular attention is paid or, equivalently, to the degree of approximation involved in obtaining such wavefunctions.

Figure 1 shows some theoretical $(\sigma_{k_2{''}}, E_{2'})$ curves where $\sigma_{k_2{''}}$ is the cross-section associated with P(2'). These are valid for general trap depths E_t, and effective masses m^*. The cross-sections $\sigma_{k_2}(i)$ ($i = a_1, a_2, b_1, b_2$) are divided by

$$4\pi\, a_0^{*2} = \frac{\pi e^4}{\epsilon^2 E_t^2}$$

where a_0^* is the effective Bohr radius. This yields the solid state cross-section as it would be found in the atomic case, for one definite spin assignment. Experimental results for the atomic case are available, and are also shown for comparison in the form of 'measured cross-section divided by πa_0^2'. The horizontal co-ordinate is the energy $E_{2'}$ in units of (E_t/E_0) eV, where $E_0 = m_0 e^4/2\hbar^2 = 13.6$ eV. In the atomic case, $E_t = E_0$. This is a most valuable check, as the experimental solid-state studies yield only total impact ionization rates, or total cross-sections, and not P(2') (or $\sigma_{k_2{''}}$) as far as the authors are aware. Figure 1 shows that the use of *orthogonal* wavefunctions in model (b_1), which is equivalent to the Born approximation of atomic physics, significantly reduces the cross-section for higher energies of the ionizing carriers in state 2'. These results are highly model-sensitive. When the Auger recombination coefficient is inferred from P(2') by integration the results obtained are, however, much less model-dependent. Thus, although serious doubts must be raised concerning the earlier P(2') (curve a_1), the earlier T_1 is surprisingly reliable. The new models (b_1) and (b_2) discussed in this paper are, however, preferable in all cases.

The calculations of P(2'), and hence T_1 and X_1, just described have been treated purely on the basis of an electronic transition. The absence of an activation energy for T_1 leads to virtual independence of temperature. An improved calculation for each of the four models which takes into account the effect of lattice coupling to the bound state via the static coupling scheme has also been performed. A detailed analysis will be the subject of a future publication. The resulting multiphonon broadening of the Auger transition leads to a slightly enhanced temperature dependence, and is closely analogous to the usual theory of radiative transitions involving a bound state. The lack of orthogonality between the wavefunctions in models (a_1), (a_2) and

(b_2), which previously appeared relatively unimportant, is now crucial however, and can lead to a large overestimation of T_1 even for moderate coupling strengths. Thus only model (b_1) ought to be relied upon. The participation of the lattice leads to an increase (over the purely electronic estimate) in the calculated reaction constant by a factor of ~2 for moderate electron-phonon coupling, at 300 K in model (b_1). A similar effect is to be expected in models (a_1), (a_2) and (b_2). So, at room temperature at least, the purely electronic estimates can still provide a guide to the likely values of the reaction constants.

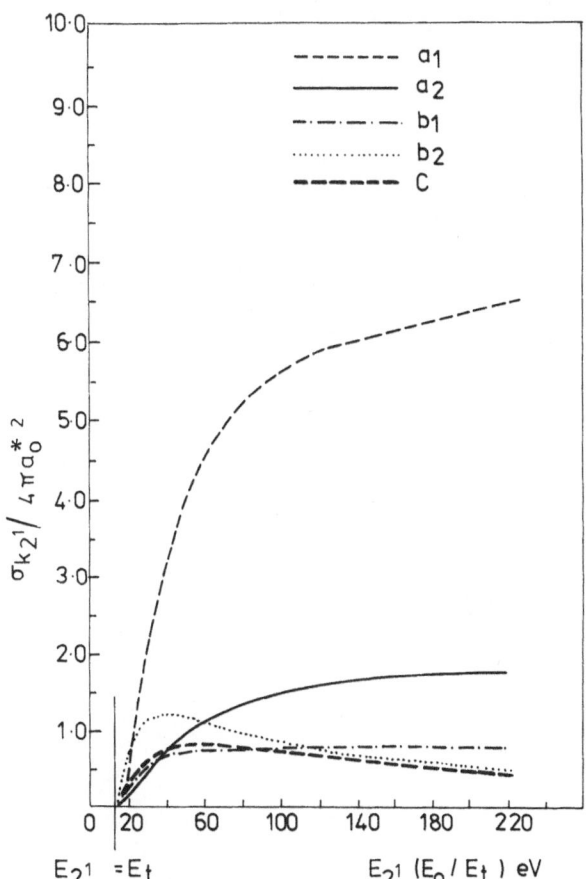

Figure 1 Impact ionization cross-sections for an electron of kinetic energy $E_{2'}$. The experimental results of [2] are also shown (curve (c)).

References

1. Landsberg, P.T., Rhys-Roberts, C. and Lal, P. (1964). *Proc. Phys. Soc.*, **84**, 915
2. Fite, W.L. and Brackman, R.D. (1958). *Phys. Rev.*, **112**, 1141

Section 5:
Influence of deep levels on device properties and material characterization

ELECTRON MOBILITY CALCULATIONS FOR Fe-DOPED InP

B.T. DEBNEY and P.R. JAY*
Plessey Research (Caswell) Limited, Allen Clark Research Centre,
Caswell, Towcester, Northants, UK

Abstract

Calculations are presented of Hall mobility and geometrical magnetoresistance mobility for n-type InP compensated with Fe. From an analysis of electrical measurements and a spectrochemical analysis to determine the Fe content of InP slices we conclude that in some cases only a small percentage of the Fe contained is electrically active as an electron acceptor, suggesting redundancy of the incorporated Fe. Analysis of Fe concentration estimates based on the segregation coefficient seems to give a better measure of the electrical activity of Fe than the total concentration of Fe, although the value quoted by Henry and Swiggard does not seem universally applicable to all growth conditions.

Introduction

As a dopant for producing high resistivity InP [1, 2], Fe has proved popular as a means of controlling the electrical behaviour of crystals that would otherwise possess a residual donor concentration of between 10^{15} and 10^{16} cm^{-3}. Semi-insulating substrates prepared in this way are used for epitaxial layer growth and as host material for ion-implanted structures in InP.

The influence of Fe impurities on the electrical behaviour of InP is of considerable interest for various reasons:

1. The unavoidable presence of Fe in ion-implanted structures.
2. Possible Fe intrusion into epilayers grown on semi-insulating substrates.
3. The risk of contamination of grown layers by transport of Fe from the starting materials (In and PCl_3) where it is frequently detectable.
4. Recent reports suggest that Fe, unintentionally present in Sn-doped conducting substrates [3], is capable of out-diffusing into epilayers grown thereupon.

One of the major device applications of InP lies in microwave transferred-electron oscillator structures which involve thin (~ 2–$5\,\mu m$) low doped ($\sim 10^{15}$ cm^{-3}) epilayers, usually prepared on Sn-doped ($\sim 10^{18}$ cm^{-3}) substrates. Since

---

*Now with Thomson-CSF, Domaine de Corbeville, 91401 Orsay, France.

Hall effect measurements are not possible in this situation, it has been found convenient to use geometrical magnetoresistance (GMR) as a method of determining mobilities of carriers in the low doped layers.

Using a theoretical representation of the InP:Fe system, the electrical properties (in particular the drift, Hall and GMR mobility) have been examined for varying degrees of Fe compensation in n-type material. By comparing the predicted quantities with published data and electrical and chemical assessments of this material, we demonstrate that Fe doping of InP is not always 100% electrically active and that the generally accepted value of the effective segregation coefficient is not universally applicable.

Theory and calculations

Calculations have been made of drift mobility, Hall mobility and geometrical magnetoresistance (GMR) mobility [4] for n-type InP compensated with Fe. The calculations are based on a direct numerical solution of the Boltzmann equation to obtain the electron probability distribution function in k-space, from which the various transport properties can be obtained [5, 6]. The conduction band in InP is sufficiently non-parabolic to justify taking account of the effect, and this is included using $k \cdot p$ band theory in the manner described by Rode [6].

The model that we have adopted for Fe in InP is the three charge state model suggested by Fung, Nicholas and Stradling [7], in which Fe behaves as a double acceptor. Fe is thought to substitute for In in the lattice with its neutral state, denoted Fe^{3+}, possessing a $3d^5$ electronic core configuration. At 300 K the occupation state of the first electron (Fe^{2+}) produces a level 0.66 eV below the conduction band edge, and the level associated with the ionization of the second electron from the doubly occupied Fe^{1+} state is 0.28 eV below the conduction band. The Fe^{2+} state has been identified by many authors [1, 7—10] with a good degree of agreement as to the energy of the level produced. The situation with regard to the second electron state is less certain. Although Fung et al. clearly see a transition corresponding to a level 0.28 eV below the conduction band, its identification with the Fe^{1+} state is tentative.

Calculations of the Hall mobility at 300 and 77 K are shown in Figures 1 and 2. The range of electron concentration for which the results are shown represents the approximate validity for the treatment of impurity scattering used; the difficulties that arise from encountering a multiple scattering regime at low carrier concentration have been discussed elsewhere [11]. For these calculations the Fermi level lies no deeper than 200 meV below the conduction band edge so that the results do not depend sensitively on the energies of the Fe acceptor levels. However, they do depend on whether Fe in InP is a single or a double acceptor. If the assumption that the level observed at $E_c - 0.28$ eV is associated with a second occupation state of the same centre is wrong, then the results shown in Figures 1 to 3 can be made valid for a single acceptor model by multiplying the Fe concentrations by a factor of 3. This takes account of the fact that in Born approximation the scattering rate is proportional to the square of the charge state.

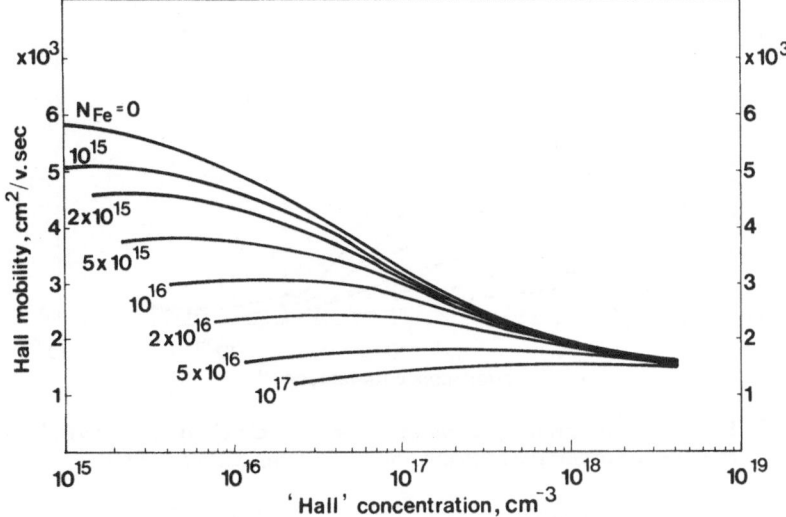

Figure 1 Calculated Hall mobility at 300 K for n-type InP compensated with Fe (concentrations are in units of cm^{-3}). The results shown here are for the model where Fe accepts two electrons. The calculations can be made applicable for a single acceptor model by multiplying the Fe concentrations by a factor of 3.

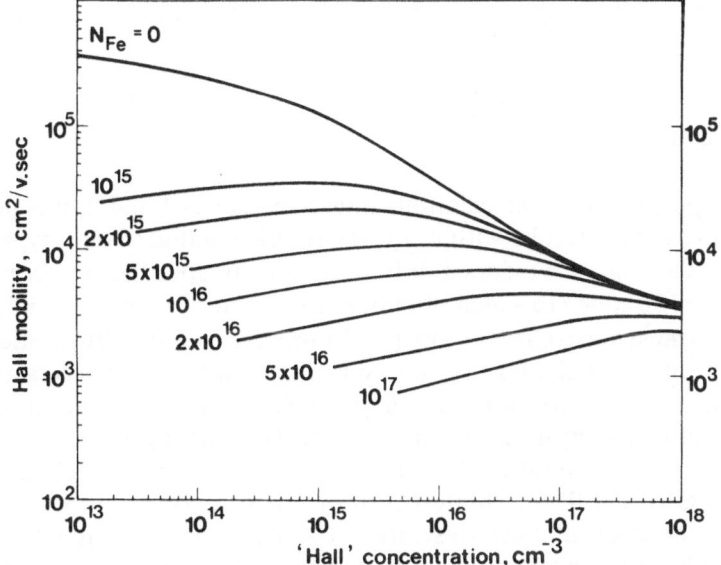

Figure 2 Calculated Hall mobility at 77 K for n-type InP compensated with Fe.

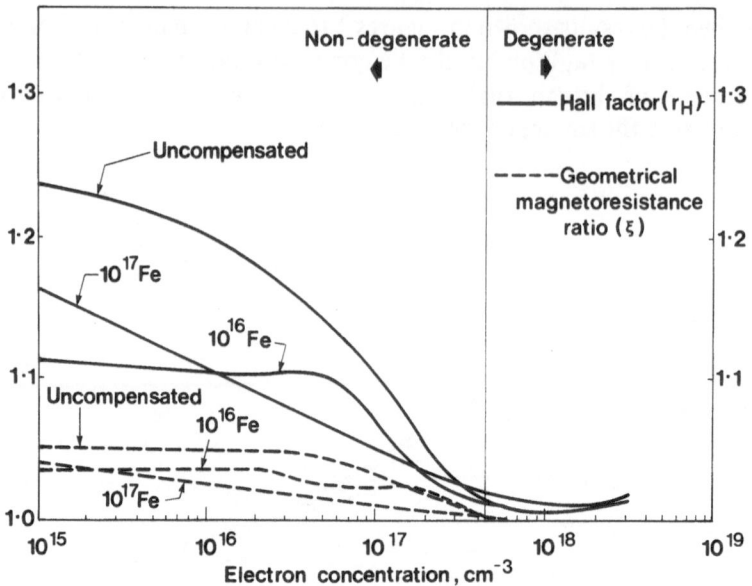

Figure 3   Hall factors and geometrical magnetoresistance ratio (GMR mobility/Hall mobility) as a function of electron concentration and Fe compensation at 300 K.

The calculations of GMR mobility have predicted that this quantity is very close to the magnitude of Hall mobility, particularly near room temperature. At 300 K the ratio, ξ, of GMR mobility to Hall mobility exceeds unity by no more than 5% under all conditions of compensation, and this bound rises to 15% at 77 K. Figure 3 illustrates the variation of Hall factor and ξ for several Fe concentrations at room temperature. The feature that ξ is very near unity has been confirmed experimentally by measurements on a range of InP material [12].

Discussion

Having performed the calculations shown in Figures 1 and 2, it is important to compare these results with available experimental data on n-type InP:Fe. The production of Fe-doped InP is frequently specified in terms of the amount of Fe added to a melt of undoped InP before an ingot is pulled by the liquid encapsulated Czochralski (LEC) process. As the distribution of Fe from the melt into the ingot is not well defined, it has been necessary for the purpose of a comparison to obtain direct chemical determinations of the Fe content of n-type crystals for which electrical data are available.

Samples of n-type InP:Fe have been made available from an ingot (L 781), LEC — grown at RSRE (Malvern) from a melt charged with insufficient Fe to make the ingot fully semi-insulating. Using the curves of Figures 1 and 2, it is possible to estimate the Fe concentration required to achieve particular measured values of Hall mobility and carrier concentration.

Table 1 shows Hall data and the associated predictions of the electrically significant Fe concentration, [Fe] 'elec.', assuming that the three charge state model prevails. This ingot was grown from a melt charged with 0.005 wt %Fe, and using a segregation coefficient K, as defined in [13], of 1.6×10^{-3} [2] we can estimate the values of Fe concentration shown in the right-hand column of the table, labelled [Fe] 'seg.'. It is interesting to observe that [Fe] 'elec.' and [Fe] 'seg.' are of the same order, but it is rather surprising when these quantities are compared with direct chemical determinations of the Fe concentration by atomic absorption measurements. For example, the chemically determined Fe concentration of slice L 781/41 is 1.3×10^{17} cm^{-3}, which represents more than an order of magnitude discrepancy from the 'expected' Fe content. As the starting material used for L 781 was n-type $<10^{15}$ cm^{-3}, it would appear that the [Fe] 'elec.' and [Fe] 'seg.' values are more compatible with the electrical behaviour than the chemical estimates, which would imply a sufficient excess of Fe to make the crystal semi-insulating, and this was not the case.

Table 1   **A comparison of estimates of the Fe impurity density in n-type InP slices**

Slice ingot / distance from seed (mm)	μ (Hall) $(cm^2V^{-1}s^{-1})$ 300/77 K	$n_{(Hall)}$ (cm^{-3}) 300/77 K	[Fe] 'elec.' (cm^{-3}) 300/77 K	[Fe] 'seg.' (cm^{-3})
L 781/'TOP'	4621/23,211	$2.3 \times 10^{15}/1.8 \times 10^{15}$	$2.0 \times 10^{15}/2.0 \times 10^{15}$	4.0×10^{15}
L 781/16	4684/23,500	$2.0 \times 10^{15}/1.6 \times 10^{15}$	$2.0 \times 10^{15}/2.0 \times 10^{15}$	5.0×10^{15}
L 781/18	3497/19,990	$3.0 \times 10^{15}/2.0 \times 10^{15}$	$6.0 \times 10^{15}/2.0 \times 10^{15}$	5.0×10^{15}
L 781/41	4265/17,944	$1.4 \times 10^{15}/0.9 \times 10^{15}$	$3.0 \times 10^{15}/2.5 \times 10^{15}$	7.0×10^{15}
L 781/50	4157/16,889	$1.2 \times 10^{15}/0.82 \times 10^{15}$	$3.0 \times 10^{15}/2.5 \times 10^{15}$	9.0×10^{15}

[Fe] 'elec.' denotes the Fe concentration required to explain the electrical measurements shown in columns 2 and 3. The Fe concentrations are obtained from calculations such as presented in Figures 1 and 2. [Fe] 'seg.' represents the estimated Fe concentration in the slices as determined from the Fe content of the melt and the segregation coefficient quoted by Henry and Swiggard [2].

Taking note of this anomaly, it is worthwhile to remember that the reported value of $K=1.6 \times 10^{-3}$ was determined chemically, but on the basis of the Fe^{3+} spin concentration in a p-type ingot, which should realistically represent the total substitutional Fe present. Henry and Swiggard [2] also reported an ingot doped with 0.0075 wt % Fe in the melt, for which the seed end of the crystal showed n-type electrical behaviour consistent with their value of effective segregation coefficient.

Iseler [10], however, found difficulty in reconciling the electrical data for his (unusually high mobility) semi-insulating ingots, with the anticipated Fe distribution and the purity of undoped boules. He suggested that either the segregation coefficient was wrong, or that unknown extra donors were associated with the Fe presence.

Mizuno and Watanabe [1] found that a large charge of 0.15 wt % Fe was necessary to obtain 10^7 Ω cm ingots from their 3—5×10^{16} cm^{-3} starting material and the associated low mobility of this material may be a consequence of the high Fe doping. In contrast, Ippolitova et al. [8] used 1—3×10^{16} cm^{-3} starting material, and achieved relatively high mobility, despite detecting a chemical concentration of 0.5—2.0×10^{17} cm^{-3} Fe.

In the light of this conflicting evidence, we can propose two new conclusions concerning Fe doping of InP. Firstly, it would appear that in some cases the Fe present in the crystal is not all electrically active, and that substantial redundancy is suggested by the analyses reported here. Secondly, the evidence, and in particular the chemical data, points heavily to the fact that the effective segregation coefficient $K = 1.6 \times 10^{-3}$ is not universally applicable, and suggests that the solubility of Fe as a deep acceptor may well be a function of growth conditions. This last fact should not be too surprising when considering the incorporation of a dopant into a binary compound which has a significant finite range of solid non-stoichiometry at growth temperatures.

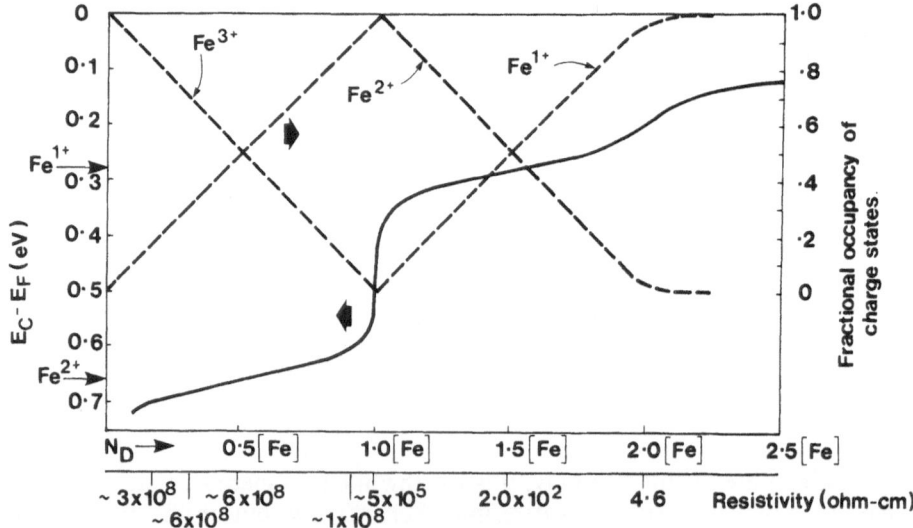

Figure 4   The variation of Fermi level and the relative occupancies of the different charge states of Fe in InP (three charge state model) as a function of donor concentration for Fe-doped InP at 300 K. Also shown on the bottom scale are the approximate magnitudes of resistivity against Fermi level position for InP containing 10^{16} cm^{-3} Fe impurities.

Figure 4 shows how the position of the Fermi level and the associated relative populations of the Fe charge states vary as a function of doping for the double acceptor model of InP:Fe. Also shown as a function of Fermi level position are the approximate magnitudes of room temperature resistivity for 10^{16} cm^{-3} Fe doping. As the Fermi energy approaches the Fe^{2+} level the density of holes in the valence band becomes significant and in the estimate of resistivity we have adopted a hole mobility of 100 cm^2 V^{-1} s^{-1}. An interesting feature of these calculations is the transition from n- to p-type conduction and a resulting limiting resistivity value of about 6×10^8 Ω-cm. This suggests that the maximum resistivity obtainable in Fe-doped InP is of the order of 10^8 Ω-cm and this magnitude could be achieved over a range of compensation ratios covering both n- and p-type conduction.

The Fermi energy calculations have been performed for a relative degeneracy of the Fe^{3+} to Fe^{2+} states of unity. The effect of using a value greater than unity would be to increase the population of the Fe^{3+} state when the Fermi level lies above the acceptor level. A point worth commenting on here is that when the Fermi level lies close to the Fe^{2+} level the relative occupancies of the charge states change rapidly with small changes in Fermi energy and we would caution against the identification of the total Fe concentration with the Fe^{3+} concentration estimated from ESR measurements on high resistivity samples [2].

In conclusion, we have demonstrated that mobility is a sensitive parameter with which to observe the presence of Fe in n-type InP, and have shown that the electrically measured Fe concentration does not necessarily correspond to the amount determined chemically. Bearing in mind reservations about the charge state distributions in high resistivity material, it could be valuable to combine chemical, electrical and ESR data to identify the Fe situation in various ingots.

ACKNOWLEDGEMENTS

The authors wish to thank Mr D.A. Stewart for the atomic absorption measurements. They also wish to thank Dr B. Cockayne, Mr W.R. McEwan and Mr W.E. Willgoss of RSRE (Malvern) for supply of samples and electrical measurements, and for valuable discussions. This work has been carried out with the support of the Procurement Executive of the Ministry of Defence; it is sponsored by DCVD and is published with the permission of the Directors of Plessey Research (Caswell) Limited.

References

1. Mizuno, O. and Watanabe, H. (1975). *Electronics Lett.*, **11**, 118
2. Henry, R.L. and Swiggard, E.M. (1978). *J. Elec. Mat.*, **7**, 647
3. Chevrier, J., Armand, M., Huber, A.M. and Linh, N.T. (1979). 21st Electronics Materials Conf. (Boulder). To be published in *J. Elec Mat.*

4. Jervis, T.R. and Johnson, E.F. (1970). *Solid State Elec.*, **13**, 181
5. Rode, D.L. (1973). *Phys. Stat. Sol.(b)*, **55**, 687
6. Rode, D.L. (1975). *Semiconductors and Semimetals* (Eds R.K. Willardsen and A.C. Beer), vol. 10, ch. 1. London; Academic Press
7. Fung, S., Nicholas, R.J. and Stradling, R.A. (1979). *J. Phys. C: Solid St. Phys.*, **12**, 5145
8. Ippolitova, G.K., Omel'yanovskii, E.M., Pavlov, N.M., Nashel'skii, A.Ya and Yakobson, S.V. (1977). *Sov. Phys. Semicond.*, **11**, 773
9. Pande, K.P and Roberts, G.G. (1976). *J. Phys. C: Solid St. Phys.*, **9**, 2899
10. Iseler, G.W. (1979). *Inst. of Phys. Conf. Ser.*, **45**, 144
11. Debney, B.T. and Jay, P.R. To be published in *Solid State Electronics*
12. Brookbanks, D.M. Private communication
13. Parr, N.L. (1960). *Zone Refining and Allied Techniques*, p. 8. London; Whitefriars Press

MATERIAL AND STRUCTURE FACTORS AFFECTING THE LARGE-SIGNAL OPERATION OF GaAs MESFETs

P.H. LADBROOKE and A.L. MARTIN*
Department of Solid-State Electronics,
University of New South Wales, Kensington, NSW 2033, Australia

Abstract

For typical recessed-gate FET structures, the drain characteristics at high drain voltages exhibit, to one degree or another, an increase in drain current for gate voltages where the channel would ordinarily be pinched off. In some devices, this 'soft' current is observed to depend on the speed of the bias sweep or pulse ($80\,\mu s$, $50\,Hz$) used to measure it. One possible consequence of soft breakdown is that it may restrict the available voltage swing across the device, and hence the r.f. power available in microwave circuit applications. It is shown that some features of the breakdown characteristic are consistent with the existence of deep levels in the n-channel and buffer epitaxial layers. Capacitance transient measurements upon finished FET structures reveal the presence of deep levels at concentrations above $10^{15}\,cm^{-3}$.

1 Introduction

It has been shown by a number of workers [1—3] that the characteristics of Schottky-barrier gate GaAs FETs (MESFETs) are affected by deep trapping levels at the substrate/n-layer interface. A common observation has been that the introduction of a semi-insulating buffer layer between the substrate and the channel layer markedly reduces such unwanted effects as looping in the drain characteristics, low drain source voltage breakdown, low transconductance, low power gain and large noise figure variations [3]. A further observation is that the drain characteristics measured under slow sweep conditions can be different from the pulsed characteristics [4], even with a buffer layer.

In large signal (power) applications of GaAs FETs, the drain characteristics provide a useful guide to the output power capability of the device. If the operating frequency is below the current cut-off frequency, the output power available can be estimated from the drain characteristics [5] as:

$$P \simeq I_F (V_{GD} - V_P)/8 > I_{DSS} (V_{GD} - V_P)/8 \tag{1}$$

*Solid-State and Quantum Electronics Group, Telecom Australia Research Laboratory, Clayton 3168, Australia.

where I_F is the channel current when the gate is driven into forward bias, V_{GD} is the gate-drain breakdown voltage, and V_P is the pinch-off voltage. A factor of major practical concern, therefore, is to establish some rules for maximizing I_{DSS} and V_{GD} by optimizing both material and processing technologies, and the device structure. The fact that experimentally the relationship between the pulsed and slow sweep characteristics is varied complicates this task.

In order to try and correlate differences in the drain characteristics with the existence of deep levels, this paper considers, first, capacitance transient measurements to determine the emission parameters of levels in finished FET structures and, second, a FET model with deep levels, which is consistent with the specific type of behaviour shown in Figure 1, i.e. where the pulsed characteristics for the most part lie below the d.c. characteristics.

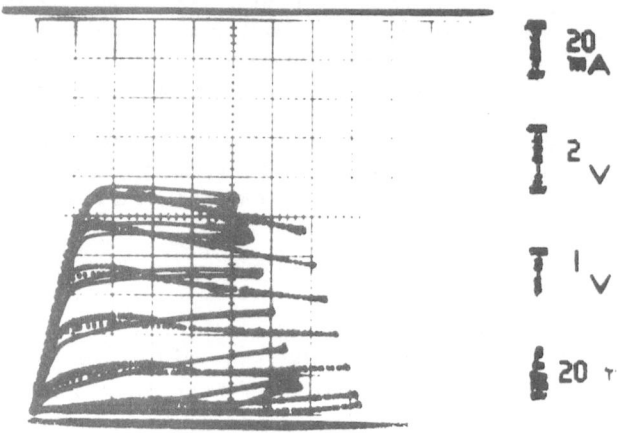

Figure 1   Slow-sweep (full) and pulsed (broken) characteristics of shallow recess FET.

2   Device technology

The materials technology and characterization procedures are as described by Crossley *et al.* [6]. Typical parameters for a power FET are an as-grown n-channel layer thickness of 0.6 μm, doped to 10^{17} cm^{-3}, with an undoped buffer layer of thickness >3 μm on a Cr-doped substrate (see Figure 4). For some devices a gate recess is etched in the as-grown layer, while for others the layer is first thinned using anodization [7] before etching the gate recess. Aluminium gate metallization and In-Ge-Au alloyed source/drain contacts as described in [8] are used.

Figure 2 shows the drain characteristics of a shallow recess power FET cell measured with 80 μs pulses. Under pulsed conditions these devices typically withstand drain-source voltages of 35 V before failure occurs. Using a slow bias sweep, however, the drain characteristics are quite different (Figure 1);

in particular, there appears an increase in drain current, for gate voltages where the channel would ordinarily be pinched off, which is often accompanied by light emission close to the edge of the gate nearest the drain [4].

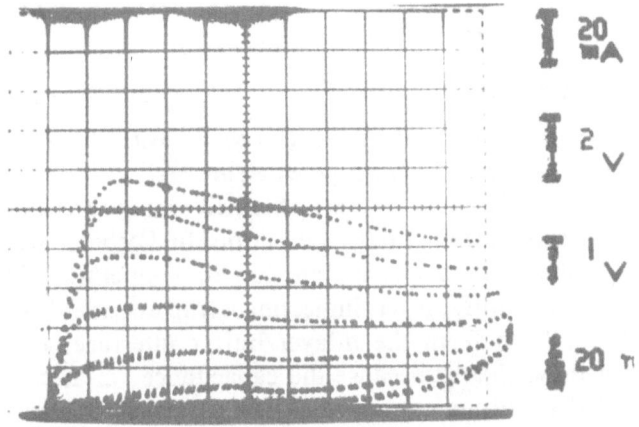

Figure 2   Extended drain characteristics (pulsed) of shallow recess FET.

It has been argued [9] that the above breakdown phenomenon and associated light emission is the result of avalanche multiplication somewhere in the gate-drain region. Given the exponential dependence of ionization rate upon the electric field, it would then be reasonable to expect that relaxation effects in the field profile within the device, due to charge exchange with deep levels, should lead to large changes in the terminal currents which are rate-dependent. The same charge relaxation mechanism would also give rise to smaller rate-dependent current changes higher up on the characteristics away from the breakdown region.

3   Transient capacitance measurements

Transient capacitance and capacitance-voltage measurements have been made on finished FET chips in an effort to identify those deep levels that exist after FET processing. Several approaches appear possible [10, 11]. Presuming the gate pad to be located on the buffer layer, slow bias-sweep C(V) measurements, made at gate voltages beyond pinch-off at temperatures high enough for the deep levels to respond, yield the deep level density at some distance into the buffer by regarding the gate pad as the profiling contact, with the gate strips acting as a parasitic element whose effect can approximately be accounted for. Further, the free carrier profile in the channel may be measured by cooling the device to freeze out the traps under the gate pad. In this measurement the gate strips form the profiling contact, with the gate pad behaving as a shunt parasitic capacitance. The profile thus obtained is a useful (approximate) check on the profile measured using an Hg-probe as part of the materials characterization procedure [6]. Both profiles can be derived from a single set of $1/C^2$ versus V plots made at different temperatures in the range $100\,K \lesssim T \lesssim 300\,K$ by applying the usual depletion-approximation relationship

$$N^{-1} = A^2 q \epsilon (dC^{-2}/dV)/2 \qquad (2)$$

to the appropriate segments of the $C^{-2}(V)$ curves with the area A set equal either to the gate strip area A_S or the gate pad area A_P.

Figure 3 is a family of curves obtained using a Hewlett-Packard instrumentation system similar to that described by Forbes and Kaempf [12]. The $C^{-2}(V,T)$ curves are a strong function of temperature in the region corresponding to the buffer layer, indicating a trap density of $\sim 10^{16}\ cm^{-3}$ at a depth of about 2 μm from the n-layer/buffer interface. (There is some uncertainty in these figures because the capacitance variation with voltage may be affected to an undetermined extent by series resistance associated with the current path followed by carriers through the buffer, between the gate and drain/source electrodes, during the measurement.)

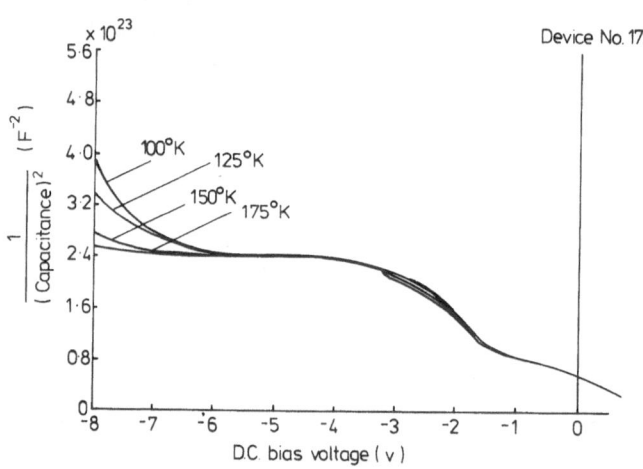

Figure 3 Plots of $1/C^2$ versus V for a FET gate at various temperatures.

The existence of deep levels has been confirmed from thermal scans of the gate capacitance transient response. An exponential function is fitted to the transient after sampling and averaging at 20 or more points. The characteristic time τ and the relative capacitance change $\Delta C/C$ are obtained as parameters of this fitting procedure. The variation of τ with temperature T allows the effective energy E_{na} and cross-section σ_{na} of the trap to be determined. For the sample in Figure 3 a prominent level exists with emission parameters

$$E_{na} = 0.43\,\text{eV}$$

$$\sigma_{na} = 1.3 \times 10^{-13}\;\text{cm}^2$$

which closely resemble the trap designated EL 16 in [13].

4   Effect of deep levels on breakdown

An idea of the effect deep levels may have on breakdown can be had by making the assumption that, for large source-drain voltages and gate voltages at or close to pinch-off, a significant fraction of the avalanche generated majority carriers follow a path such as (2) in Figure 4. A rectilinear approximation to the equipotential distribution along this path enables the physics of breakdown to be examined quasi two-dimensionally by evaluating numerically the ionization integral over the approximate field distribution shown in Figure 5 (with this approximation the calculation may be done using only a small computing facility).

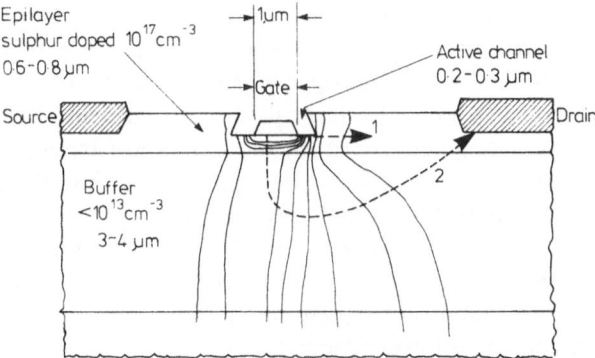

Figure 4   Sketch of recessed-gate structure showing general shape of equipotentials and possible breakdown paths.

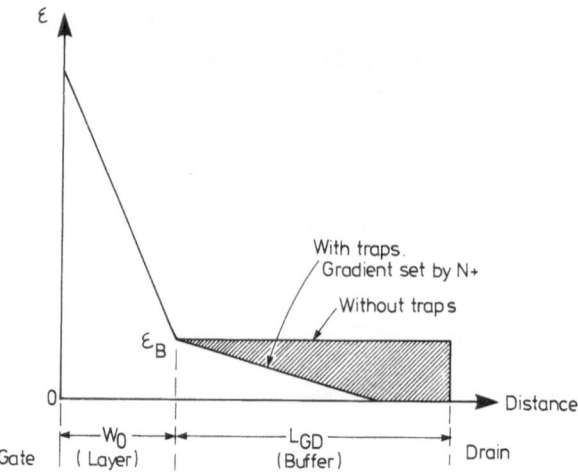

Figure 5   Approximate field profile along a path through the buffer with and without traps.

For a 10^{17} cm^{-3} uniformly doped channel layer and gate-drain spacing of $L_{GD}=2\,\mu m$, such calculations in the absence of traps yield an estimate for the gate-drain breakdown voltage of 60 V with a maximum output power of $P_{max}\cong1.8$ watts per mm gate width, corresponding to a channel thickness of $\sim0.3\,\mu m$ and an open channel current of $I_{DSS}\cong240$ mA mm^{-1}. It should be noted from Figure 5 that P_{max} depends on L_{GD} in this model by virtue of the fact that the electric field penetrates to the drain electrode. An important point is that a major contribution to the ionization integral comes from the region where the field ε is highest (i.e. close to the gate), whereas a major contribution to the gate-drain voltage comes from the field in the buffer.

When there are deep acceptor states in the buffer, it is helpful to consider bulk effects separately from charge exchange at the n-layer/buffer interface. Given that the application of a large gate-drain voltage which reverse-biases the Schottky barrier gives rise to electron emission from deep levels in the bulk of the buffer layer, the electric field penetrating into the buffer from the channel will terminate on the nett positive ionic space charge thus created instead of in the drain contact, as shown in Figure 5. For a given channel field the area under the total field profile is now much reduced or, equivalently, the gate-drain potential at which a given level of current multiplication occurs is lower, depending upon the trap density. For $N_T=10^{16}$ cm^{-3}, the theoretical limiting output power is reduced to 1 W mm^{-1} at a gate-drain voltage of 40 V, in reasonable agreement with the performance of good X-band power FETs [5, 14].

A similar conclusion is reached for deep levels at the n-layer/buffer interface which empty of electron charge under normal operating bias (which is in the nature of a forward bias at the n-layer/substrate junction). The associated contraction of the secondary depletion region in the n-layer adjacent to the interface [1] has the effect of absorbing any increase in buffer field ε_B in Figure 5, again resulting in a lower gate-drain voltage at a given level of current multiplication in the channel. The possible importance of this effect can be assessed by considering the (hypothetical) case of an interface state density large compared with the n-layer doping density. The theoretical limiting output power is in this case reduced to approximately 1 W mm^{-1} at 40 V gate-drain breakdown for an interface state energy 0.8 eV below the conduction band edge. For a state with $E_{na}=0.43$ eV, as found in Section 3, the degradation would be less serious.

5 Conclusions

It has been found that differences between the pulsed and slow sweep drain characteristics of GaAs FETs in the breakdown region can be interpreted using a simple model of current multiplication in the channel which is sensitive to the presence of deep levels in the buffer layer and at the n-layer/buffer interface. From transient capacitance measurements on FET chips it is known that electron traps characteristic of vapour-phase epitaxial material exist at densities up to 10^{16} cm^{-3} in the buffer layer of some devices. On the basis of the simple model a trap density of 10^{16} cm^{-3} reconciles reasonably well with the practically observed gate-drain breakdown voltages of ~40 V and an output power per unit gate width of 1 W mm^{-1}.

ACKNOWLEDGEMENTS

The work described in this paper was in part done while one of us (P.H.L.) was on leave at Plessey Research (Caswell) Ltd. The significant contribution of Dr P.M. White during this period is gratefully acknowledged. The paper is published by permission of the Directors, Plessey Research (Caswell) Ltd, and by permission of the Chief General Manager, Telecom Australia.

References

1. Hower, P.L., Hooper, W.W., Tremere, D.A., Lehrer, W. and Bittman, C.A. (1969). *Inst. Phys. Conf. Ser.*, **7**, 187
2. Yokoyama, N., Shibatomi, A., Ohkawa, S., Fukuta, M. and Ishikawa, H. (1977). *Inst. Phys. Conf. Ser.*, **33b**, 201
3. Houng, Y.M. and Pearson, G.L. (1978). *J. Appl. Phys.*, **49**, 3348
4. Yamamoto, R., Higashisaka, A. and Hasegawa, F. (1978). *IEEE Trans. Electron. Dev.*, **ED—25**, 567
5. Dilorenzo, J.V. and Wisseman, W.R. (1979). *IEEE Trans. Microwave Theory and Techniques*, **MTT—27**, 367
6. Crossley, I., Goodridge, I.H., Cardwell, M.J. and Butlin, R.S. (1977). *Inst. Phys. Conf. Ser.*, **33b**, 289
7. Rode, D.L., Schwartz, B. and Dilorenzo, J.V. (1974). *Solid St. Electron.*, **17**, 1119

8. Abbott, D.A. and Turner, J.A. (1976). *IEEE Trans. Microwave Theory and Techniques,* **MTT—24,** 317

9. Furutsuka, T., Tsuji, T. and Hasegawa, F. (1978). *Trans. IEEE Electron. Dev.,* **ED—25,** 563

10. Senechal, R.R. and Basinski, J. (1968). *J. Appl. Phys.,* **39,** 4581

11. Lang, D.V. (1974). *J. Appl. Phys.,* **45,** 3023

12. Forbes, L. and Kaempf, U. (1979). *Hewlett Packard Journal,* **30,** 29

13. Martin, G.M., Mitonneau, A. and Mircea, A. (1977). *Electron. Lett.,* **13,** 192

14. White, P.M. Private communication

DEEP LEVEL MEASUREMENTS OF LAYERS ON SEMI-INSULATING GaAs SUBSTRATES BY MEANS OF THE PHOTOFET METHOD

F.J. TEGUDE and K. HEIME
Duisburg University, FB 9, Solid State Electronics Laboratory,
Kommandantenstrasse 60, D—4100 Duisburg 1, West Germany

Abstract

Deep level measurements have been performed on three different types of sample: (1) Sn-doped layers diffused into SI-GaAs:Cr; (2) Sn-doped LPE-GaAs layers on SI-GaAs:Cr; and (3) Sn-doped layers diffused into high resistivity undoped LPE layers on SI-GaAs:Cr. Measurement technique was the PHOTOFET method [1], which monitors changes in I_{ds} of a MESFET with respect to the energy of monochromatic light. The characteristics of the three types of sample are presented and related to trap behaviour, especially near interfaces. Deep traps at the interfaces between active layers and substrates may cause serious degradation of device performance. It appears that a buffer layer (sample type 3) is not always a good way of improving device behaviour.

Introduction

Long time constant effects have often been observed in planar devices such as transferred electron devices and field-effect transistors (FET). Deep traps in the active layer, and especially at the layer-substrate interface, are commonly related to these troublesome effects [2]. The deep levels can be produced or introduced during the technological processes. For example, heat treatment during epitaxial growth, diffusion processes or annealing after ion implantation can produce defects or can drive defects towards surfaces or interfaces where they accumulate. Stress due to thermal expansion mismatch between encapsulant and GaAs produces defects which act as deep levels themselves or which getter contaminants. Chromium and/or oxygen commonly used for compensation of semi-insulating substrates are a particular problem if they are present in or near the active layers of devices and are thought to impair their performance [3—6]. We used the PHOTOFET method [1] with MESFETs. It makes use of the fact that trap charging in the space-charge region beneath the Schottky gate or in the dipole layer at the active layer-substrate interface varies the thickness of the conductive channel of the MESFET. In n-type material, an increasing positive trap charge results in an increase in I_{ds} while an increasing negative trap charge results in a decrease in I_{ds}. Thus electron and hole traps can be distinguished, as in photocapacitance measurements. Trap charging is governed by illumination with chopped monochromatic light. If variations of I_{ds} are plotted against the energy $\hbar\omega$ of the incident light, activation energies of trap levels can be deduced from thresholds of ΔI_{ds}. Measurements are done at room temperature. The main advantages of the PHOTOFET method are high sensitivity and high spatial resolution.

Sample preparation

SAMPLE TYPE 1

The conductive layers were diffused into SI-GaAs from spun-on emulsion films [7]. After drying in air, the films decompose to SiO_2:Sn, forming the Sn diffusion source and encapsulant simultaneously. The diffusion is performed at 900°C for 10 minutes in an open tube system with N_2 ambient. The resulting diffusion profile is nearly rectangular with an electron concentration of about $n=3\times10^{18}$ cm^{-3}.

SAMPLE TYPE 2

This is an Sn-doped GaAs LPE layer on SI-GaAs substrate. Epitaxial growth is performed in a graphite boat in an H_2 ambient after a melt-back of the substrate surface. Growth temperature is 700°C and electron concentration is $n=1.5\times10^{17}$ cm^{-3}.

SAMPLE TYPE 3

In this type the active layer is Sn diffused as with sample type 1; however, diffusion is now performed into high resistivity undoped LPE material. Preparation and diffusion processes are the same as for sample type 1. The diffusion profile again is nearly rectangular with an electron concentration of $n=3\times10^{18}$ cm^{-3}. The LPE buffer layer is grown in a system similar to the one in which the conductive layers are grown. LPE layer thickness is about 1 μm and $n<10^{14}$ cm^{-3}.

The desired channel thicknesses and the mesas for device isolation are obtained by wet chemical etching. All samples are provided with Ni/AuGe/Ni alloyed ohmic contacts and with Cr/Au Schottky gate contacts. Gate length is L=1.3 μm and gate width is W=300 μm.

Experimental results

SAMPLE TYPE 1

In FETs of this type we find a decrease of I_{ds} at energies 0.7 and 0.9 eV, indicating hole emission from acceptor-like traps 0.7 and 0.9 eV, respectively, above the valance band edge. An increase of I_{ds} at 1.05 eV indicates electron emission from a donor-like level 1.05 eV beneath the conduction band edge.

From the FET characteristics displayed on a curve tracer we obtain a transconductance of $g_m = 120$ mS per mm gate width at $U_g = 0$ V. The conductive channel can be pinched off completely.

SAMPLE TYPE 2

In FETS of this type we find the acceptor-like level at 0.9 eV and the donor-like one at 1.05 eV but not the acceptor-like trap at 0.7 eV. Instead, there is another donor-like level at an energy less than 0.65 eV. Since our optical equipment operates at light energies of $\hbar\omega \gtrsim 0.65$ eV, the activation energy of this donor cannot be determined.

The FET characteristics displayed on the curve tracer show a transconductance of $g_m = 80$ mS/mm gate width. Again, I_{ds} can be pinched off completely.

SAMPLE TYPE 3

In FETs with the LPE buffer layer the PHOTOFET technique reveals the same levels as in sample type 1: 0.7 and 0.9 eV acceptor-like and 1.05 eV donor-like. The variation of I_{ds} due to the 0.9 eV level is much smaller than in samples of types 1 and 2, and sometimes it cannot be detected at all. We observe an excellent transconductance of $g_m = 150$ mS per mm gate width at $U_g = 0$ V but I_{ds} cannot be pinched off completely.

One interesting fact with this sample type is that the strong dependence of I_{ds} on light energy as observed at $U_g \approx 0$ V (cf. Figure 1, trace for $U_g = -0.3$ V) dissappears almost completely for $U_g < -0.8$ V (cf. Figure 1, trace for $U_g = -1.2$ V). Only a continuous rise in I_{ds} with $\hbar\omega$ and a small additional increase at 1.05 eV is detected. This behaviour was not observed with sample types 1 and 2.

Another remarkable fact is observed in the d.c. characteristics when displayed on the curve tracer. In Figure 2 U_g becomes more negative from the upper left to the lower right, according to an increasing number of traces. The current I_{ds} ($U_g = 0$ V) significantly increases by 20% or more after applying gate voltages U_g more negative than -2.5 V (lower right in Figure 2). In sample types 1 and 2 the d.c. characteristics do not show the memory effect of the gate voltages applied before.

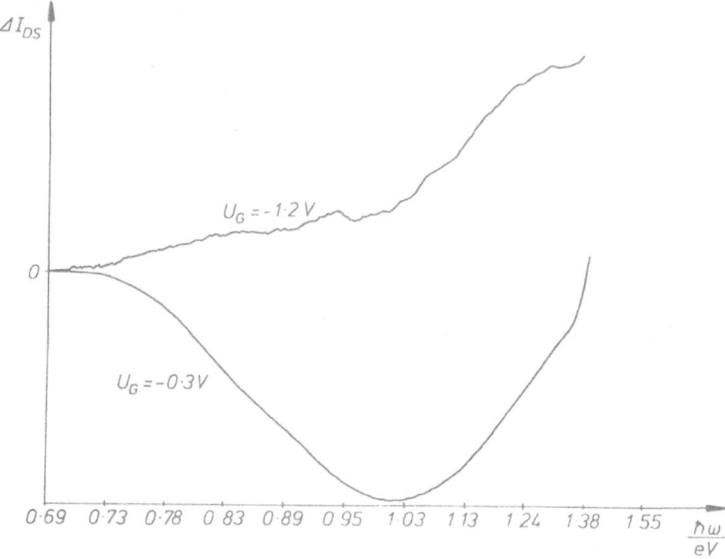

Figure 1   PHOTOFET spectra of sample type 3 typical for small (−0.3 V) and large (−1:2 V) negative gate voltages.

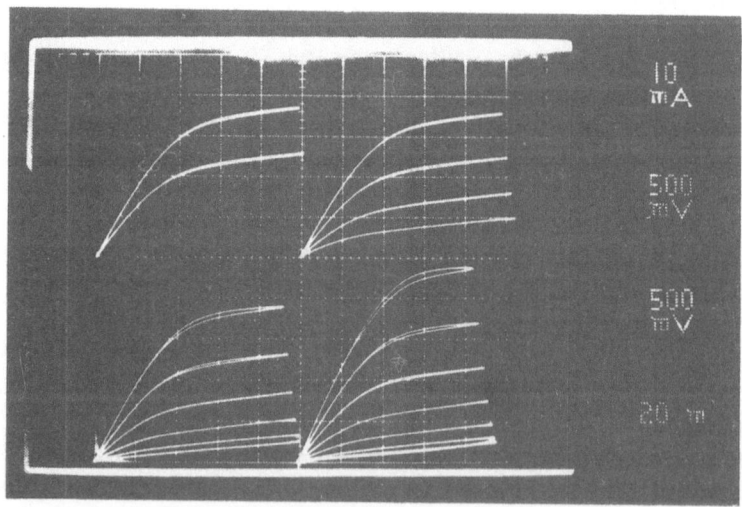

Figure 2   D.c. characteristics of a sample type 3 FET with increasing negative gate voltage from upper left to lower right. Uppermost trace is always for $U_g=0$ V! Numerical values displayed are in turn: I_{ds}, U_{ds}, U_g per step, g; per section, respectively.

Table 1 Summary of sample preparation and experimental results

	Type 1	Type 2	Type 3
		Sn-doped active layer	
FET channel preparation	Diffused into SI-GaAs:Cr	LPE layer on SI-GaAs:Cr	Diffused into undoped LPE buffer layer on SI-GaAs:Cr
	900°C, 10 min	700°C, 2 min	900°C, 10 min + 700°C, 2 min
Deep levels, ΔE			
Acceptor-like	0.7 eV; 0.9 eV	0.9 eV	0.7 eV; 0.9 eV (weak)
Donor-like	1.05 eV	0.65 eV; 1.05 eV	1.05 eV
U_g-memory effect	None	None	Approx. 20% I_{ds} increase

Experimental results of all three types of sample are summarized in Table 1.

Discussion

The trap levels revealed in our investigations are commonly found in SI-GaAs:Cr or GaAs:O and in connection with active layers on these substrates [6, 8, 9, 10]. The 1.05 eV level is usually observed with optical excitation techniques [8, 11, 12]. The fact that we find these levels not only in samples of types 1 and 2 but also in or close to the active layer of sample type 3 with the LPE buffer layer, suggests an out-diffusion of Cr from the substrate into the buffer layer where it generates Cr-related levels. Assuming a diffusion coefficient $D \approx 10^{-11}$ cm^2 s^{-1} [13] for Cr in GaAs, we estimate a diffusion depth of the order of 1 μm or more during the diffusion process of Sn from the SiO$_2$ film. It was demonstrated [3, 4] that stress due to thermal expansion mismatch between GaAs and a surface protective layer generates defects which can getter and accumulate Cr near the surface. Additional stress is introduced at the interface between the high n-doped and very pure buffer layer because of the extremely high gradient of dopant material in type 3 samples. Furthermore, only this sample type suffers two high temperature processes (cf. Table 1) for which the above-mentioned considerations are applicable.

It was shown [2] that trapping of negative charge at the substrate or buffer layer side of the FET channel decreases I_{ds}. The following model is in agreement with experimental results and thus can explain the d.c. characteristics of the sample type 3 (Figure 2). Due to the pile-up of defects and Cr near the diffused layer-buffer layer interface, negatively charged traps are present and partially pinch the channel at moderately low gate voltages (Figure 3). With increasing negative gate voltage the electric field in the channel rises to such values that optically assisted field emission of electrons from deep levels can occur. This results in a widening of the conductive channel towards the substrate and an increase of I_{ds} with increasing $\hbar\omega$ (cf. Figure 1, trace for $U_g = -1.2$ V). Without optical excitation the anomalous increase of I_{ds} is observed at more negative gate voltages ($-2 \ldots -3$ V) since now the electric field alone has to free the electrons from the deep levels.

The retrapping of electrons is slow [6]. Therefore the $U_g = 0$ V trace, displayed approximately 30 ms after the $U_g = -2 \ldots -3$ V trace (lower right in Figure 2), shows the current through a channel the thickness of which has not yet relaxed to the equilibrium condition of $U_g = 0$ V (upper left in Figure 2).

A similar model was used in [14] to calculate the d.c. characteristics of MESFETs.

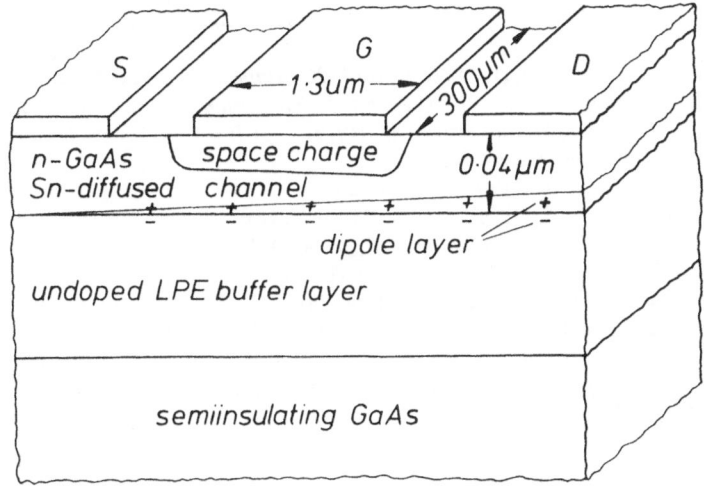

Figure 3 Cross-sectional view of sample type 3 MESFET with interface charges.

Conclusion

It was shown by the PHOTOFET method that deep traps commonly related to Cr are present in all our samples. The samples with buffer layer show high transconductances ($g_m = 150$ mS/mm). For these devices there is strong evidence that traps accumulated at the interface between active layer-buffer layer are responsible for undesirable high-field effects which are not observed in our samples without buffer layer. Significant improvements are possible if the concentration of Cr is lowered by growing thicker buffer layers and/or by reducing the diffusion temperature.

ACKNOWLEDGEMENTS

We would like to thank N. Arnold and H. Dämbkes for helpful discussions and W. Schmitz, G. Howahl and G. Schulzek for preparation of samples. Thanks are also due to U. Ringer for typing the manuscript. This work was supported by the Deutsche Forschungsgemeinschaft.

References

1. Tegude, F.J. and Heime, K. (1980). *Electron. Lett.*, **16**, 22
2. Tanimoto, M., Suzuki, K., Itoh, T., Yanai, H., Kaufmann, L.M.F., Nievendick, W. and Heime, K. (1977). *Trans. IECE:* **60-C**, 698
3. Asbeck, P., Tandon, J., Sin, D., Fairman, R. and Welch, B. (1979). *IEEE GaAs IC Symposium,* Research Abstracts No. 14

4. Evans, C.A. and Deline, V.R. (1979). *IEEE GaAs IC Symposium,* Research Abstracts No. 15
5. Forbes, L. and Chang, C.D. (1979). *IEEE GaAs IC Symposium,* Research Abstract No. 16
6. Simons, M. and King, E.E. (1979). *IEEE GaAs IC Symposium,* Research Abstract No. 35
7. Arnold, N., Dämbkes, H. and Heime, K. (1979). *11th Int. Conf. on Solid State Devices,* Tokyo. Digest of Techn. Papers, No. 89
8. Look, D.C. (1977). *Solid St. Commun.,* **24,** 825
9. Lin, A.L. and Bube, R.H. (1976). *J. Appl. Phys.,* **47,** 1859
10. Lin, A.L. and Bube, R.H. (1976). *J. Appl. Phys.,* **47,** 1852
11. Grimmeis,H.G. (1977). *Ann. Rev. Mater. Sci.,* **7,** 341
12. Bois, D. and Pinard, P. (1973). *Japan. J. Appl. Phys.,* **12,** 936
13. Asbeck, P., Tandon, J., Babcock, E. and Welch, B. (1979). *37th Annual Device Research Conference,* Boulder, WP-A2
14. Bonjour, P., Castagne, R. and Pone, J.F. (1979). *IEEE GaAs IC Symposium,* Research Abstract No. 34

DARK CAPACITANCE, PHOTOCAPACITANCE, DARK CONDUCTANCE AND PHOTOCONDUCTANCE TRANSIENTS ON GaAs MESFETS

C.D. CHANG and L. FORBES
Department of Electrical and Computer Engineering,
University of California at Davis, Davis, California 95616, USA

Abstract

The techniques of capacitance and conductance transients have been utilized in the identification of chromium and oxygen in GaAs. A good correspondence has been obtained between thermal emission rate measurements and optical emission rate, or photo-ionization cross-section measurements.

The technique of capacitance transients [1], as applied in the characterization of imperfection centres, is by now well known; these techniques were subsequently automated by Lang [2] as deep level transient spectroscopy (DLTS) measurements. Capacitance transient measurements have been applied to a wide range of silicon and compound semiconductor devices [3, 4, 5, 6].

The results described here utilize the older techniques [1], since these provide an optical verification, by photocapacitance measurements, of the ionization energies found from the thermal emission rate measurements in dark capacitance transient measurements. In addition, the original techniques have been expanded to include dark conductance transients on FETs, which will be discussed in more detail later.

Figure 1 shows the energy level assignments for oxygen and chromium in GaAs based on our dark capacitance and photocapacitance transient measurements. We have been able to observe two thermal emission processes and the threshold energies in three photo-ionization cross-sections. These results can be described in terms of simple single level models for the oxygen and chromium centres, without any lattice relaxation or additional levels.

Figure 2 shows the traditional plot of thermal emission time constants versus $1000\,T^{-1}$; the simplest is seen following the zero bias initial condition where electron emission is observed. The second main transient is a decay down following illumination with a red light emitting diode or red light from a monochromator and corresponds to hole emission [1]. These measurements are made on the gate capacitance of an n-channel MESFET with an aluminium gate, which is fabricated by direct implantation into Crystal

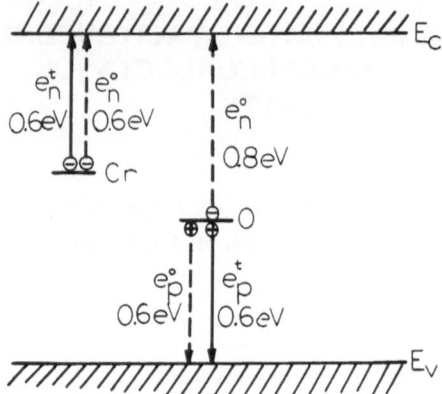

Figure 1   Energy level model for oxygen and chromium in GaAs, showing observed thermal and optical emission processes.

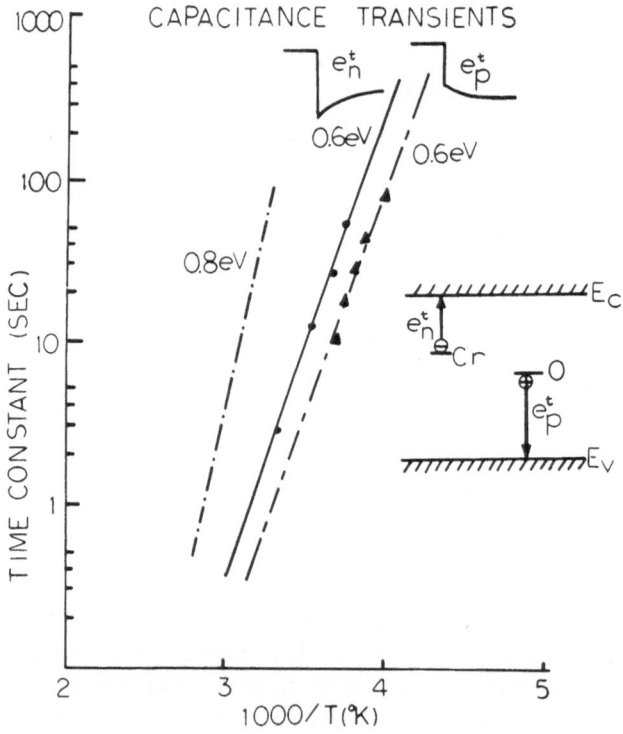

Figure 2   Time constants for the observed thermal emission processes.

Specialties chromium- or oxygen-doped substrate. While these capacitance transients are small they are, however, free of series impedance effects, the drain-source impedance being about 5 ohms. On large area Schottky barrier diodes fabricated on implanted layers it has been found that many capacitance transient and DLTS systems can give erroneous data because of the large series impedance which, as we shall see later, can in itself be time-dependent.

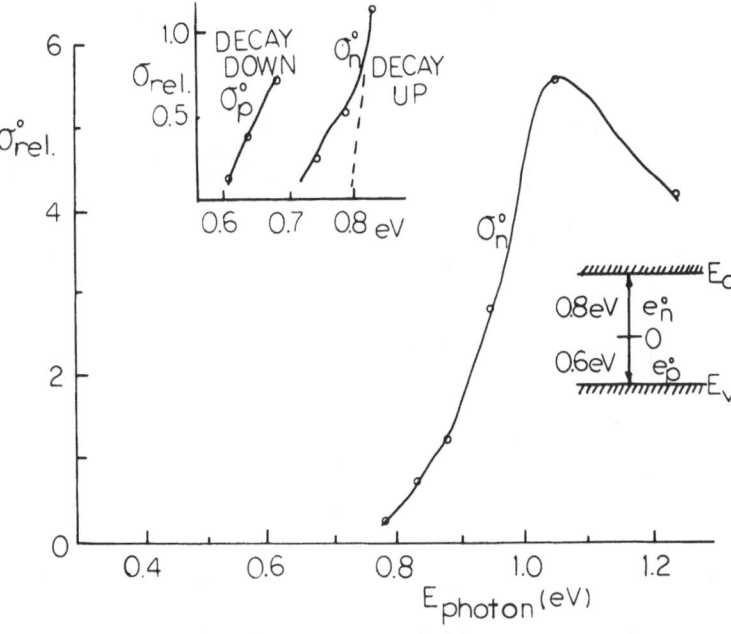

Figure 3 Photo-ionization processes and relative cross-sections observed on oxygen-doped material.

Both of the thermal activation energies are around 0.6 eV without any corrections for power laws of temperature, and without any lattice relaxation correction. In order to verify these results since they may, in general, be different from others in the literature, we have made measurements of the photo-ionization cross-sections, on both chromium-doped and oxygen-doped substrates, placing particular emphasis on the threshold energies. Figure 3 shows the results on oxygen-doped substrates where two distinctly different processes can be observed in the threshold. For photon energies larger than 0.8 eV, only a decay up, optical emission of electrons can be observed. However, for photon energies in the range 0.6—0.8 eV, a decay down, optical emission of holes [1] can be observed with a threshold of around 0.6 eV. This then constitutes the reason for the association of the level at $E_c - 0.8$ eV with the oxygen level. These results correspond fairly well with photocapacitance [7] and photoconductivity measurements [8] by other authors for electron emission from the oxygen donor centre; in those cases, however, some other levels were observed so their results are not quite as clear as they might otherwise have been. Figure 4 shows the results of photocapacitance measurements on chromium-doped substrates where the electron emission or photoionization cross-sections suggests two thresholds: one at 0.8 eV and one at 0.6 eV. These are believed to be associated with electron emission from the oxygen and chromium levels, respectively, the threshold energies corresponding then to the optical emission processes for electrons, as shown in Figure 1.

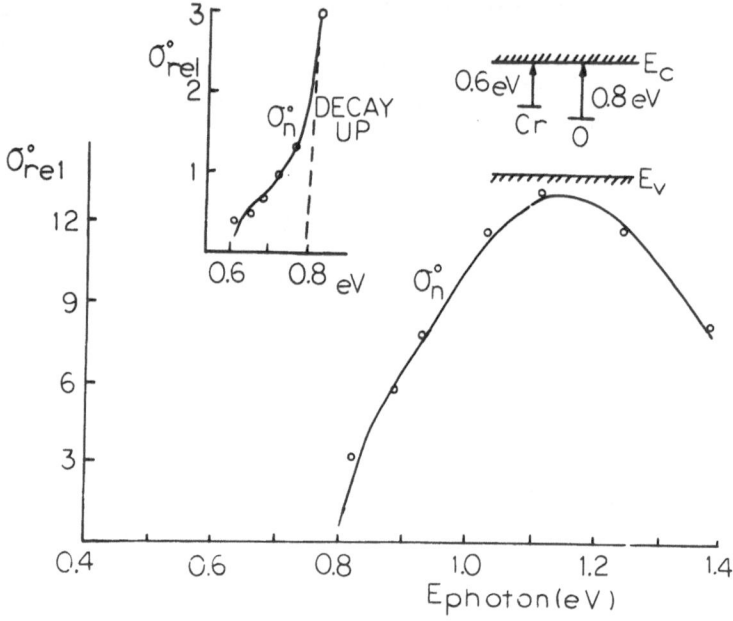

Figure 4   Relative photo-ionization cross-sections observed on chromium-doped material.

Clearly, if the emission of carriers from impurities in the space charge region of a FET results in capacitance transients, the modulation of this space charge will equally well result in a change or transient in the drain to source conductance. These can, in fact, be utilized in so-called conductance DLTS techniques [9, 10, 11].

The theory for these conductance transients and experimental results on GaAs MESFETs has been presented previously [11]. For a uniform doping profile the conductance transient is given by,

$$\Delta g_d(t)/g_d = (N_{TT}/2N_D)(a^2/(a-x_d)^2)\exp(-t/\tau)$$

where g_d is the drain conductance in the linear region, N_{TT} the deep level impurity concentration, N_D the donor concentration, a the channel depth and

x_d the gate junction depletion region width. As shown in Figure 5, two transients, B and C, are seen which correspond to the capacitance transient data for thermal electron and hole emission. The transient labelled A is believed to be due to a backgating effect [11]. Photoconductance transients can also be observed.

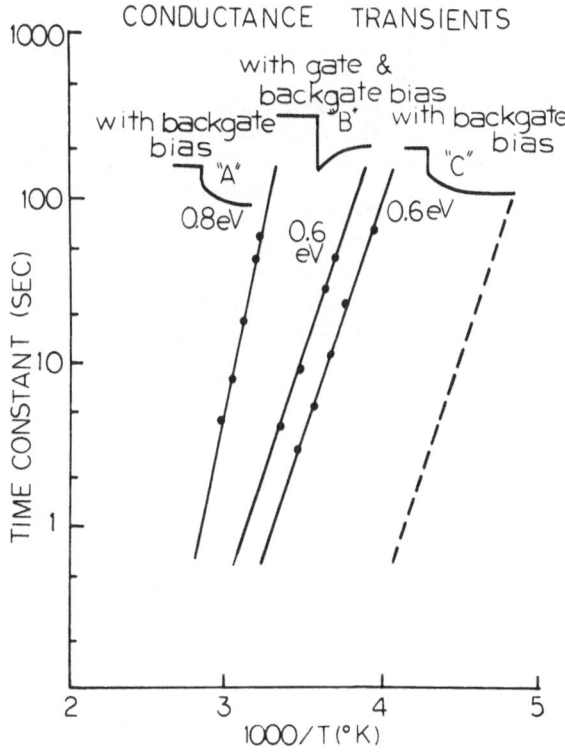

Figure 5   Time constants for conductance transients.

In conclusion, we have been able to demonstrate a correspondence between the thermal and optical emission processes at the chromium and oxygen centres in GaAs and capacitance and conductance transients on FETs. Our results suggest a simple single level model for both chromium and oxygen in GaAs [7, 8]. The assignments of chromium and oxygen levels are only by association with the device material — their origin has not been absolutely identified.

ACKNOWLEDGEMENTS

The authors would like to acknowledge helpful discussions and help with materials/devices by E.M. Swiggard and S.H. Lee of the Naval Research

Laboratory, R. Zuleeg of McDonnell-Douglas, A. Chu of MIT Lincoln Laboratories, and P. Wang of Hewlett-Packard. This work was supported by the National Science Foundation Grant ENG76-80128 and Grant ENG79-09955.

References

1. Sah, C.T., Forbes, L., Rosier, L.L., and Tasch, A.F. Jr., (1970). *Solid St. Electr.*, **13**, 759
2. Lang, D.V. (1974). *J. Appl. Phys.*, **45**, 3023
3. Sakai, K. and Ikoma, T. (1974). *Appl. Phys.*, **5**, 165
4 Lang, D.V. and Logan, R.A. (1975). *J. Electr. Mat.*, **4**, 1053
5. Martin, G.M., Mitonneau, A. and Mircea, A., (1977). *Electr. Lett.*, **13**, 191
6. Mitonneau, A., Martin, G.M. and Mircea, A., (1977). *Electr. Lett.*, **13**, 667
7. Vasudev, P.K. and Bube, R.H. (1978). *Solid St. Electr.*, **21**, 1095
8. Grimmeiss, H.G. and Ledebo, L. (1976). *J. Appl. Phys.*, **46**, 255
9. Forbes, L. and Kaempf, U. (1979). *Hewlett-Packard J.*, **30**, (4), 29
10. Forbes, L., Brown, R., Sheikholesham, M. and Current, W. (1979). *Solid St. Electr.*, **22**, 391
11. Forbes, L. and Chang, C.D. (1979). *Proc. IEEE Symp. GaAs Int. Ccts;* paper 16

SEMI-INSULATING GaAs — A USER'S VIEW

R. ZUCCA, B.M. WELCH, P.M. ASBECK, R.C. EDEN and S.I. LONG
Rockwell International Electronics Research Center,
1049 Camino dos Rios, Thousand Oaks, California 91360, USA

Abstract

The extraordinary progress which has taken place in GaAs digital IC technology over the last two years is reviewed, highlighting the role of the semi-insulating substrate. The most critical requirements of the material are discussed in the context of current circuit fabrication techniques. At the same time, recent advances in the fundamental knowledge of the material are briefly reviewed and weighted according to their impact on the technology.

Introduction

The ever increasing demands for high speed in large scale integrated (LSI) circuits has provided a unique opportunity for developing a GaAs integrated circuit technology [1]. The higher electron mobility of GaAs (compared to Si) results in higher device transconductances, which can be exploited for higher speed. Some speed may be traded off for lower power dissipation (required for large scale integration). The semi-insulating GaAs substrate provides a 'natural' way of reducing circuit parasitics without resorting to hybrid structures such as silicon on sapphire. Recent progress in the control and development of semi-insulating bulk GaAs crystals, and the development of reliable GaAs processing techniques (ion implantation, vapour phase epitaxy, ohmic contacts, etc.) has led to full development of low-noise microwave GaAs MESFETs [2] and to great progress in power MESFETs [3]. At the same time, we are witnessing spectacular advances in digital integrated circuits [4, 5] coupled with the appearance of a technology for monolithic microwave integrated circuits [6], while the highest speed CCDs have been demonstrated with GaAs devices. In this paper we will focus on just one of these exciting GaAs applications: digital integrated circuits. We will discuss the dependence of past and future achievements on our understanding of semi-insulating GaAs.

Digital GaAs IC technology status

The planar fabrication technology developed at Rockwell International for high-speed low-power digital integrated circuits will be used as an example [8] because it relies more than any other GaAs IC technology on good control of the semi-insulating substrate. The scanning electron micrograph in Figure 1 exemplifies the circuit complexity already reached with this technology. The figure shows a 3×3 bit parallel multiplier employing 75 logic NOR gates (225 FET structures and approximately 150 diodes). This medium scale level of integration (MSI) has been reached very quickly, only two and a half years after fabrication of the first planar FET.

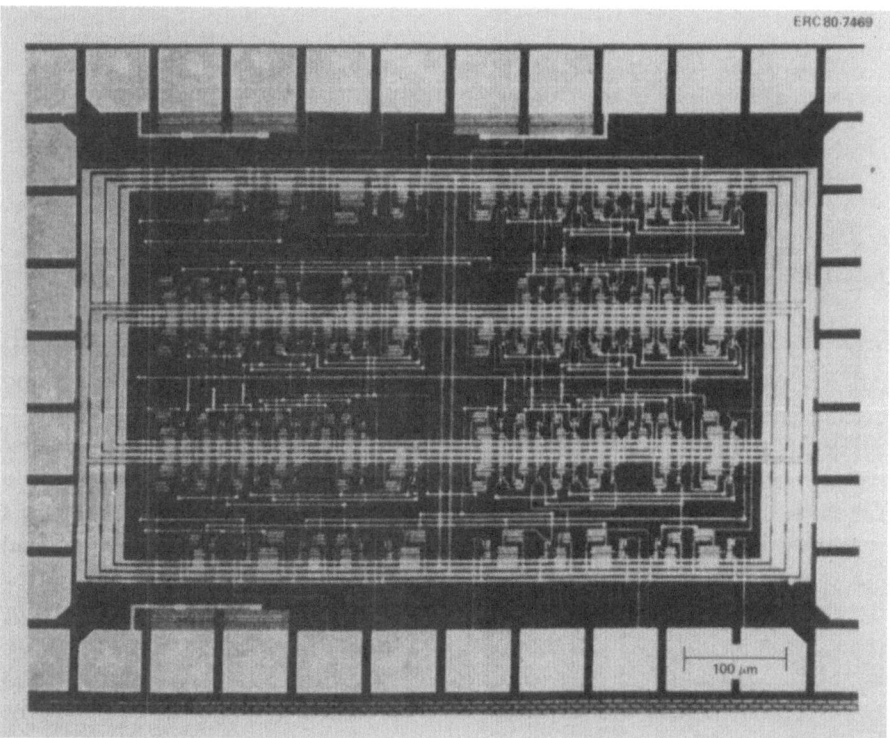

Figure 1   Scanning electron micrograph of a 3×3 bit parallel multiplexer using 75 SDFL gates. The circuit is capable of operating at a rate of 320 MHz (τ_d = 170 ps/gate) while dissipating only 56 mW (750 μW/gate).

The multiplier circuit shown in Figure 1 operated at a rate as high as 323 MHz, corresponding to a propagation delay τ_d = 172 ps/gate, while the power dissipation was only 56 mW (750 μW/gate) [5]. These figures are representative of the performance demonstrated by other circuits fabricated with this technology, ranging from simple ring oscillators to frequency dividers, shift registers, and multiplexers/demultiplexers [5]. Perhaps a better grasp of the significance of the above figures can be obtained by projecting the performance of an 8×8 bit parallel multiplier. For a $\tau_d = 172$ ps, an 8×8 bit multiplication could be completed in approximately 6 ns with a power dissipation of 300 mW, while the fastest commercially available 8×8 silicon multipliers have 45 ns internal multiply time and dissipate 1.2 W [9].

The performance figures are quite promising. The success obtained at MSI (~100 gates) suggests that the planar GaAs digital IC technology can be implemented at the LSI level (~1000 gates) provided that sufficient yields can be reached. Although the most ambitious long range applications are expected to take place in fast arithmetic processors, earlier applications are expected in communications for high speed signal processors, beginning with simple signal conditioners such as frequency dividers.

Description of the planar GaAs IC technology

The building block for the circuits discussed above is the Schottky diode-FET logic (SDFL) NOR gate [10]. As shown in the diagram and photograph in Figure 2, a logic OR operation is performed by very small (typically 1×2 μm) high speed Schottky barrier diodes, while a depletion mode FET (gate length 1 μm, channel width between 5 and 20 μm, typically 10 μm) provides inversion and gain. The circuit requires a fabrication process capable of optimizing the carrier concentration profiles of the diodes and the FET channels independently from one another. This is accomplished by selective implantations into the semi-insulating substrate using photoresistant masks to cover the unimplanted areas [8].

A schematic representation of the planar fabrication steps is shown in Figure 3. Note that the localized implantations are done through the Si_3N_4 cap used for post-implantation annealing. This dielectric is never removed from the GaAs surface keeping it clean and, to some degree, passivated [8]. The ohmic contact and Schottky barrier metals (the latter are also used for first level interconnections) are deposited in holes opened in the dielectric by plasma etching, offering a nearly flat surface for the second level metallization [8].

The ion implantation step in our process is very critical. This is illustrated in Figure 4, where carrier concentration profiles for the FET channels (Se implanted) are displayed. Note that the active channel regions of these low pinch-off voltage FETs (V_p ~1 V) are just the 'tails' of very shallow profiles (~2000 Å), and depth variations of only 64 Å cause depletion voltage variations of 0.11 V. Although we are able to achieve good device uniformity, as shown by the histograms of pinch-off voltage of test FETs on an IC wafer (Figure 5) [11], and good reproducibility within an ingot as shown by the profiles in Figure 4, this is accomplished through a very careful process of substrate qualification, which rejects a significant fraction of the ingots sampled. Reproducibility from ingot to ingot is achieved by implant dose adjustments based on the substrate qualification tests. The weakness of such an empirical process is quite obvious. We need better understanding of the complex interplay between the semi-insulating substrate, the implanted ions and the encapsulant in order to achieve better profile predictability. Some recent studies of Cr out-diffusion during post-implantation anneal (discussed later) represent a good first step [12, 13].

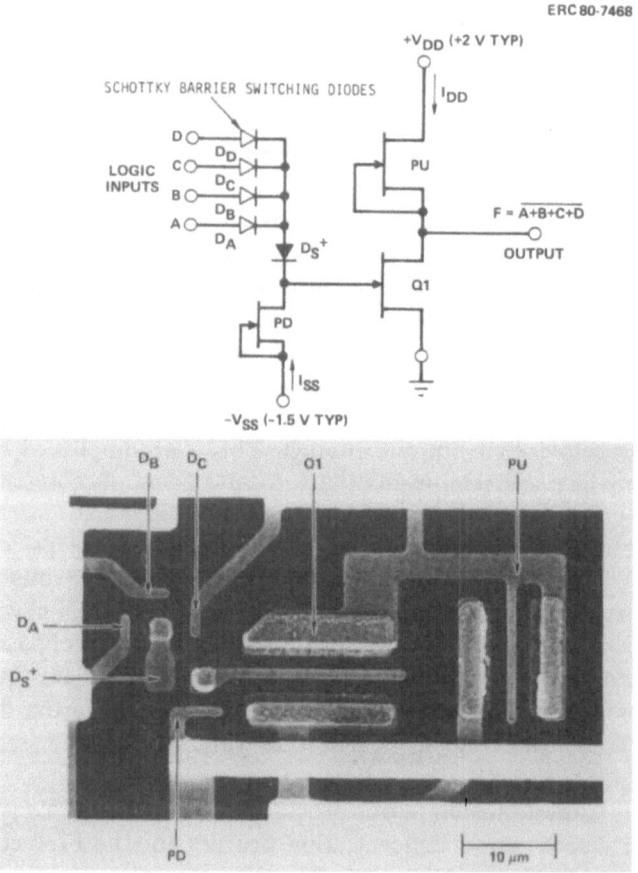

Figure 2 Schematic and scanning electron micrograph of an SDFL (Schottky diode-FET logic) gate. The main FET is 15 μm wide and has a 1 μm long gate. The dimensions of the high-speed logic diodes are 1×2 μm.

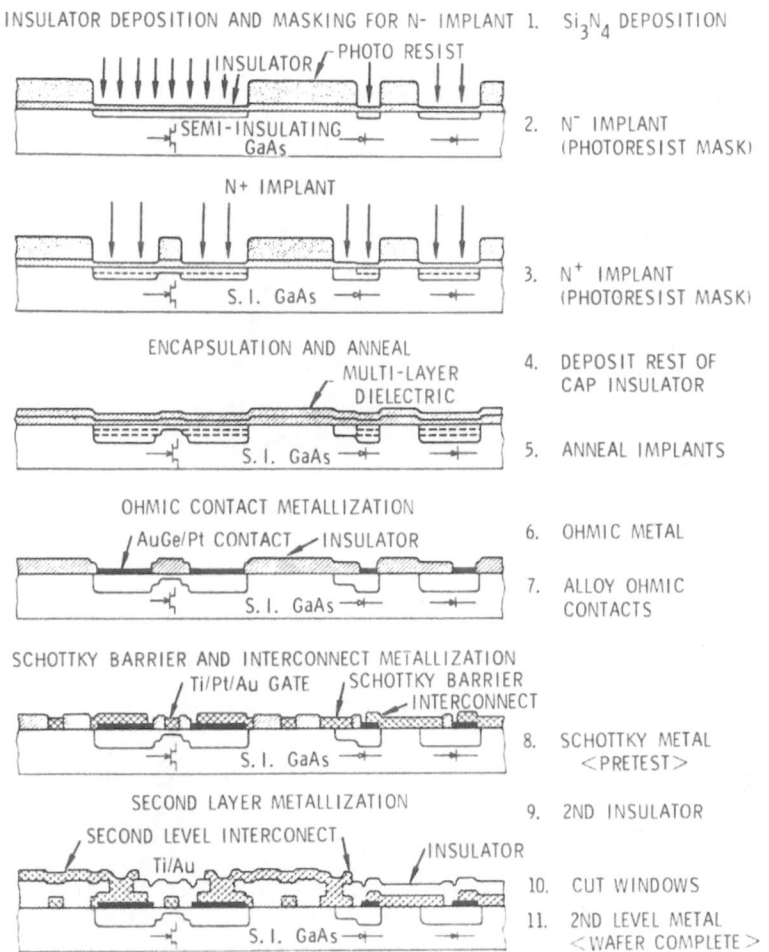

Figure 3 Fabrication steps for planar fabrication process using localized implantations into semi-insulating GaAs. Note that the bare surface of GaAs is never exposed, except for the areas where the encapsulating dielectric is briefly open for metal depositions (steps 7 and 8).

SC 79-5915

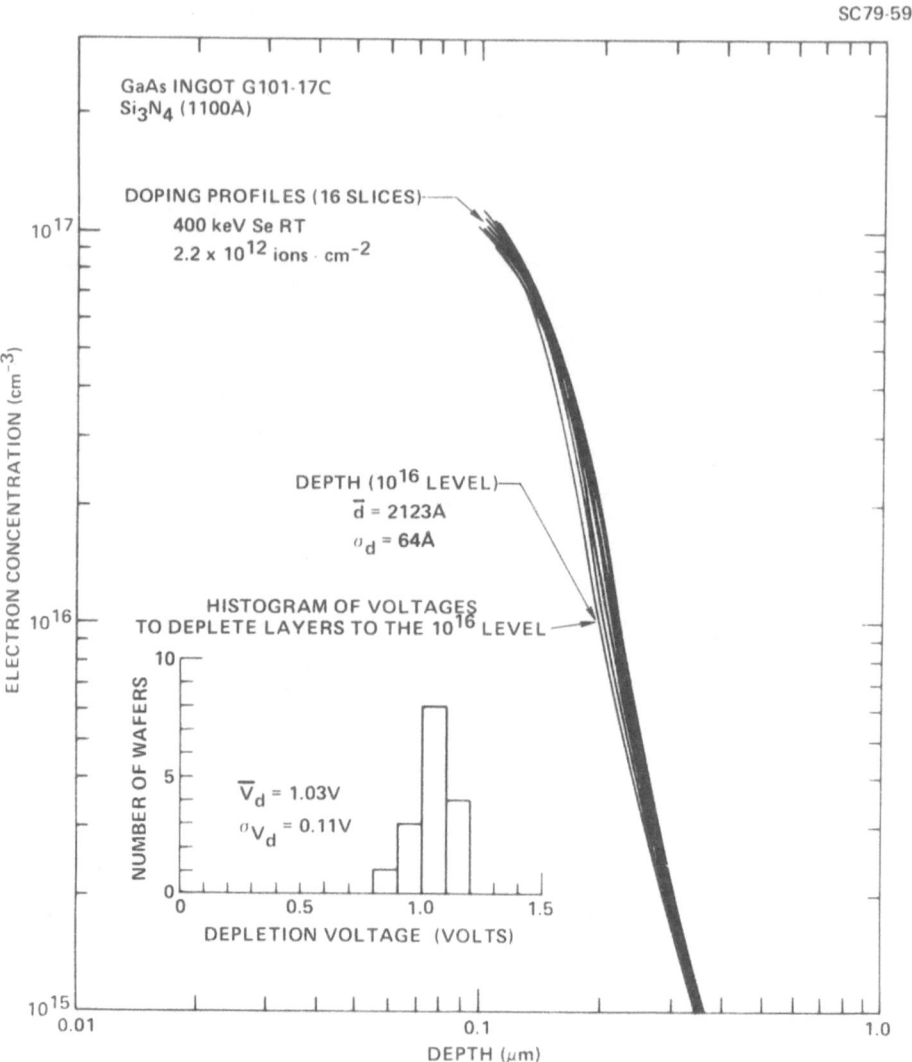

Figure 4 Carrier concentration profiles of FET channel layers prepared by Se implantation into semi-insulating GaAs. Note how sensitive the depletion voltage is to small changes in the depth of the profile.

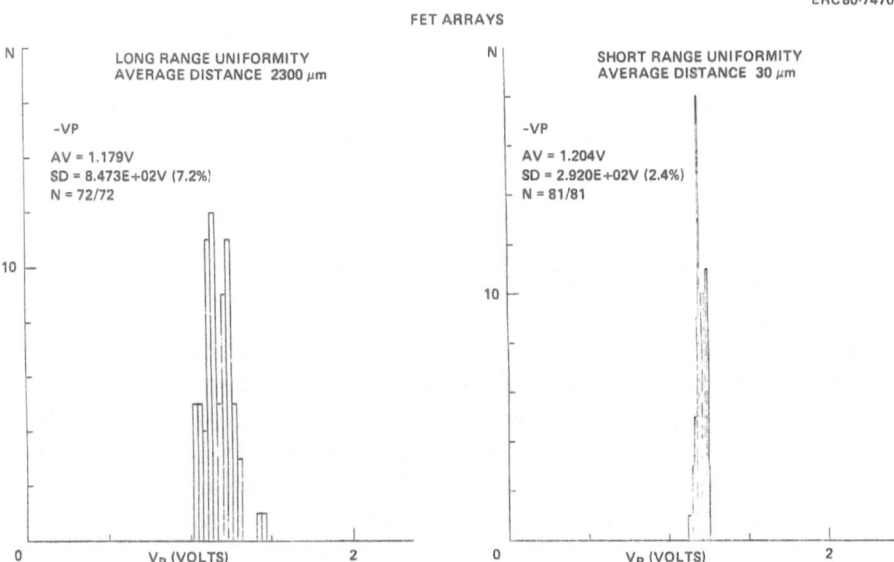

Figure 5   Histograms of pinch-off voltages of test FETs on a typical wafer fabricated with the planar IC process. The histogram to the left corresponds to any array of devices covering the whole wafer (21×21 mm), and it reflects the long range uniformity ($\bar{V}_p$=1.18 V, σ_{V_p}=85 mV). The histogram to the right corresponds to an array covering a small area (0.3×0.3 mm), and it reflects the short range uniformity ($\bar{V}_p$=1.20 V, σ_{V_p}=29 mV).

Requirements on semi-insulating GaAs substrates

Our planar fabrication process for GaAs digital ICs, as it is at present, poses severe demands on the semi-insulating substrate. This section contains a list of such requirements.

AMPLE MATERIAL SUPPLY

The supply of semi-insulating GaAs has been and still is marginal, to some extent because only a fraction of a grown ingot is usable. Pressed by needs for a continuous supply and better material control, several users have been attracted to LEC (liquid encapsulated Czochralski [14]) GaAs grown by the large Metals Research pullers now available [15]. The prospects are good in terms of quantity. Quality, measured by the qualification of the material for reproducible low dose implants, is not yet known on a sufficiently large database. The preliminary data are very promising [15].

LARGE WAFERS

As manufacture of large numbers of circuits is beginning to be considered, wafers larger than the current 1—2 inches are needed. The capability of the new LEC pullers for 3 inch ingot diameters should help respond to this requirement. Processing the large wafers will pose many practical problems (wafer flatness, for example), but not at the fundamental level, and none being insurmountable.

LOW DISLOCATION DENSITY

The current boat grown material has dislocation densities between 10^3 and 10^4 cm^{-2}. The LEC material has higher dislocation densities, between 10^4 and 5×10^4 cm^{-2}, a potential drawback for this material. If we made the very pessimistic assumption that a dislocation under a FET gate or a diode barrier causes a fatal circuit failure, we would project the yield of 1000 gate SDFL circuits (for 24 μm^2/gate of critical area) to be 79% for material with 10^3 dislocations cm^{-2}, and to drop to only 7% for a dislocation density of 10^4 cm^{-2}. However, by the same estimates, the 100 gate circuits made today on material with 10^4 dislocations cm^{-2} would have a yield ceiling of 79%, and yet we have no evidence for dislocation limited yield. It is likely that the noise margins of digital circuits absorb most of the distortions of device characteristics caused by dislocations without catastrophic failures. It is plausible that dislocations may be more critical from the process standpoint because dislocations may act as sinks for impurities and contribute to unwanted rapid diffusion processes [16]. In summary, we do not at present specify low dislocation densities, but it may become necessary to do so in the future.

HIGH BULK RESISTIVITY

The resistivity of as grown semi-insulating GaAs is always in the quite satisfactory $10^8 - 10^9$ Ω-cm range, which is near the intrinsic resistivity of GaAs at room temperature [17]. The yield of ingots with high resistivity appears to be quite high.

NO TRAP RELATED EFFECTS

Voltages applied to the implanted devices can alter the distribution of electrical potential in the substrate inducing changes of trap population. Such changes cannot follow the high speed signals ($\gtrsim$ 1 GHz) because of the long relaxation time of semi-insulating GaAs (0.1—1 ms). However, changes of trap populations induced by static (bias) voltages can result in distortion of FET I-V characteristics labelled backgating effects [18]. Zylbersztejn, Bert and Nuzillat [19] have attributed such effects to hole traps in the substrate near the interface with the active layer. It is possible that backgating may be unavoidable, to some extent, since traps are associated with the deep levels that are needed for the electrical compensation. Although undesirable, these effects can be compensated for by circuit design. However, this task is somehow hampered by variations in the magnitude of backgating effects, which appears to be ingot-dependent. We expect that better knowledge and control of the trap system will lead to material improvements.

HIGH RESISTIVITY AFTER ANNEAL AND REPRODUCIBILITY OF IMPLANTED PROFILES

The resistivity of the unimplanted substrate areas remains high after the post-implantation anneal cycle (as required for isolation between devices) provided that the substrate has passed qualification tests [20]. In these tests, ingots are sampled and those which do not conserve high resistivity after an anneal cycle (simulating the real process) or which exhibit unwanted deep 'tails' in the carrier concentration profiles of implanted layers are rejected [13]. We believe that both failure modes, which tend to appear together, are mainly due to shallow donors becoming uncompensated near the GaAs surface as a result of Cr out-diffusion [12, 13]. Asbeck *et al.* [13] showed that a carrier concentration profile with a deep tail can be constructed from the shallow donor concentration and the atomic Cr and Se profiles measured by SIMS (the Cr concentration decreases towards the surface due to out-diffusion). It appears that reduction of background impurities to a 10^{15} cm^{-3} level (a normal goal in crystal growth) may lead to better annealing behaviour of the substrate (implanted or unimplanted). If such a goal were met, much fewer implantation tests and dose adjustments would be needed. However, full control of the annealing processes will require better understanding of the role of the cap. The cap must be treated both as a chemical interface and as a source of strain influencing impurity diffusion through the substrate (suggested by gettering experiments [16]).

FURTHER PROGRESS IN ION IMPLANTATION

We would like to see ion implantation break the 2—3×10^{18} cm^{-3} 'barrier' of maximum carrier concentration [21], so that non-alloyed ohmic contacts can be developed. We hope that pulsed annealing techniques may accomplish this [22]. Better control of pn junctions formed by implantation of both the n and p layers [23] could lead to a planar IC fabrication technology using JFETs [24]. In general, ion implantation has the potential for providing the new techniques which will spur innovations in circuit and device design.

Conclusions

In the last few years, applications of semi-insulating GaAs have been blooming. The most dramatic advances were made in digital integrated circuits, to the point that we now envisage GaAs high speed signal processors and even computers. And yet, our knowledge and control of the semi-insulating substrate material is not sufficient. The most pressing substrate problems today are: a short material supply, a very painstaking and selective material qualifications process and, to a lesser degree, uncontrolled substrate trapping effects. Many of the engineering advances have been accomplished by sheer cleverness in developing processing techniques based more on good intuition than on scientific knowledge. We hope that broadening the scientific base will not only help in solving the problems, but that it will also nurture new ideas and new applications.

ACKNOWLEDGEMENTS

This work was supported, in part, by the Defense Advanced Research Projects Agency of the US Department of Defense, and monitored by the Air Force Office of Scientific Research under contract No. F49620—77—C—0087.

References

1. Eden, R.C. Welch, B.M., Zucca, R. and Long, S.I. (1979). *IEEE J. Solid St. Circ.*, **14**, 221
2. Liechti, C.A. (1976). *IEEE MTT Trans.*, **24**, 279; Higgins, J.A., Kuvas, R.L., Eisen, F.R. and Ch'en, D.R. (1978). *IEEE Trans. Electr. Dev.* **25**, 587
3. Macksey, H.M. Adams, R.L., McQuiddy, D.N., Shaw, D.W. and Wisseman, W.R. (1977). *IEEE Trans. Electr. Dev.* **24**, 113; Immorlica, A.A., Ch'en, D.R., Decker, D.R. and Fairman, R.D. (1979). *Proc. 7th Biennial Cornell Electrical Engineering Conf.*, in press
4. Van Tuyl, R.L., Liechti, C.A., Lee, R.E. and Gowen, E. (1977). *IEEE J. Solid. St. Circ.*, **12**, 485
5. Long, S.I., Lee, F.S., Zucca, R., Welch, B.M. and Eden, R.C. (1980). *IEEE MTT Trans.*, June issue, in press
6. Higgins, J.A., Gupta, A., Robinson, J. and Ch'en, D.R. (1979). *IEEE Int. Solid State Conf.*, Digest Tech. Papers, 120; Gupta, A.K., Higgins, J.A. and Decker, D.R. (1979). *IEDM*, Digest Tech. Papers, 269
7. Deyhimy, I., Eden, R.C., Anderson, P.J. and Harris, J.F. (1980). *Appl. Phys. Lett.*, **36**, 151
8. Welch, B.M., Shen, Y.D., Zucca, R., Eden, R. C. and Long, S.I. (1980). *IEEE Trans. Electr. Dev.*, June issue, in press
9. Editorial (1980). *Electronics*, **53**, 202
10. Eden, R.C., Welch, B.M. and Zucca, R. (1978). *IEEE J. Solid St. Circ.*, **13**, 419
11. Zucca, R., Welch, B.M., Eden, R.C. and Long, S.I. (1980). *IEEE Trans. Electr. Dev.*, June issue, in press
12. Asbeck, P., Tandon, J., Babcock, E., Welch, B., Evans, C.A. and Deline, V.R. (1979). *IEEE Trans. Electr. Dev.* **26**, 1853; Favennec, P.N. and L'Haridon, H. (1979). *Appl. Phys. Lett.*, **35**, 699
13. Asbeck, P.M., Tandon, J., Welch, B.M., Evans, C.A. and Deline, V.R. (1980). *IEEE Electr. Dev. Lett.*, March issue, in press
14. Henry, R.L. and Swiggard, E.M. (1977). *Inst. Phys. Conf. Ser.*, **33b**, 29
15. Fairman, R.D. and Oliver, J.R. This volume

16. Magee, T.J., Peng, J., Hong, J.D., Evans, C.A. and Deline, V.R. (1979). *Phys. Stat. Sol. (a)*, **55**, 169
17. Zucca, R. (1977). *J. Appl. Phys.*, **48**, 1987; Look, D.C. (1977). *J. Appl. Phys.*, **48**, 5141
18. Rossel, P., Tranduc, H., Graffevil, J., Azizi, C., Nuzillat, G. and Bert, G. (1978). *Rev. Phys. Appl.*, **13**, 503
19. Zylbersztejn, A., Bert, G. and Nuzillat, G. (1979). *Inst. Phys. Conf. Ser.*, **45**, 315
20. Eisen, F.H., Zucca, R. and Welch, B.M. (1979). US Patent 4, 157, 497
21. Tandon, J.L., Nicolet, M.-A. and Eisen, F.H. (1979). *Appl. Phys. Lett.*, **34**, 165
22. Tandon, J.L., Golecki, I., Nicolet, M.-A. Sadana, D.K. and Washburn, J. (1979). *Appl. Phys. Lett.*, **35**, 867
23. Donnelly, J.P. (1977). *Inst. Phys. Conf. Ser.*, **33b**, 166
24. Troeger, G.L., Behle, A.F., Frieberthauser, P.E., Hu, K.L. and Watanabe, S.H. (1979). *IEDM*, Digest Tech. Papers, 497

INTERFACE EFFECTS ON NOISE TEMPERATURE
OF UNGATED GaAs-MESFETs

M.K. AHMED and H. BENEKING
Institute of Semiconductor Electronics, Aachen Technical University,
Templergraben 55, D-5100 Aachen, West Germany

Abstract

The noise behaviour of GaAs MESFETs depends on the microwave noise temperature of the epilayer [1, 2]. In order to investigate its dependence on:

1. the quality of the interface between active layer and substrate; and
2. the substrate preparation before the epitaxial process,

four layers were suitably prepared and ungated GaAs FETs (channels) were fabricated on them. We measured the noise temperature of these devices at frequencies between 0.1 and 8 GHz. The measurements showed that the insertion of an undoped (buffer) layer between substrate and epilayer improves the noise behaviour of the epilayer in the MHz and GHz range. Etching the substrate surface before the epitaxial process also improves the noise temperature of the active layer with and without an inserted buffer layer.

As a basis of our investigation the most accepted empirical relationship [3, 4] between electron noise temperature T_n and electric field E

$$\frac{T_n}{T_0} = 1 + \delta \left(\frac{E}{E_{sat}}\right)^3 \tag{1}$$

is used, where T_0 is the lattice temperature, E_{sat} the saturation field and δ the noise temperature coefficient. The lower δ, the better will be the layer for low noise FET fabrication.

This relationship has also been derived theoretically [5]. To avoid measuring errors due to correlation and field inhomogeneities, ungated MESFET structures have been tested. The noise temperature measurements were made between 100 MHz and 8 GHz using the substitution method (the noise temperature was compared to that of a calibrated noise source). To avoid heating, the saturation current should be less than 150 mA mm^{-1}; on the other hand, the thickness of the channel should be large in comparison to the mean free path of electrons in GaAs to avoid the influence of the layer thickness on the physical parameters, mobility and E_{sat}, that would affect our measurement. Saturation currents of approximately 10 mA for 200 μm wide channels fulfil these conditions (layer thickness approx. 0.04 μm). The test devices were fabricated on four epitaxial layers grown under the same conditions on parts of the same Monsanto SI substrate wafer but after different surface treatment. The following abbreviations are used for the different surface treatments:

ME: epitaxy directly on SI substrate,
MEE: epitaxy after etching of the SI substrate surface for 1 min,
MBE: epitaxial growth of an undoped (n~10^{15} cm^{-3}) buffer layer followed by the active layer,
MEBE: epitaxy after etching the substrate surface and growing an undoped buffer layer.

Figure 1 shows the variation of δ depending on the substrate preparation before the epitaxial process. These results prove that the insertion of a buffer layer and also the etching before the epitaxy of the active layer improve the noise temperature coefficient to one third of the untreated value. Figure 2 indicates the variation of δ as a function of frequency for two series of devices. We observe that δ decreases with frequency up to 1.5 GHz in the case of the unbuffered and unetched devices (ME) and up to 200 MHz in the case of MEBE. This result suggests that the low frequency noise at frequencies lower than 1.5 GHz is greatly influenced by G-R noise in the case of ME which is a consequence of the deep traps with different time constants due to the interface between the active layer and the substrate. The second observation is that δ is lower for MEBE devices than for ME devices in the range above 1.5 GHz. This result may be due to a noise centre with a short time constant in the interface region.

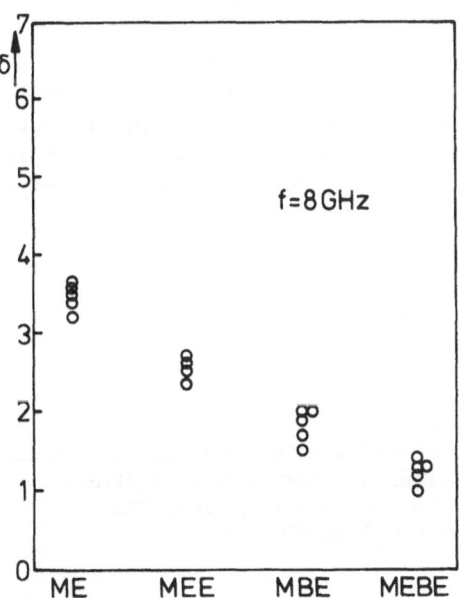

Figure 1   The dependence of the noise temperature coefficient on the insertion of a buffer layer and substrate preparation.

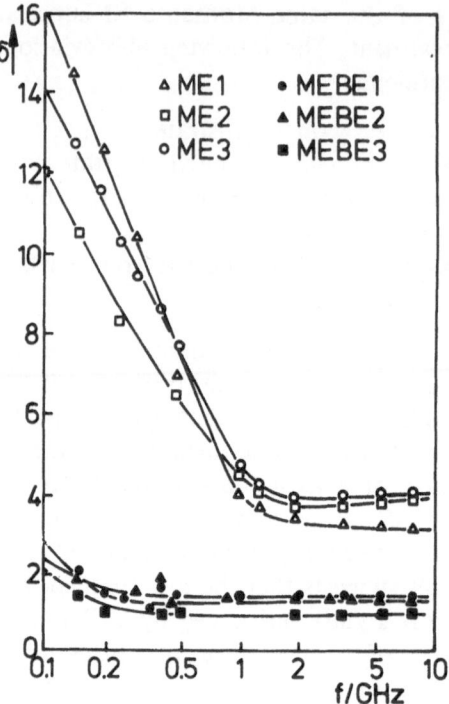

Figure 2   The dependence of the noise temperature coefficient on the frequency.

Conclusion

The influence of technological fabrication and substrate preparation of vapour phase GaAs epitaxy layers on noise temperature up to 8 GHz has been shown. Etching the substrate surface and inserting a buffer layer reduces the overall noise of the layers and should enable the fabrication of lower noise GaAs FETs.

References

1. Tsironis, C. (1979). *AEU*, **33**, 32
2. Statz, H., Haus, H. und Pucel, R. (1974), *IEEE Trans. Electron. Dev.*, **ED-21**, 549
3. Bächtold, W. (1972). *IEEE Trans. Electron. Dev.*, **ED-19**, 674
4. Frey, J. (1976). *IEEE Trans. Electron. Dev.*, **ED-23**, 1298
5. Ahmed, M.K., Beneking, H. To be published

SELENIUM ION IMPLANTED GaAs MESFETs

I.R. SANDERS, A.H. PEAKE and R.K. SURRIDGE*

Plessey Research (Caswell) Ltd, Caswell, Towcester, Northants, UK

Abstract

Schottky barrier gated FETs have been fabricated by selenium ion implantation at room temperature and 200°C into Cr-doped semi-insulating GaAs and into undoped epitaxial buffer layers. Dual dose and energy implants were investigated, the depth variation of carrier concentration and mobility in the implanted layers being determined by the differential Hall technique. Devices with the best microwave performance were produced on buffer layers implanted at room temperature with 5×10^{13} Se cm^{-2} at 100 keV plus 7.5×10^{12} Se cm^{-2} at 800 keV. The noise figure and associated gain of these devices at 12.75 GHz were 2.2 dB and 6.6 dB, respectively. The transfer characteristics and d.c. parameter uniformity of devices produced on the two materials are compared for samples implanted at both temperatures investigated.

Introduction

The potential advantage of ion implantation over conventional doped epitaxial layers for GaAs MESFET production is the ability to introduce uniformly donor impurities into large numbers of large area semi-insulating substrates. In this investigation selenium ions were implanted at room temperature and 200°C into Cr-doped semi-insulating material and undoped epitaxial buffer layers of GaAs on semi-insulating substrates. Post-implantation annealing was performed at 900°C for 30 s with a protective layer of CVD Si_3N_4.

A dual dose and energy implantation schedule was adopted in order to produce devices with low source to drain resistance. A highly doped n$^+$ near surface region was produced by implantation at 100 keV and a deeper n layer suitable for the channel region of the devices was achieved by implantation at 800 keV (Se^{2+} at 400 keV). The depth variations of electron concentration and Hall mobility were determined by a differential Hall technique in conjunction with chemical layer stripping.

MESFET devices were manufactured by the etched channel technique [1]. The depth of the channel etch was sufficient to remove the n$^+$ surface layer and situate the gate at or near the peak of the higher energy implant.

* Formerly of the University of Surrey, now at
Philips Research Laboratories, Redhill, Surrey, U.K.

Results and discussion

The variations of carrier concentration and Hall mobility with depth, for bulk SI material samples from the same ingot, implanted with two dose combinations, are shown in Figure 1. Similar results were obtained for

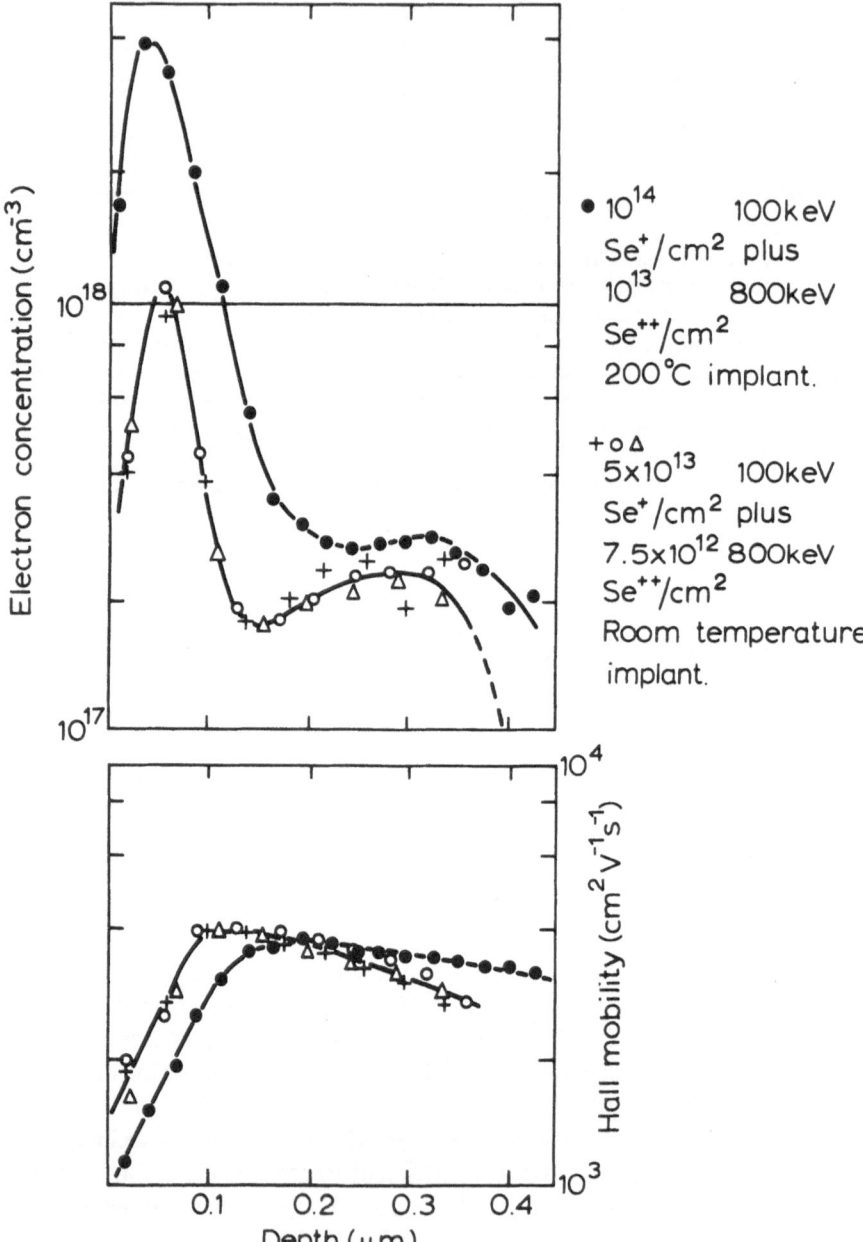

● 10^{14} 100keV
Se$^+$/cm^2 plus
10^{13} 800keV
Se^{++}/cm^2
200°C implant.

+ o △
5×10^{13} 100keV
Se$^+$/cm^2 plus
7.5×10^{12} 800keV
Se^{++}/cm^2
Room temperature
implant.

Figure 1   Carrier concentration and Hall mobility profiles.

corresponding implants into buffer layers. The symbols +, 0 and △ show measurements from three samples of the same room temperature implanted slice, and show excellent reproducibility.

It can be seen that the low energy implantation was successful in increasing the near surface electron concentration to around 1×10^{18} cm^{-3} for the 5×10^{13} cm^{-2} room temperature implant, and about 4×10^{18} cm^{-3} for the 10^{14} cm^{-2} 200°C implant. The carrier concentration level achieved for any given dose of 800 keV ions was similar at both of the implantation temperatures studied in the dose range suitable for the channel region of the MESFET. For the samples shown in Figure 1, the doping levels measured at the projected range of the 400 keV Se^{2+} ions were approximately 3×10^{17} cm^{-3} and 2×10^{17} cm^{-3} for the 10^{13} and 7.5×10^{12} cm^{-2} implants, respectively. It was observed that the carrier concentration levels decreased more rapidly with depth for all the specimens implanted at room temperature compared to samples implanted with similar doses at 200°C.

The one micron gate length devices with the best r.f. performance were fabricated on buffer layers implanted with 5×10^{13} 100 keV Se^{1+}cm^{-2} plus 7.5×10^{12} 400 keV Se^{2+} cm^{-2} at room temperature. The noise figures and associated gain measured at 12.75 GHz of these devices were 2.2 dB and 6.6 dB, respectively. In comparison, the best results from devices on bulk semi-

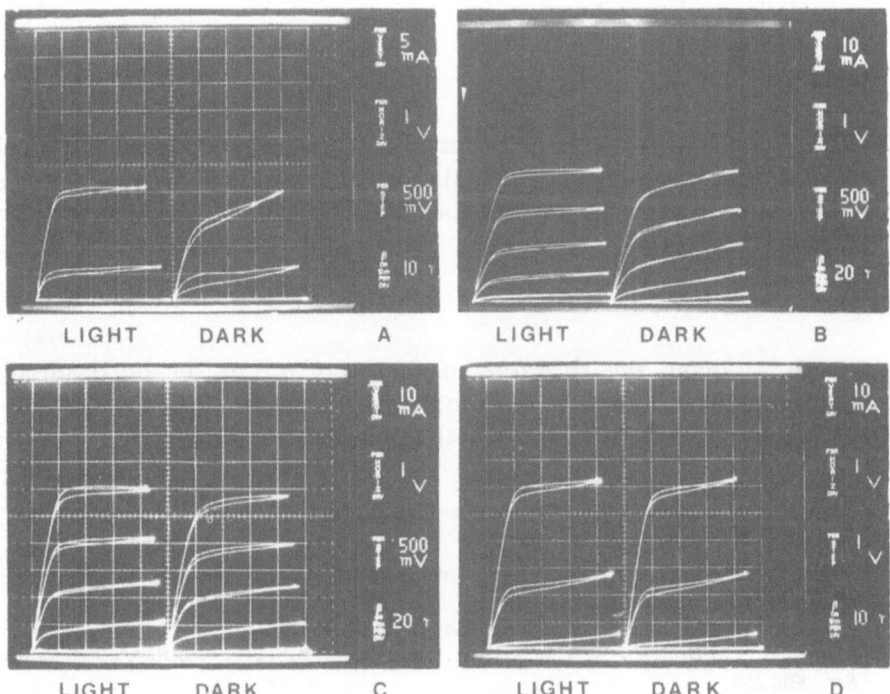

Figure 2 IV characteristics of implanted devices from implantation into semi-insulating material at: (*a*) room temperature and (*b*) 200°C; and from implantation into buffer layers at (*c*) room temperature and (*d*) 200°C.

insulating material were 2.9 dB and 5.5 dB for samples implanted at 200°C with 10^{14} 100 keV Se^{1+} cm^{-2} plus 10^{13} 400 keV Se^{2+} cm^{-2}.

The IV characteristics shown in Figure 2 compare the light sensitivity and hysteresis of selenium implanted devices fabricated on buffer layers and semi-insulating material at room temperature and 200°C. Light sensitivity and hysteresis of device characteristics have been attributed to the existence of Cr [2] and electron traps in the active region [3]. It can be seen that significantly less light sensitivity is exhibited by those devices on implanted buffer layers. While the reason for this is not clear, it is consistent with an expected lower Cr concentration in the top micron of the buffer layer than in a similar region of bulk Cr-doped substrate material [4].

The d.c. parameters of these devices are comparable to implanted FET results quoted in the literature [5, 6] and also to results achieved on conventional epitaxial material. In general, better d.c. parameter uniformity over a wafer is obtained from implantation into buffer layers than Cr-doped substrates, and results are similar to uniformities reported by other workers [5]. This superior uniformity on buffer material indicates that uniformity is dominated by variations in the substrates rather than by spatial variation of doping due to the implanted species. The reasons for variations in substrate properties are not fully understood but may reflect variation in Cr-doping levels either in the 'as grown' material or due to out-diffusion during encapsulation and annealing.

Conclusions

MESFETs with good operating characteristics can be produced by the dual implantation of selenium ions either at room temperature or at 200°C. Device parameters and uniformity are superior where implantation is performed into epitaxial buffer layers.

ACKNOWLEDGEMENTS

We wish to thank the Directors of Plessey Co. Ltd for provision of laboratory facilities and permission to publish this work. R.K. Surridge was supported by an SRC CASE award at the University of Surrey.

References

1. Butlin, R.S., Parker, O., Walker, A.J. and Turner, J.A. (1976). *Proc. 6th European Microwave Conf.*, Rome, Sept., p. 606
2. Houng, Y.M. and Pearson, G.L. (1978). *J. Appl. Phys.*, **44**, 3348
3. Crossley, I., Goodridge, I.H., Cardwell, M.J. and Butlin, R.S. (1977). *Inst. of Phys. Conf. Ser.*, **33b**, 289
4. Tuck, B., Adegboyega, G.A., Jay, P.R. and Cardwell, M.J. (1979). *Inst. of Phys. Conf. Ser.*, **45**, 114
5. Higgins, J.A., Eisen, F., Welch, B., Robinson, G. and Hill, W. (1977). *Inst. of Phys. Conf. Ser.*, **33b**, 236
6. King, J.K., Malbon, R.M. and Lee, D.H. (1977). *Elec. Lett.*, **13**, 187

HIGH RESISTIVITY LAYERS IN GaP, GaAs AND GaAsP USING ELECTRON BOMBARDMENT

TAKAO WADA, SASHIRO UEMURA* and NOBORU KITAMURA
Department of Electrical Engineering, Mie University,
Kamihama, Tsu, 514, Japan

High energy (1 and 7 MeV) electron bombardment produced high resistivity layers in GaP, GaAs and GaAsP crystals, and changed the GaP crystal from orange to black. The Laue pattern of the wafer remained unchanged by bombardment. The process is useful for electrically and optically isolating GaP junction devices.

Electron beams of 1 MeV with a spot size of 5—35 μm were scanned to demonstrate the separation of an array of p-n junction diodes. The spatial distributions of red luminescence and high resistivity layers near the irradiated region were investigated using a scanning electron microscope (SEM).

The samples were irradiated with electrons of 7 MeV from an electron linear accelerator, and 1 MeV from an ultra-high voltage electron microscope. In the case of 7 MeV, they were placed in a water-bath to inhibit temperature rise and at 1 MeV they were exposed to an electron beam with a spot size at room temperature.

The resistivities, ρ, of the n-GaP and n-GaAsP wafer irradiated with 7 MeV electrons [1], and n-GaAs crystal for 2 MeV [2] are shown by dotted lines in Figures 1a, 1b and 1c, respectively, as a function of total flux, ϕ. The measured value of ρ is expressed as $\rho=(e\mu_n n+e\mu_p p)^{-1}$ in the usual notation. The relationship between n and the incident electron flux follows from the condition of charge neutrality for the irradiated sample by [3]:

$$\phi\left(\sum_{E_A} \eta_{E_A}\, f_{Ae} - \sum_{E_D} \eta_{E_D}\, f_{Dp}\right)= p + N_d^h - n$$

where N_d^h is the hole concentration on the donor impurities present in GaP before irradiation, η is the production rate of each defect, N_d is the impurity donor density, E_A and E_D are the acceptor and donor levels of defects (six levels [4]), respectively, the Fermi level is approximately obtained by $E_F=kT \ln(n/N_C)+E_C$, and $f_{Ae}(f_{Dp})$ is the probability of an electron (hole) occupying the defect acceptor (donor) levels.

*Ise Electronics Central Laboratory, Tsumura, Ise, 516—11, Japan

354

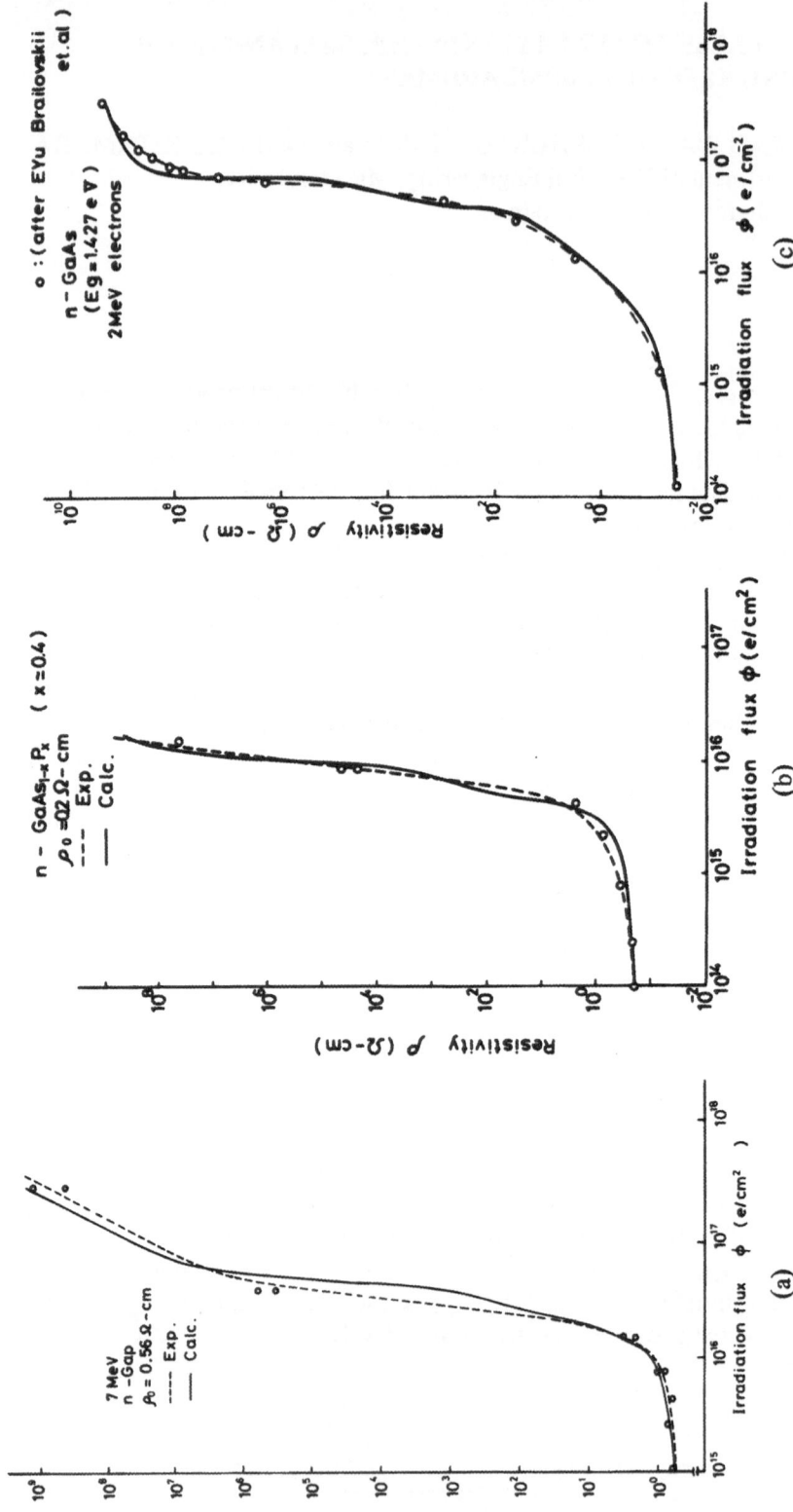

Figure 1 Resistivities for the wafers of (*a*) n-GaP, (*b*) n-GaAsP, and (*c*) n-GaAs as a function of irradiation flux.

Assuming that only one acceptor level of the defect in irradiated samples is introduced in the forbidden gap, a curve of ρ versus the defect density N_D is obtained from $N_D=(p+N_D^h-n)/f_{Ae}$. The full theoretical curve is a result of six such curves and is shifted so as to superimpose some points of the experimental curve by multiplying the flux scale by a constant factor. These factors represent approximately the production rate of the respective defect states.

Further details of the methods of fitting the $\rho-\phi$ curves are given in a previous report [3].

The resultant $\rho-\phi$ curves calculated by using the appropriate values of η agree satisfactorily with the experimental results, as shown by solid lines in Figure 1. This calculation simultaneously determines the η of the respective defects.

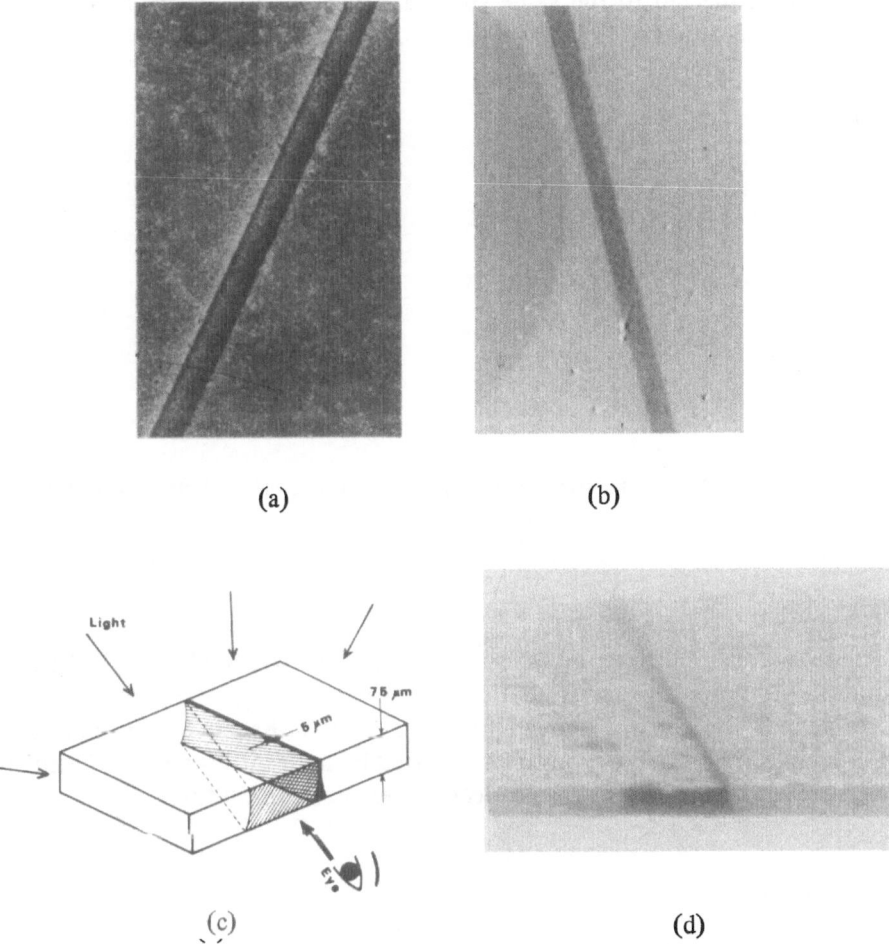

(a) (b)

(c) (d)

Figure 2 (*a*) Electron beam induced current image of the irradiated GaP LED. (*b*) Cathodoluminescent image of the irradiated GaAsP LED. (*c*) Schematic representation of a photomicrograph (*d*) of an irradiated GaP LED.

Figure 2a shows an electron beam induced current image by SEM for the surface of a GaP LED, irradiated at a total flux of $\sim 2 \times 10^{20}$ electrons cm^{-2} by a scanned electron beam (1 MeV) with ~ 32 μm width. It indicates a very marked line (~ 32 μm width) produced by the high resistivity stripe. An example of a cathodoluminescent image for the surface of a GaAsP LED irradiated by a scanned electron beam is shown in Figure 2b. The well-defined stripe corresponds to the disappearance of luminescence from the irradiated region. Figure 2c is a schematic representation of the observed spatial distribution of the changed colour in the irradiated GaP wafer (t$\sim$75 μm) and Figure 2d shows its photomicrograph. In this case, the thickness of the high resistivity layer is estimated to be greater than 75 μm.

Figure 3 shows how the XY addressable LED arrays are fabricated using the present method. The column and row leads are connected to the front and rear faces of the wafer, respectively; in a conventional array both X and Y leads are connected to the same surface. High density matrix LED arrays become available using this technique.

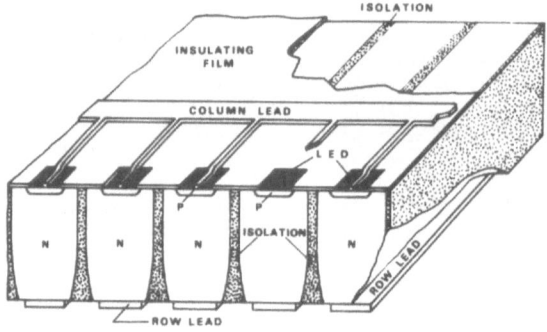

Figure 3   Matrix LED display structure fabricated by using high energy electron bombardment.

ACKNOWLEDGEMENTS

The authors would like to express their thanks to Drs S. Okabe, T. Tabata and K. Tsumori of the Radiation Center of Osaka prefecture, and to Professor S. Maruse of Nagoya University, for their help in the irradiation of the sample. The work was partly supported by a Grant-in-Aid for Scientific Research from the Ministry of Education, Science and Culture.

References

1. Wada, T. and Uemura, S. (1975). *Tech. of Int. Electron Devices Meeting,* Washington, DC, p. 192; also (1977) p. 486, and Wada, T., Uemura, S., Kakehi, M. and Kitamura, N. (1978) p. 638

2. Brailovskii, E.Yu, Braundnyi, V.N. and Groza, A.A. (1972). *Inst. Phys. Conf. Ser.,* **16,** 121
3. Wada, T., Yasuda, K., Ikuta, S., Takeda, M. and Masuda, H. (1977). *J. Appl. Phys.,* **48,** 2145
4. Lang, D.V. and Kimerling, L.C. (1976). *Appl. Phys. Lett.,* **28,** 248
5. Lang, D.V. and Kimerling, L.C. (1975). *Inst. Phys. Conf. Ser.,* **23,** 581

CONCLUDING REMARKS

RICARDO ZUCCA
Rockwell International Electronic Research Center,
Thousand Oaks, California 91360, USA

The recent advances in GaAs high speed technologies have been paralleled by an increased level of interest in the physics and chemistry of semi-insulating III—V materials. This conference brought together, for the first time, specialists on these subjects from all over the world. Such an opportunity to meet and discuss problems of common interest, both formally and informally, will encourage and hasten progress. The meeting was very successful and I hope that this success will stimulate the organization of further meetings.

Despite the general title of *Semi-Insulating III—V Materials,* semi-insulating GaAs was the dominant topic. Several papers on semi-insulating InP reveal, however, sustained interest in the material. We hope that this interest will expand, as InP holds promise to become a valuable device material.

In the area of growth of bulk, semi-insulating GaAs crystals, results from the new LEC systems are beginning to emerge with promise for relief from the chronic shortages of good semi-insulating GaAs substrates to which we are accustomed. Growth of epitaxial high resistivity (buffer) layers by vapour transport appears to be understood and well controlled, while organometallic growth is actively being pursued. Great emphasis is being placed on understanding the heat treatment behaviour of semi-insulating GaAs, as evidenced by a number of papers on this subject; Cr redistribution appears to be the main reason for the observed effects. Although the key role of Cr in the electrical compensation of the material is universally accepted, the mechanism by which the compensation takes place is still under scrutiny. This investigation is being aided by progress in determining the electronic states of Cr using EPR techniques and a variety of optical and transport techniques. And yet, a consensus has not been reached, as shown by the discussions in the rump session. In the field of deep level spectroscopy techniques, a very complex trap system continues to emerge. More work is needed to associate deep levels with transients, hysteresis and backgating effects in FET characteristics. Interesting work on the relationship between material properties and device behaviour is also being pursued.

A general conclusion from this conference is that the investigation of III—V semi-insulating materials is very active and is attracting great interest; but a change has occurred in this field over the last few years. In the past, the main driving force in this research used to be scientific curiosity and search for technological excellence, attracting interest only among investigators in this field. Today there is considerable outside interest in the scientific and technological developments in semi-insulating III—V semiconductors. It began when the microwave MESFET caught the interest of microwave circuit designers. Today a number of high speed circuits and devices, such as microwave integrated circuits, high speed digital integrated circuits and GaAs CCDs are emerging, with potential for some of these devices and circuits to be integrated with opto-electronic devices. This is happening at a time when system needs have grown to the point that system designers are prepared to take advantage of this new generation of high speed devices and circuits in a variety of communications and signal processing applications. Since most of the new devices and circuits depend very critically on semi-insulating substrates, the researchers find themselves in a new position, where their work is looked upon with great interest. This new status brings better financial support for their projects — a very helpful development. But, it also adds a new limitation to their projects: time. Better understanding and control of the semi-insulating substrates is crucial for the success of the new generation of GaAs high speed technology. There is a limited time, one to two years, for this to happen. This is a formidable challenge. If, as I hope, enough progress is made, a solid basis for a new family of technologies will be established.

As a final remark, I would like to congratulate the organizing committee for their remarkable work, which resulted in a very effective and pleasant meeting. I extend my thanks to the sponsoring organizations which made this meeting possible. And finally, a special word of appreciation is deserved by the University of Nottingham and the local committee for their hospitality, which made it possible for all of us to enjoy our stay in that beautiful place.

AUTHOR INDEX